INTERNATIONAL CENTRE FOR MECHANICAL SCIENCES

COURSES AND LECTURES - No. 292

APPLICATIONS OF TENSOR FUNCTIONS IN SOLID MECHANICS

EDITED BY

J.P. BOEHLER

UNIVERSITY OF GRENOBLE

SPRINGER-VERLAG WIEN GMBH

Le spese di stampa di questo volume sono in parte coperte da contributi
del Consiglio Nazionale delle Ricerche.

This volume contains 36 illustrations.

Originally published by Springer-Verlag Wien New York in 1987

ISBN 978-3-211-81975-3 ISBN 978-3-7091-2810-7 (eBook)
DOI 10.1007/978-3-7091-2810-7

PREFACE

The mechanical behavior of materials with oriented internal structures, produced by formation processes and manufacturing procedures (crystal arrangements, stratification, fibrosity, porosity, etc.) or induced by permanent deformation (anisotropic hardening, softening, creep internal damage growth, etc.) requires a suitable mathematical modelling. The properties of tensor valued functions of tensor variables constitute a rational basis for a consistent mathematical modelling of complex material behavior.

This book, which contains lectures presented at a CISM Advanced School, presents the principles, methods and results of applications in solid mechanics of the general laws governing tensor functions. The principles of mathematical techniques employed to derive representations of tensor functions are explained. The rules of specifying irreducible sets of tensor invariants and tensor generators for various classes of material symmetries are discussed. Representations of isotropic and anisotropic tensor functions are derived, in order to develop the general invariant forms of non-linear constitutive laws in mechanics of solids.

Within this approach, the mathematical modellization of the materials' mechanical response is explained and specific models are presented in elasticity, plasticity, hardening, internal damage and failure, for materials such as metals, composites, stratified rocks, consolidated soils and granular materials. The approach specifies a rational way to develop approximate theories and gives the necessary precision as to the number and the type of independent variables entering the mechanical laws to be used in engineering applications.

Experimental justifications as to the pertinence of the approach are given on examples of composite materials, rolled sheet-steel and stratified rocks. Information concerning proper experimental setting of tests for materials with oriented internal structures is developed.

This book is addressed to specialists in solid mechanics, both theoretical and applied, material scientists concerned with metals and composites, specialists in soil and rock mechanics and to structural engineers facing problems involving anisotropic and inelastic solids at various environments, nonlinearity and couplings.

I wish here to pay homage to the memory of my deeply missed friend Professor Antoni Sawczuk, co-coordinator of the CISM Advanced School. His untimely death did not allow him to see the final fulfilment of our shared project ; nevertheless, his help was invaluable in the preparatory phase of the Advanced School.

I would like to take this opportunity to thank CISM for having provided lecturers and participants with a chance to work together on Applications of Tensor Functions in Solid Mechanics. I am indebted to Professor Giovanni Bianchi, Secretary General of CISM, for his help with the organization of the School and to Miss Elsa Venir for her kindness and efficiency.

Jean-Paul Boehler

CONTENTS

LIST OF AUTHORS

J. BETTEN — Rheinisch - Westfälische Technische Hochschule Aachen
Templergraben 55
D-5100 AACHEN
Federal Republic of Germany

J. P. BOEHLER — Université Scientifique Technologique et Médicale de Grenoble
Institut de Mécanique de Grenoble
B.P. n° 68
38402 SAINT MARTIN D'HERES CEDEX
France

A. J. M. SPENCER — The University of Nottingham
Department of Theoretical Mechanics
NOTTINGHAM NG7 2RD, England
United Kingdom

APPLICATIONS OF TENSOR FUNCTIONS

IN SOLID MECHANICS

Chapter 1

PHYSICAL MOTIVATION

J.P. Boehler
University of Grenoble, France

1. INTRODUCTION

Theorems of representations for tensor functions are valuable for modelling non-linear constitutive laws, particularly when the mechanical response of the material depends on more than one tensor agency. It is an approach that leads to the general invariant forms of the non-linear constitutive equations and gives the number and type of the scalar variables involved.

These representations for tensor functions have proved to be even more pertinent in attempts to model the mechanical behavior of anisotropic materials, since here invariance conditions predominate and the number of independent scalar variables cannot be found by simple arguments.

In this Chapter, we present experimental evidence of anisotropic response of materials and show the complexity of phenomena observed, which indicates the need for a rational and unified formulation of anisotropic constitutive laws.

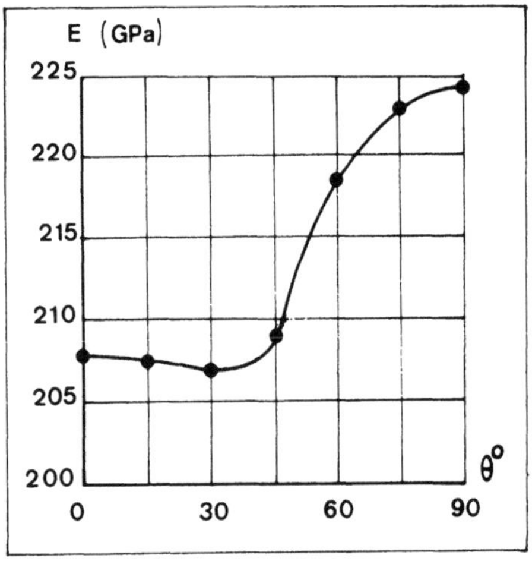

Fig.1 - Anisotropy of the elastic modulus of rolled sheet-steel (after [1]).

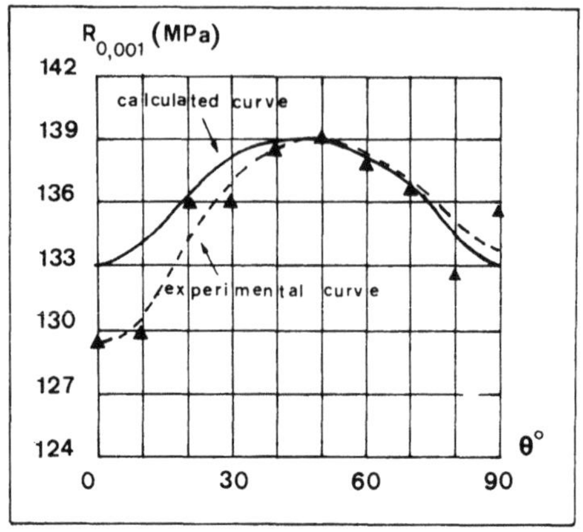

Fig.2 - Anisotropy of the elastic limit of rolled sheet-steel (after [2]).

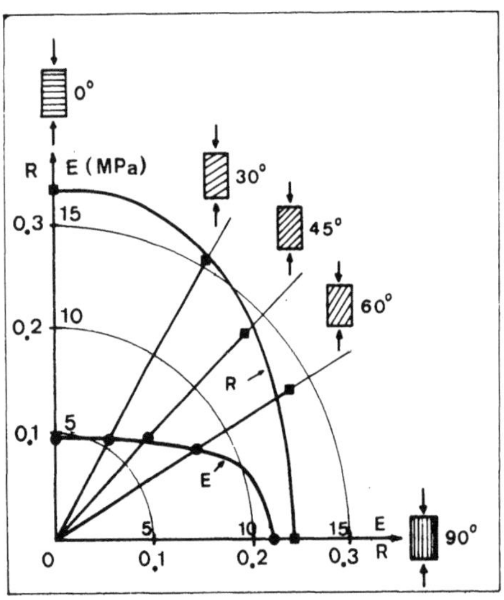

Fig.3 - Anisotropy of the elastic modulus and compressive strength of a natural clay (after [3]).

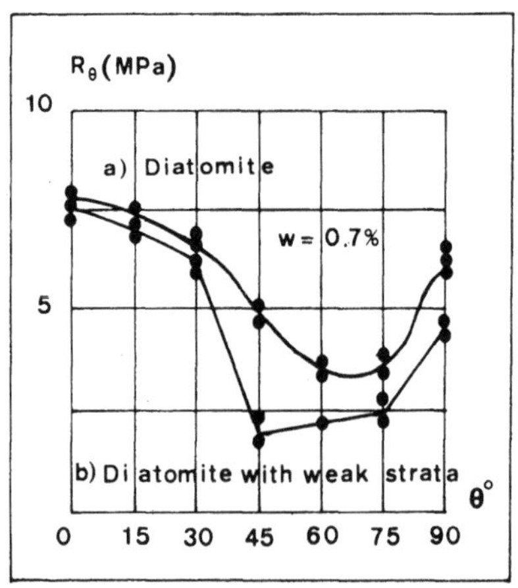

Fig.4 - Strength in simple compression for uniform and periodically non-homogeneous structure of diatomite (after [4]).

2. DIFFERENT DOMAINS OF MECHANICAL ANISOTROPY

Oriented internal structures of solids, such as oriented crystallographic axes, grains, particles, fissuration, cracks, cavities, etc..., result on the macroscopic level in a directional mechanical response to applied agencies. Different domains of the mechanical behavior can be influenced.

Fig.1 shows the variation of the elastic modulus E of rolled sheet-steel with respect to the angle θ between the direction of the tensile stress and the rolling direction (after [1]), whereas the anisotropy of the elastic limit R is presented in Fig.2 (after [2]). Anisotropy of the elastic modulus E and strength R of a consolidated clay subjected to oriented compressions is given in Fig.3 (after [3]).

According to the type of the oriented internal structure, the variation of mechanical characteristics, with respect to the orientation of the material, may be continuous or discontinuous. This is shown in Fig.4 (after [4]), where the uniaxial compressive strength of diatomite, a stratified soft rock, is traced versus the orientation θ of the specimens with respect to the normal of the strata. It is seen that the standard compressive strength possesses a minimum within the range of inclination of the strata. For a diatomite with marked layers of weaker strata, there appears a sudden drop in the strength for inclinations ranging between 30° and 45°.

In Fig.5, results regarding the strength of different consolidated clays in compression are presented (after [5, 6]) : a) range of variation for London clay ; b) Little Belt clay ; c) Vienna clay ; d) Welland clay ; e) experimental points for Grenoble clay. It is seen that the variation of strength with the orientation of the privileged direction of transverse isotropy is quite irregular. The strength either decreases, increases or passes through an extremum when the inclination of the privileged direction changes with respect to the principal stress direction. For triaxial tests on shales, Fig.6, the strength is plotted against this inclination angle for several values of the confining pressure (after [7]). Two remarks are appropriate in connection with the experimental results presented. In the first place a diversity of mechanical response with the orientation of the privileged plane of transverse isotropy can be seen in the figures. Secondly, a quite irregular dependence of the strength on the hydrostatic pressure is noticed. This indicates that proposed criteria of failure should possess a sufficient flexibility to permit an explanation of experimentally observed behavior of solids with directional mechanical properties. In view of the complexity of mechanical behavior, a study of yielding and failure of oriented solids requires an adequate and unified approach.

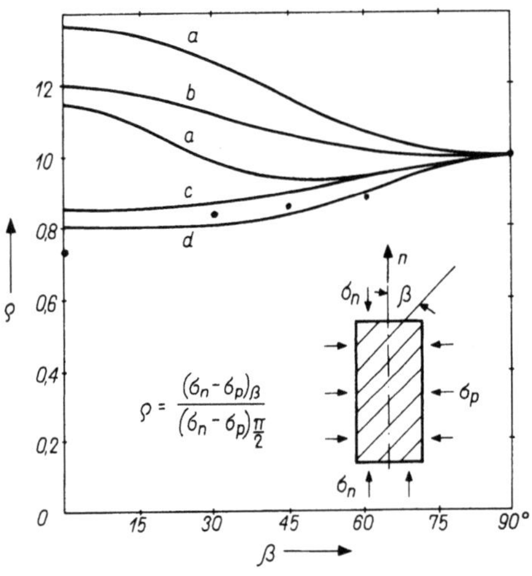

Fig.5 - Directional strength of clays (after [5, 6]).

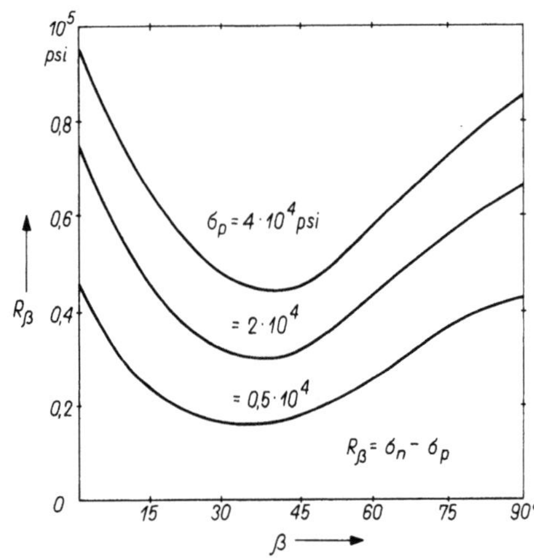

Fig.6 - Directional strength of shales under confining pressure (after [7]).

3. ESSENTIAL FEATURES OF THE ANISOTROPIC MECHANICAL BEHAVIOR OF ROLLED SHEET-STEEL

In this Section, we present some essential features of the anisotropic mechanical behavior of rolled sheet-steel, as well as the evolution of the anisotropic behavior during irreversible deformations (anisotropic hardening). Three questions are considered.

* First question : what is mechanical anisotropy ?

The definition is : "Variation of the mechanical response with the orientation of the agencies with respect to the material". This is sketched in Fig.7, which presents stress-strain curves for different oriented specimens of rolled soft steel when subjected to simple tensile tests; L is the rolling direction and θ the angle between L and the direction of the applied tensile stress. The figure shows the generally obtained stress-strain curves in the rolling direction (θ = 0), the transverse direction (θ = 90°) and at 45° from the rolling direction. The essential features are:

- The yield stress in the transverse direction is higher than in the rolling direction; at 45°, it is higher than in the transverse direction.
- The deformations corresponding to the different yield stresses vary with the orientation of the specimens.
- For the elastic behavior, we remark that the material is stiffer for the orientations of lower yield stress.

We can already notice that the description of these anisotropic phenomena is not simple: there exists no a priori rule for the variation of the stiffness, the yield stress and the corresponding deformation; moreover, these three variations are generally independent.

Fig.8 shows the evolution of the yield stress with the orientation of the specimens. The more the curve moves away from the horizontal line, the more the degree of anisotropy increases. In the case of isotropy, we have a horizontal straight line.

* Second question : how does the anisotropic mechanical behavior evolve during cold rolling ?

Fig.9 represents experimental results [8] concerning the yield criterion in simple tensile tests performed on oriented specimens of the same steel initially subjected to three different rolling reductions: r = 5 %, 10 % and 15 %. We observe that the anisotropy increases with the rolling reduction, but no simple relation exists between the three dif-

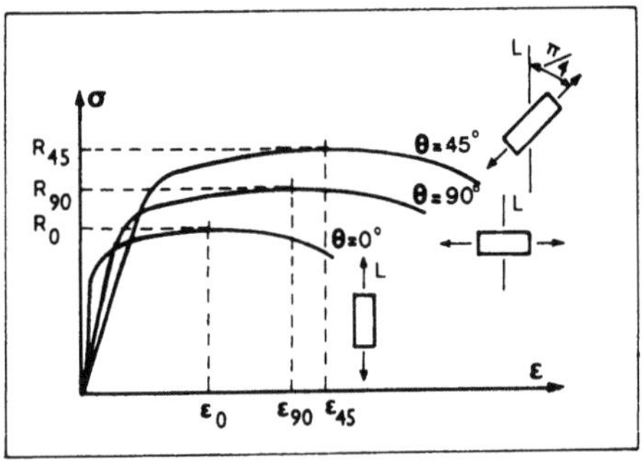

Fig.7 - Anisotropic stress-strain relations.

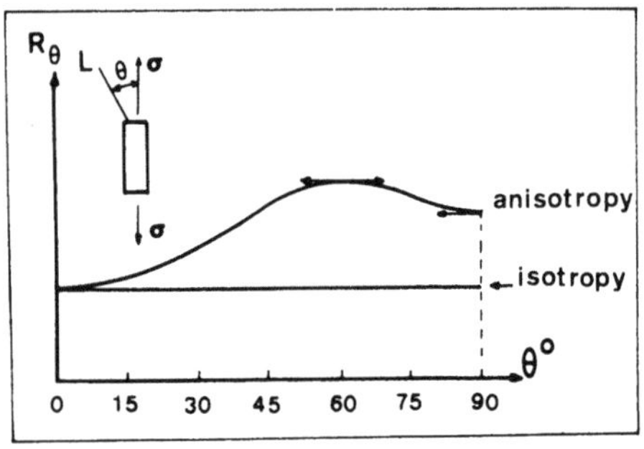

Fig.8 - Anisotropic yield criterion in simple tension.

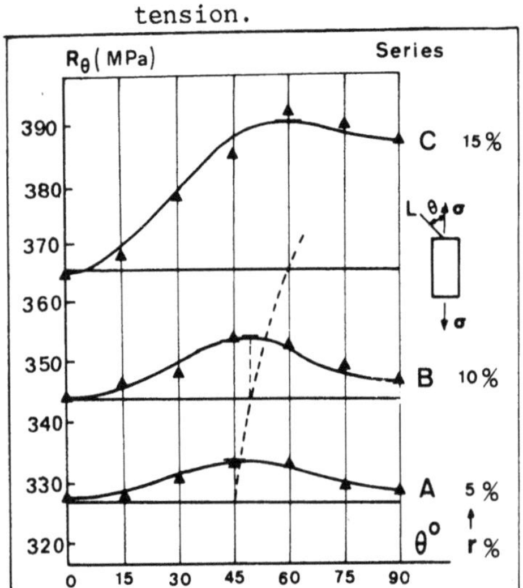

Fig.9 - Evolution of the yield criterion with the rolling reduction.

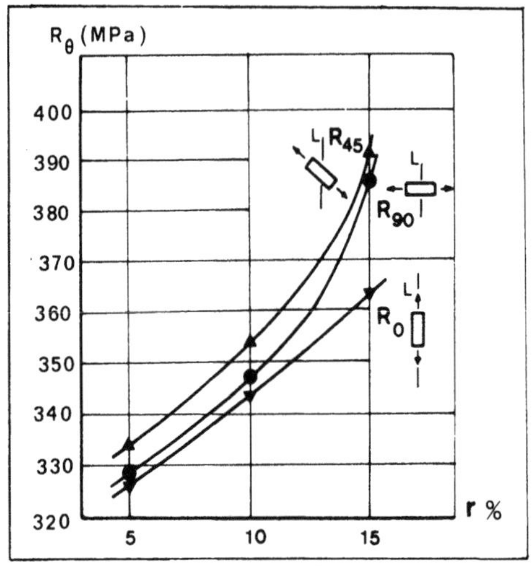

Fig.10 - Evolution of oriented yield stresses with the rolling reduction.

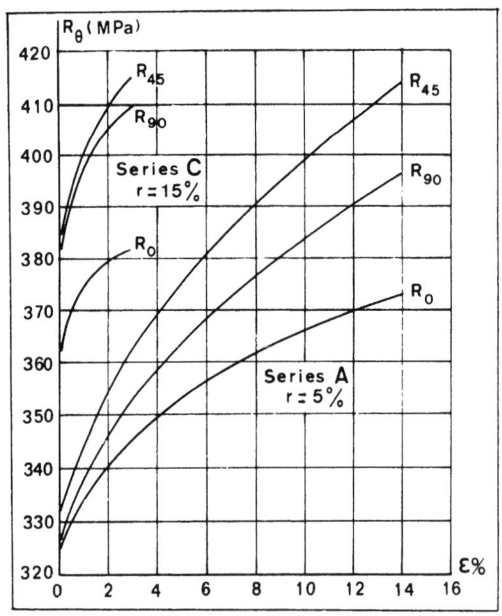

Fig.11 - Evolution of oriented yield stresses of rolled sheet-steel with further applied irreversible prestrains.

ferent yield curves. The evolution of the criterion is non-linear. This is also the case for the evolution of the orientation corresponding to the maximum yield stress.

Another means of stuyding the evolution of the anisotropy of the criterion is presented in Fig.10, which shows the evolution of the yield stress in the rolling direction, transverse direction and at 45° with respect to the rolling direction. In the isotropic case, the three curves would coincide. We observe that in the anisotropic case, the curves are strongly different and that there exists no simple relation between them.

* Third question : how does the anisotropic mechanical behavior of rolled sheet-steel evolve when subjected to further irreversible deformatio: (prestrainings) ?

Fig.11 represents the evolution of the yield stress in the rolling and transverse directions and at 45°, with respect to the prestrainings ε applied to steel initially rolled at r = 5 % and 15 % reduction. We can observed two facts:

- For a given rolling reduction, the three curves are not parallel; thus, the anisotropy changes with the prestrain: this is the phenomenon of anisotropic hardening.
- The respective evolutions of the three curves are not the same for two different rolling reductions.

So we see here further the complexity of anisotropic phenomena.

So far, we were concerned with the anisotropy of the yield criterion. We shall now analyse the anisotropy of the flow law, not directly, but through one of its consequences, which is the fact that the two necking lines are no longer symmetric with respect to the axis of the specimens, as it is the case for isotropic steel. Fig.12 (after [8]) shows the evolution of the orientation of the two necking lines with the orientation of the specimens. For each value of the irreversible prestrain, the two curves ϕ_1 and ϕ_2 together form the shape of an eight. We observe that the figures evolve in a complex manner with the prestrain.

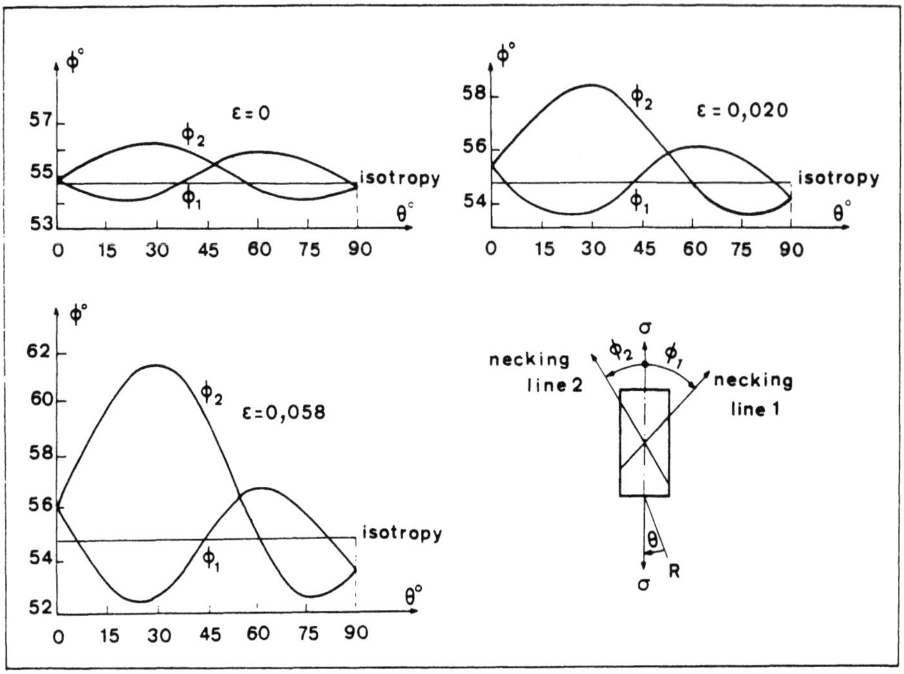

Fig.12 - Evolution of the orientation of the necking lines (after [8]).

From these experimental facts, we can conclude that the anisotropic mechanical behavior concerns very different phenomena, which are very complex and strongly non-linear. There appears to be a need to develop an objective and unified theory able to describe these different phenomena. In such a development, two main problems arise:

- What is the general invariant form of a constitutive law able to describe the anisotropic phenomena ?
- What are the essential variables, i.e. the number and type of variables which have to be measured during experiments in order to identify the specific constitutive equations ?

We shall see in the next Chapters that the theorems of representations for anisotropic tensor functions allow these two requirements to be fulfilled.

REFERENCES

1. PARNIERE, P., Relations entre Textures et Comportement Mécanique des Métaux, in: *Mechanical Behavior of Anisotropic Solids*, Proc. Euromech Colloquium 115 - Colloque International 295 du CNRS, J.P.BOEHLER ed., Editions du CNRS (Paris) and Martinus Nijhoff Publishers (The Hague), 1982, 303-332.

2. BUNGE, H.J. and ROBERT, W.T., *J. Applied Cryst.*, 9 (1969) : 116.

3. BOEHLER, J.P., Contribution à l'Etude de l'Equilibre Limite des Sols Anisotropes, *Thèse de Doctorat de Spécialité*, Grenoble, 1968.

4. ALLIROT, D., BOEHLER, J.P. and SAWCZUK, A., Pressure-Induced Evolution of Anisotropies in Stratified Rock, *Studia Geotechnica et Mechanica*, Vol. III, n°2-4, 1981, 59-73.

5. DUNCAN, J.M. and SHEED, H.B., Anisotropy and Stress-Reorientation in Clay, *Proc. ASCE 92*, SM5, 1956, 21-50.

6. BOEHLER, J.P., Contributions Théoriques et Expérimentales à l'Etude des Milieux Plastiques Anisotropes, *Thèse de Doctorat ès Sciences*, Grenoble, 1975.

7. GRAY, K.E., Some Rock Mechanics Problems of Petroleum Engineering, *Proc. 9th. Symp. Rock Mechanics*, The Colorado School of Mines, 1967, 405-433.

8. BOEHLER, J.P. and RACLIN, J., Anisotropic Hardening of Prestrained Rolled Sheet-Steel, in: *Current Advances in Mechanical Design and Production*, Proc. Second Cairo University MDP Conference, Dpt. of MDP, Faculty of Engineering, Cairo University, 1982, 483-492.

Chapter 2

INTRODUCTION TO THE INVARIANT FORMULATION OF ANISOTROPIC CONSTITUTIVE EQUATIONS

J.P. Boehler
University of Grenoble, France

1. INTRODUCTION

The aim of this Chapter is to present some elementary notion for non-specialists in the invariant formulation of anisotropic constitutive equations. Much of this Chapter is taken from [1].

We consider the very simple case of a two-dimensional anisotropic material of constitutive equation specifying a symmetric 2nd order tensor $\underset{\sim}{T}$, which is a function of two symmetric second order tensors $\underset{\sim}{D}$ and $\underset{\sim}{\xi}$:

$$\underset{\sim}{T} = \underset{\sim}{F}(\underset{\sim}{D}, \underset{\sim}{\xi}) \qquad (1)$$

In (1), $\underset{\sim}{\xi}$ is the structural tensor, taking into account the symmetries of the material internal structure, $\underset{\sim}{D}$ is the mechanical agency and $\underset{\sim}{T}$ the mechanical response of the material. We use the absolute notation, not to be related to coordinates at this point. For definiteness, $\underset{\sim}{T}$ is the stress tensor, $\underset{\sim}{D}$ is the strain tensor in the case of elastic behavior or the rate of strain tensor in the case of plastic behavior. In general, the principal directions ($\underset{\sim}{e}_1$, $\underset{\sim}{e}_2$) of the response $\underset{\sim}{T}$ do not coincide with the principal directions ($\underset{\sim}{E}_1$, $\underset{\sim}{E}_2$) of the agency $\underset{\sim}{D}$.

The constitutive equation (1) is sketched in the Figure below, where the material, in its initial configuration, is represented by a circle and the attached orthonormal frame ($\underset{\sim}{v}_1$, $\underset{\sim}{v}_2$); the tensors $\underset{\sim}{T}$ and $\underset{\sim}{D}$ are represented by their associated ellipses. Similar sketches will be used throughout this Chapter.

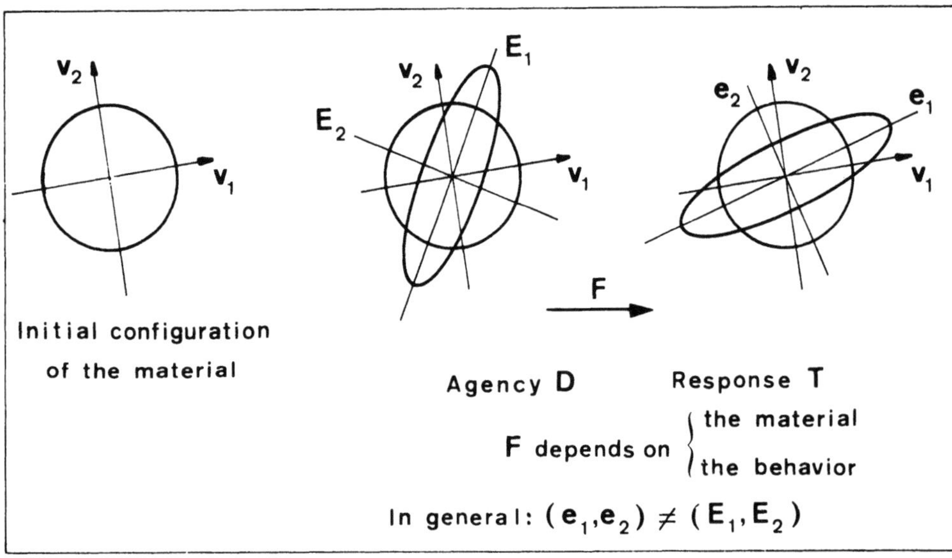

Constitutive relation

2. PRINCIPLE OF ISOTROPY OF SPACE

2.1 Physical concept

Constitutive equations are subjected to the invariance requirements of the Principle of Isotropy of Space (or Principle of Material Indifference [2]). We shall apply it here to the simple situation involved in the equation (1).

A consequence of this Principle is that an arbitrary transformation $\underset{\sim}{Q}$ of the orthogonal group O and applied to both the material and the agency $\underset{\sim}{D}$, results in the same orthogonal transformation of the material response $\underset{\sim}{T}$. A naive statement of this is that the orientation of the material in the space has no effect on its constitutive relation. This is sketched in the Figure below.

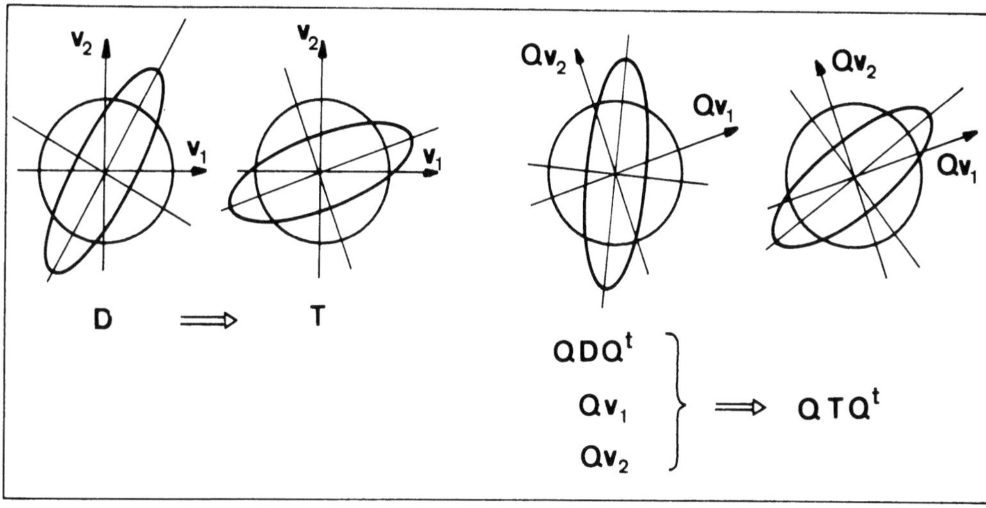

Consequence of the Principle of Isotropy of Space

2.2 Mathematical concept

The transformation of the body : $\underset{\sim}{\xi} \Rightarrow \underset{\sim}{Q}\underset{\sim}{\xi}\underset{\sim}{Q}^t = \bar{\underset{\sim}{\xi}}$

and the transformation of the agency : $\underset{\sim}{D} \Rightarrow \underset{\sim}{Q}\underset{\sim}{D}\underset{\sim}{Q}^t = \bar{\underset{\sim}{D}}$

result in the same transformation of the response: $\underset{\sim}{T} \Rightarrow \underset{\sim}{Q}\underset{\sim}{T}\underset{\sim}{Q}^t = \bar{\underset{\sim}{T}}$

Finally we obtain:

$$\forall \underset{\sim}{Q} \in O : \underset{\sim}{F}(\underset{\sim}{Q}\underset{\sim}{D}\underset{\sim}{Q}^t, \underset{\sim}{Q}\underset{\sim}{\xi}\underset{\sim}{Q}^t) = \underset{\sim}{Q}\underset{\sim}{F}(\underset{\sim}{D}, \underset{\sim}{\xi})\underset{\sim}{Q}^t$$

or:
$$\forall \underset{\sim}{Q} \in O : \underset{\sim}{F}(\bar{\underset{\sim}{D}}, \bar{\underset{\sim}{\xi}}) = \bar{\underset{\sim}{F}}(\underset{\sim}{D}, \underset{\sim}{\xi}) \qquad (2)$$

Relation (2) indicates the invariance conditions to which the constitutive equation (1) is subjected by the application of the Principle of Isotropy of Space.

The tensor-valued function $\underset{\sim}{F}$, subjected to the invariance conditions (2), is called isotropic with respect to its two arguments $(\underset{\sim}{D}, \underset{\sim}{\xi})$.

3. ISOTROPIC MATERIALS

3.1 Physical concept

A material is called isotropic if an arbitrary transformation $\underset{\sim}{Q}$ of the orthogonal group O and applied to the material, but not to the agency $\underset{\sim}{D}$, results in the same response $\underset{\sim}{T}$. This is sketched in the Figure below.

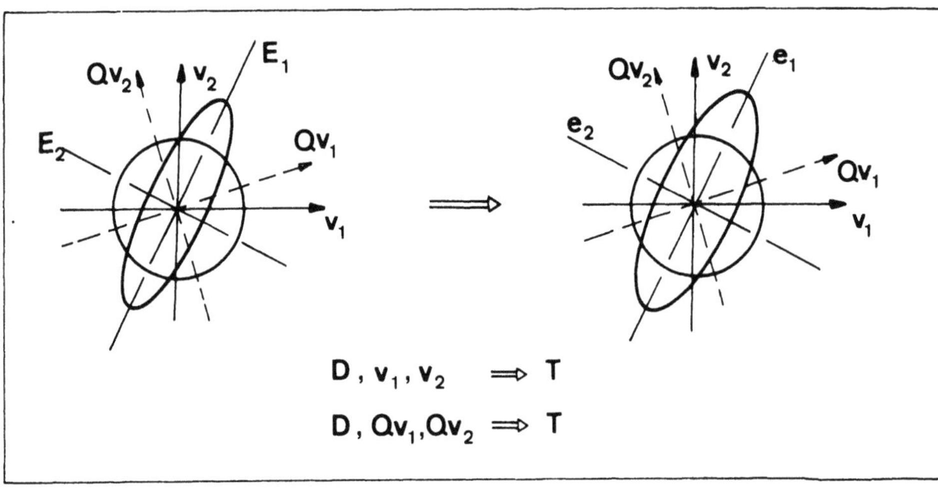

Definition of an isotropic material

A direct consequence of this definition is that the principal directions $\underset{\sim}{E}_1$, $\underset{\sim}{E}_2$) of the agency $\underset{\sim}{D}$ and the principal directions $(\underset{\sim}{e}_1, \underset{\sim}{e}_2)$ of the response $\underset{\sim}{T}$ are the same. To see that, we apply the reflection $\underset{\sim}{S}$ with respect to the direction $\underset{\sim}{E}_1$ to both the material and the agency. We obtain obviously:

$$\underset{\sim}{D} \Rightarrow \bar{\underset{\sim}{D}} = \underset{\sim}{S}\underset{\sim}{D}\underset{\sim}{S}^t = \underset{\sim}{D} \tag{3}$$

The Principle of Isotropy of Space requires that:

$$\underset{\sim}{T} \Rightarrow \bar{\underset{\sim}{T}} = \underset{\sim}{S}\underset{\sim}{T}\underset{\sim}{S}^t \tag{4}$$

But, in view of (3), the agency is unchanged by the transformation $\underset{\sim}{S}$. Thus, the isotropy of the material requires that:

$$\bar{\underset{\sim}{T}} = \underset{\sim}{T} \tag{5}$$

Introducing (5) in (4), we obtain:

$$\underset{\sim}{S}\underset{\sim}{T}\underset{\sim}{S}^t = \underset{\sim}{T} \tag{6}$$

The relation (6) implies that $\underset{\sim}{E}_1$ is a principal direction of $\underset{\sim}{T}$; finally, the principal directions of $\underset{\sim}{T}$ and $\underset{\sim}{D}$ are the same.

Thus, the mechanical behavior of an isotropic material can be sketched by the following Figure, where the isotropic material is represented by a circle and the agency $\underset{\sim}{D}$ and the response $\underset{\sim}{T}$ by their associated ellipses, which admit the same principal axes.

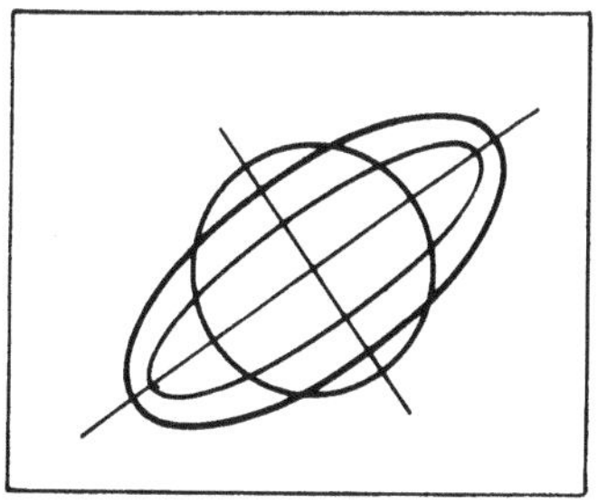

Isotropic material

3.2 Mathematical concept

The isotropy of the material implies that if we:

transform the body : $\underset{\sim}{\xi} \Rightarrow \underset{\sim}{Q}\underset{\sim}{\xi}\underset{\sim}{Q}^t = \bar{\underset{\sim}{\xi}}$

and keep the same agency : $\underset{\sim}{D} \Rightarrow \underset{\sim}{D}$

then the response is the same: $\underset{\sim}{T} \Rightarrow \underset{\sim}{T}$

Finally, we obtain:

$$\forall \underset{\sim}{Q} \in \mathcal{O} : \underset{\sim}{F}(\underset{\sim}{D}, \underset{\sim}{Q}\underset{\sim}{\xi}\underset{\sim}{Q}^t) = \underset{\sim}{F}(\underset{\sim}{D}, \underset{\sim}{\xi}) \tag{7}$$

It is easy to see that the condition (7) means that the structural tensor $\underset{\sim}{\xi}$ is an isotropic tensor:

$$\underset{\sim}{\xi} = \lambda \underset{\sim}{I}$$

Then,for isotropic materials, the constitutive equation (1) reduces to:

$$\underset{\sim}{T} = \underset{\sim}{F}(\underset{\sim}{D}, \lambda \underset{\sim}{I}) = \underset{\sim}{F}(\underset{\sim}{D}) \tag{8}$$

If we now apply the Principle of Isotropy of Space, we obtain:

$$\forall \underset{\sim}{Q} \in \mathcal{O} : \underset{\sim}{F}(\overline{\underset{\sim}{D}}) = \overline{\underset{\sim}{F}}(\underset{\sim}{D}) \tag{9}$$

which means that for isotropic materials, the constitutive function $\underset{\sim}{F}$ is isotropic with respect to the agency $\underset{\sim}{D}$.

4. ANISOTROPIC MATERIAL

4.1 Physical concept

A material is called anisotropic if when we transform the material by an arbitrary orthogonal transformation $\underset{\sim}{Q}$ and keep the same agency $\underset{\sim}{D}$, then, in general, the response $\underset{\sim}{T}^*$ is different from the response $\underset{\sim}{T}$ obtained before the transformation. This is sketched in the Figure below. The following question arises immediately: are there orthogonal transformations $\underset{\sim}{Q}$ such that the response $\underset{\sim}{T}$ is the same:

$$\exists\ \underset{\sim}{Q} \in \mathcal{O} : \underset{\sim}{T}^* = \underset{\sim}{T}\ ? \tag{10}$$

If only the transformations $(\underset{\sim}{I}, -\underset{\sim}{I})$ satisfy (10), then the material is called "general anisotropic material". If other orthogonal transformations $\underset{\sim}{Q}$ satisfy (10), then they constitute a subgroup $\mathcal{S}$ of the orthogonal group $\mathcal{O}$. The subgroup $\mathcal{S}$ is called the material symmetry group.

For example, the material symmetry group for a two-dimensional orthotropic material, with privileged directions $(\underset{\sim}{v}_1, \underset{\sim}{v}_2)$, is given by:

$$S = \left\{ \pm \underset{\sim}{I}, \underset{\sim}{S}_1, \underset{\sim}{S}_2 \right\} \tag{11}$$

where $\underset{\sim}{S}_1$ and $\underset{\sim}{S}_2$ are the reflections with respect to the directions $\underset{\sim}{v}_1$ and $\underset{\sim}{v}_2$.

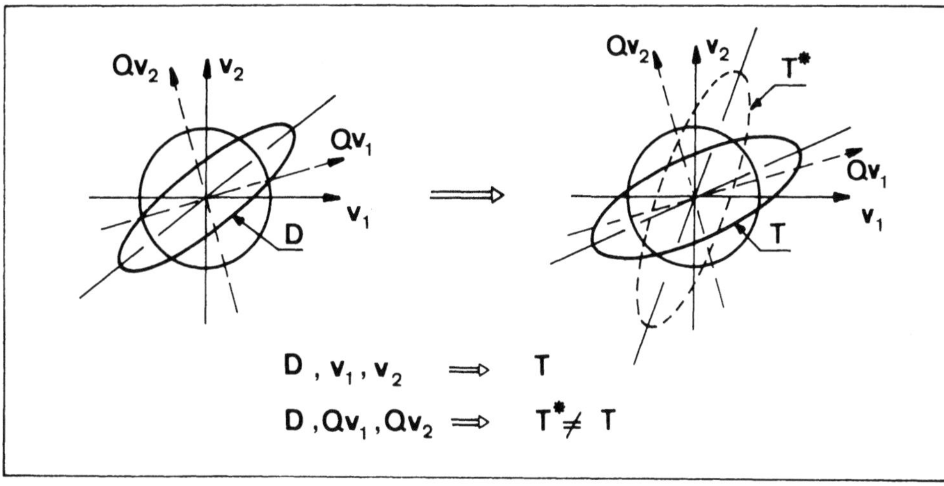

Anisotropic material

4.2 Mathematical concept

Suppose that S, the symmetry group of the material, is also the "invariance group" of the structural tensor $\underset{\sim}{\xi}$:

$$\underset{\sim}{Q} \in S \iff \bar{\underset{\sim}{\xi}} = \underset{\sim}{Q}\underset{\sim}{\xi}\underset{\sim}{Q}^t = \underset{\sim}{\xi} \tag{12}$$

It follows that:

$$\underset{\sim}{Q} \in S \Rightarrow \underset{\sim}{F}(\bar{\underset{\sim}{D}}, \bar{\underset{\sim}{\xi}}) = \underset{\sim}{F}(\bar{\underset{\sim}{D}}, \underset{\sim}{\xi}) \tag{13}$$

The Principle of Isotropy of Space requires that:

$$\forall \underset{\sim}{Q} \in O : \underset{\sim}{F}(\bar{\underset{\sim}{D}}, \bar{\underset{\sim}{\xi}}) = \bar{\underset{\sim}{F}}(\underset{\sim}{D}, \underset{\sim}{\xi}) \tag{14}$$

Employing (14) in (13), we arrive at:

$$\underset{\sim}{Q} \in S \Rightarrow \underset{\sim}{F}(\bar{\underset{\sim}{D}}, \underset{\sim}{\xi}) = \bar{\underset{\sim}{F}}(\underset{\sim}{D}, \underset{\sim}{\xi}) \tag{15}$$

Conversely, suppose that the relation (15) holds for a given transformation $\mathbf{Q}^*$. Employing the Principle of Isotropy of Space (14), we obtain:

$$\exists\ \mathbf{Q}^* \in \mathcal{O} : \mathbf{F}(\bar{\mathbf{D}}, \boldsymbol{\xi}) = \mathbf{F}(\bar{\mathbf{D}}, \bar{\boldsymbol{\xi}}) \tag{16}$$

for an arbitrary agency $\bar{\mathbf{D}}$. It is easy to see that for a general function $\mathbf{F}$, the condition (16) leads to:

$$\bar{\boldsymbol{\xi}} = \boldsymbol{\xi} \quad \Rightarrow \quad \mathbf{Q}^* \in \mathcal{S} \tag{17}$$

Finally, considering the relations (13), (16) and (17), we obtain the following converse implication:

$$\mathbf{Q} \in \mathcal{S} \iff \mathbf{F}(\bar{\mathbf{D}}, \boldsymbol{\xi}) = \bar{\mathbf{F}}(\mathbf{D}, \boldsymbol{\xi}) \tag{18}$$

The implication (18) means that, with respect to $\mathbf{D}$, the function $\mathbf{F}$ is invariant in the group $\mathcal{S}$ and only in this group. Thus, the function $\mathbf{F}$ is anisotropic with respect to $\mathbf{D}$, the type of anisotropy being characteristized by the invariance group $\mathcal{S}$ of the structural tensor $\boldsymbol{\xi}$. This leads to the following fundamental result :

$\mathbf{T} = \mathbf{F}(\mathbf{D}, \boldsymbol{\xi})$ + Isotropy of Space
⇓
$\mathbf{F}$ isotropic / $(\mathbf{D}, \boldsymbol{\xi})$
⇓
$\mathbf{F}$ anisotropic / $(\mathbf{D})$
the material symmetry group S
being the invariance group of the structural tensor $\boldsymbol{\xi}$

Finally, for an anisotropic material, the invariance group of the structural tensor $\boldsymbol{\xi}$ involved in the constitutive equation (1) is identical to the group of transformations characterizing the material symmetries of the body. The problem now is to find a structural tensor for each type of anistropy.

In the constitutive equation (1) for the considered two-dimensional anisotropic material, we supposed that the internal structure can be taken into account by a single symmetric second order tensor ; this is indeed the case for a two-dimensional orthotropic material (cf. Section 5 below). For other types of anisotropy and in the three-dimensional case, several structural tensors may be necessary. For example, the internal structure of a three-dimensional orthotropic material is represented by three symmetric second order tensors ([3] ; see also Chapter 3, Section 3-4) ; in this case, the group S is the invariance group of the set of structural tensors.

5. ORTHOTROPIC MATERIALS

Consider a two-dimensional orthotropic material, with privileged direction $(\underset{\sim}{v}_1, \underset{\sim}{v}_2)$. The material symmetry group S is given by (11).We introduce the following tensor $\underset{\sim}{M}$:

$$\underset{\sim}{M} = \underset{\sim}{v}_1 \otimes \underset{\sim}{v}_1 \qquad (19)$$

The tensor $\underset{\sim}{M}$ is a very particular symmetric second order tensor, which transforms the unit vectors $\underset{\sim}{n}_\alpha$ into their projections upon the $\underset{\sim}{v}_1$ axis:

$$\underset{\sim}{n}_\alpha \left| \begin{matrix} \cos\alpha \\ \sin\alpha \end{matrix} \right. \quad \overset{\underset{\sim}{M}}{=>} \quad \underset{\sim}{d}_\alpha \left| \begin{matrix} \cos\alpha \\ 0 \end{matrix} \right.$$

M : 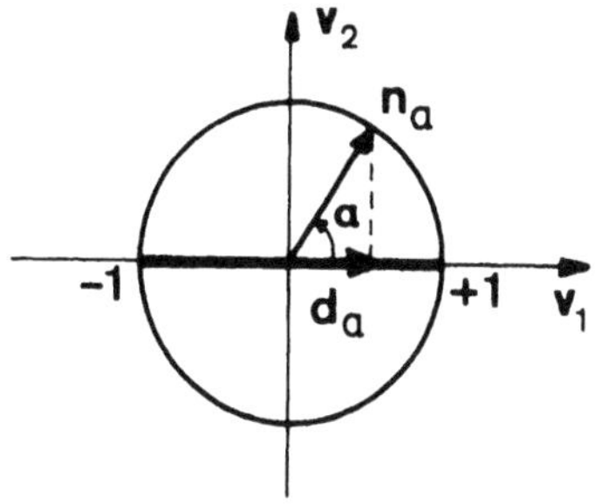

Thus, the tensor $\underset{\sim}{M}$ transforms the unit circle into the segment [-1, +1] lying on the $\underset{\sim}{v}_1$ axis. The principal directions of $\underset{\sim}{M}$ are the privileged directions $(\underset{\sim}{v}_1, \underset{\sim}{v}_2)$ of the orthotropic material. In the $(\underset{\sim}{v}_1, \underset{\sim}{v}_2)$ axes and in the $(\underset{\sim}{x}, \underset{\sim}{y})$ axes deduced from $(\underset{\sim}{v}_1, \underset{\sim}{v}_2)$ by a rotation of angle θ, the tensor $\underset{\sim}{M}$ is expressed by :

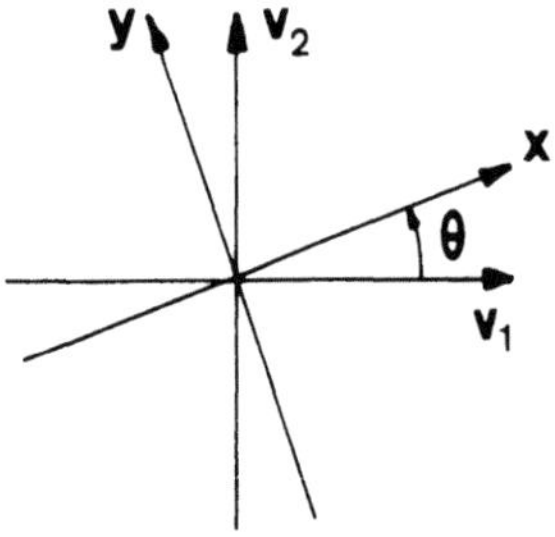

$$\underset{\sim}{M} = \begin{vmatrix} 1 & 0 \\ 0 & 0 \end{vmatrix}_{(\underset{\sim}{v}_1, \underset{\sim}{v}_2)} = \begin{vmatrix} \cos^2\theta & \sin\theta\cos\theta \\ \sin\theta\cos\theta & \sin^2\theta \end{vmatrix}_{(\underset{\sim}{x}, \underset{\sim}{y})} \tag{20}$$

It is easy to see that the tensor $\underset{\sim}{M}$ verifies the following converse implication:

$$\underset{\sim}{Q} \in \left\{ \pm \underset{\sim}{I},\ \underset{\sim}{S}_1,\ \underset{\sim}{S}_2 \right\} \iff \underset{\sim}{Q}\underset{\sim}{M}\underset{\sim}{Q}^t = \underset{\sim}{M} \tag{21}$$

Thus, the invariance group of $\underset{\sim}{M}$ is the material symmetry group (11), i.e. $\underset{\sim}{M}$ is the structural tensor for a two-dimensional orthotropic material.

We now apply the fundamental result of Section 4. Consider the constitutive equation:

$$\underset{\sim}{T} = \underset{\sim}{F}(\underset{\sim}{D}, \underset{\sim}{M}) \tag{22}$$

The Principle of Isotropy of Space implies that the function $\underset{\sim}{F}$ is isotropic with respect to $(\underset{\sim}{D}, \underset{\sim}{M})$, thus anisotropic with respect to $(\underset{\sim}{D})$, the type of anisotropy being characterized by the invariance group of the structural tensor $\underset{\sim}{M}$. Finally, the function $\underset{\sim}{F}$ is orthotropic with respect to $\underset{\sim}{D}$:

$$\text{Isotropy of Space} \Rightarrow \forall \underset{\sim}{Q} \in \mathcal{O} : \underset{\sim}{F}(\underset{\sim}{\bar{D}}, \underset{\sim}{\bar{M}}) = \underset{\sim}{\bar{F}}(\underset{\sim}{D}, \underset{\sim}{M}) \tag{23}$$

$$\downarrow$$

$$\forall \underset{\sim}{Q} \in \mathcal{S} : \underset{\sim}{F}(\underset{\sim}{\bar{D}}, \underset{\sim}{M}) = \underset{\sim}{\bar{F}}(\underset{\sim}{D}, \underset{\sim}{M}) \tag{24}$$

$$\downarrow$$

$$\underset{\sim}{F} \text{ orthotropic } / \ (\underset{\sim}{D})$$

Finally, for orthotropic materials, the constitutive equation (1) takes the particular form (22) which is subjected to the invariance condidition (23) given by the Principle of Isotropy of Space. Equations (22) and (23) constitute the general form of a two-dimensional orthotropic constitutive law relating two symmetric second order tensors.

6. REPRESENTATION OF THE FUNCTION $\underset{\sim}{F}$

6.1 Introduction

In Mechanics of Anisotropic Solids, it appears necessary to establish invariant forms of the constitutive laws, i.e. forms which verify automatically the material symmetries of the considered body and are thus automatically invariant under the material symmetry group. This is not the case of the constitutive equation (22), because it is subjected to the invariance condition (23). It is also necessary to derive the essential variables, i.e. the type and the minimal number of independent scalar variables which must appear in the constitutive law and which constitute the set of independent anisotropic mechanical variables.

The theorems of representations for anisotropic tensor functions allow canonical forms which fulfil these two requirements to be developed.

6.2 Theorem of representation

The theorem of representation for orthotropic tensor functions ([3]) indicates that the relation (22), together with the invariance condition (23), admits the following irreducible canonical form:

$$\left.\begin{aligned} \underset{\sim}{T} &= \alpha_o \underset{\sim}{I} + \alpha_1 \underset{\sim}{M} + \alpha_2 \underset{\sim}{D} \\ \alpha_i &= \alpha_i(\mathrm{tr}\underset{\sim}{D},\ \mathrm{tr}\underset{\sim}{D}^2,\ \mathrm{tr}\underset{\sim}{M}\underset{\sim}{D}) \end{aligned}\right\} \qquad (25)$$

In the representation (25), the response $\underset{\sim}{T}$ appears as a linear combination of the following three well determined tensors:

$$\underset{\sim}{I},\ \underset{\sim}{M},\ \underset{\sim}{D} \qquad (26)$$

which constitute the generating set of the representation. More precisely the generators (26) are known S - invariant symmetric second order tensors, where S is the material symmetry group (11) for two-dimensional orthotropy. The coefficients α_i are arbitrary scalar-valued functions, of arguments:

$$\mathrm{tr}D,\ \mathrm{tr}D^2,\ \mathrm{tr}MD \tag{27}$$

which are called "orthotropic invariants" of D. They constitute the "functional basis" of D for the considered two-dimensional orthotropic group of transformations. Precise definitions of the terms introduced above are given in Chapter 3.

6.3 Orthotropic invariants of D

We introduce the expressions of the agency D in its principal frame $(E_1,\ E_2)$ and in the privileged frame $(v_1,\ v_2)$ of the material:

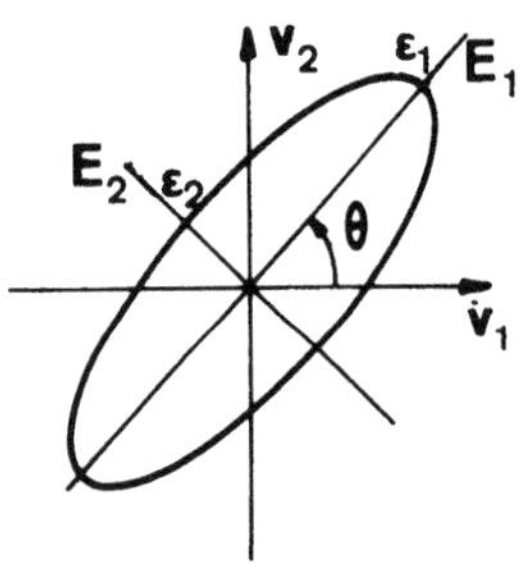

$$D = \begin{vmatrix} \varepsilon_1 & 0 \\ 0 & \varepsilon_2 \end{vmatrix}_{(E_1, E_2)} = \begin{vmatrix} \varepsilon_{11} & \varepsilon_{12} \\ \varepsilon_{12} & \varepsilon_{22} \end{vmatrix}_{(v_1,\ v_2)} \tag{28}$$

The first two invariants (27) admit the following expressions :

$$\left.\begin{aligned} \mathrm{tr}D &= \varepsilon_1 + \varepsilon_2 = \varepsilon_{11} + \varepsilon_{22} = \mathrm{tr}(QDQ^t)\ ,\ \forall Q \in \mathcal{O} \\ \mathrm{tr}D^2 &= \varepsilon_1^2 + \varepsilon_2^2 = \varepsilon_{11} + \varepsilon_{22} + 2\varepsilon_{12}^2 = \mathrm{tr}(QDQ^t)^2\ ,\ \forall Q \in \mathcal{O} \end{aligned}\right\} \tag{29}$$

Thus, $\mathrm{tr}D$ and $\mathrm{tr}D^2$ admit the same values for arbitrary orthogonal transformations of D: they are called isotropic invariants of D. As they are invariant under the full orthogonal group $\mathcal{O}$, they are invariant under the group $\mathcal{S}$ a fortiori ; thus, they constitute also orthotropic invariants of D.

The expressions of the last invariant (27) are given by:

$$\operatorname{tr}\underset{\sim}{M}\underset{\sim}{D} = \varepsilon_{11} = \varepsilon_1\cos^2\theta + \varepsilon_2\sin^2\theta \quad ; \quad \theta = (\underset{\sim}{v}_1, \underset{\sim}{E}_1) \tag{30}$$

Thus, $\operatorname{tr}\underset{\sim}{M}\underset{\sim}{D}$ is equal to the component ε_{11} of $\underset{\sim}{D}$ in the privileged frame $(\underset{\sim}{v}_1, \underset{\sim}{v}_2)$. If we transform both $\underset{\sim}{D}$ <u>and</u> $\underset{\sim}{M}$ by an arbitrary orthogonal transformation $\underset{\sim}{Q}$, the value of this component remains constant: $\operatorname{tr}\underset{\sim}{M}\underset{\sim}{D}$ is an isotropic invariant of the set $(\underset{\sim}{D}, \underset{\sim}{M})$. If we transform $\underset{\sim}{D}$, but <u>not</u> $\underset{\sim}{M}$, the value of the component ε_{11} is modified, except if, and only if, the transformation $\underset{\sim}{Q}$ belongs to the orthotropic group S given by (11) ; thus, $\operatorname{tr}\underset{\sim}{M}\underset{\sim}{D}$ is an orthotropic invariant of $\underset{\sim}{D}$. Finally, we have:

$$\forall \underset{\sim}{Q} \in O \supset S \;:\; \operatorname{tr}\overline{\underset{\sim}{D}} = \operatorname{tr}\underset{\sim}{D} \quad ; \quad \operatorname{tr}\overline{\underset{\sim}{D}}^2 = \operatorname{tr}\underset{\sim}{D}^2 \quad ; \quad \operatorname{tr}\overline{\underset{\sim}{M}\underset{\sim}{D}} = \operatorname{tr}\underset{\sim}{M}\underset{\sim}{D} \tag{31}$$

$$\forall \underset{\sim}{Q} \in S \;:\; \operatorname{tr}\underset{\sim}{M}\overline{\underset{\sim}{D}} = \operatorname{tr}\underset{\sim}{M}\underset{\sim}{D} \tag{32}$$

6.4 <u>Fundamental result</u>

The representation (25) verifies automatically the consequence (2) of the Principle of Isotropy of Space. The proof of this property is obtained in considering that, in view of (31), we have:

$$\forall \underset{\sim}{Q} \in O \;:\; \alpha_i(\operatorname{tr}\overline{\underset{\sim}{D}}, \operatorname{tr}\overline{\underset{\sim}{D}}^2, \operatorname{tr}\overline{\underset{\sim}{M}\underset{\sim}{D}}) = \alpha_i(\operatorname{tr}\underset{\sim}{D}, \operatorname{tr}\underset{\sim}{D}^2, \operatorname{tr}\underset{\sim}{M}\underset{\sim}{D}) \tag{33}$$

$$\left.\begin{aligned} \forall \underset{\sim}{Q} \in O : \underset{\sim}{F}(\overline{\underset{\sim}{D}}, \overline{\underset{\sim}{M}}) &= \alpha_o\overline{\underset{\sim}{I}} + \alpha_1\overline{\underset{\sim}{M}} + \alpha_2\overline{\underset{\sim}{D}} \\ &= \underset{\sim}{Q}(\alpha_o\underset{\sim}{I} + \alpha_1\underset{\sim}{M} + \alpha_2\underset{\sim}{D})\underset{\sim}{Q}^t \\ &= \overline{\underset{\sim}{F}}(\underset{\sim}{D}, \underset{\sim}{M}) \end{aligned}\right\} \tag{34}$$

Relation (34) signifies that the representation (25) is automatically isotropic with respect to the set of arguments $(\underset{\sim}{D}, \underset{\sim}{M})$; thus, it is automatically orthotropic with respect to the mechanical agency $\underset{\sim}{D}$. The essential variables are the arguments $(\operatorname{tr}\underset{\sim}{D}, \operatorname{tr}\underset{\sim}{D}^2, \operatorname{tr}\underset{\sim}{M}\underset{\sim}{D})$ of the arbitrary scalar-valued functions α_i ; they are the independent variables which must be observed in experimental investigations, in order to be able to specify the particular forms of the functions α_i, for each type of material and each type of mechanical behavior. Finally, both requirements mentioned in Section 6.1 for the invariant formulation of the constitutive equation (22) for a two-dimensional orthotropic material are fulfilled by the representation (25).

In representation (25):

- The structural tensor $\underset{\sim}{M}$ with principal directions $(\underset{\sim}{v}_1, \underset{\sim}{v}_2)$ characterizes the orthotropy ; i.e. it specifies the privileged directions of the material.
- The invariant $\mathrm{tr}\underset{\sim}{M}\underset{\sim}{D} = \varepsilon_1\cos^2\theta + \varepsilon_2\sin^2\theta$ characterizes the "degree" of the orthotropy, i.e. it takes into account the influence of the orientation θ of the agency $\underset{\sim}{D}$ with respect to the privileged directions $(\underset{\sim}{v}_1, \underset{\sim}{v}_2)$ onto the mechanical response $\underset{\sim}{T}$ of the orthotropic material.

6.5 Characteristic property

If the privileged directions $(\underset{\sim}{v}_1, \underset{\sim}{v}_2)$ are not the principal directions of the agency $\underset{\sim}{D}$, then the principal directions of $\underset{\sim}{D}$ and of the response $\underset{\sim}{T}$ do not coincide in general. This property is obvious, in view of the linear combination:

$$\underset{\sim}{T} = \alpha_0\underset{\sim}{I} + \alpha_1\underset{\sim}{M} + \alpha_2\underset{\sim}{D} \tag{35}$$

It is also clear in the sketch below :

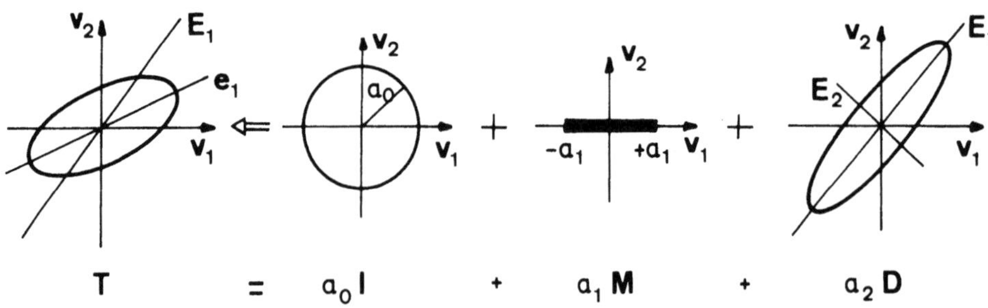

It can be proved in considering the expression of (35) in the principal frame $(\underset{\sim}{E}_1, \underset{\sim}{E}_2)$ of $\underset{\sim}{D}$:

$$\begin{vmatrix} \sigma_{11} & \sigma_{12} \\ \sigma_{12} & \sigma_{22} \end{vmatrix} = \alpha_o \begin{vmatrix} 1 & 0 \\ 0 & 1 \end{vmatrix} + \alpha_1 \begin{vmatrix} \cos^2\theta & \sin\theta\cos\theta \\ \sin\theta\cos\theta & \sin^2\theta \end{vmatrix} + \alpha_2 \begin{vmatrix} \varepsilon_1 & 0 \\ 0 & \varepsilon_2 \end{vmatrix} \quad (36)$$

where :

$$\theta = (\underset{\sim}{v}_1, \underset{\sim}{E}_1)$$

If we exclude the special case $\alpha_1 \equiv 0$, then the component:

$$\sigma_{12} = \alpha_1 \sin\theta\cos\theta \quad (37)$$

of $\underset{\sim}{T}$ vanishes if and only if: $\theta = 0, \pi/2$.

6.6 Structural interpretation of $\underset{\sim}{M}$

In Section 5, we remarked that the tensor $\underset{\sim}{M}$ transforms the unit circle into the segment [-1, +1] on the $\underset{\sim}{v}_1$ axis. The unit circle represents the isotropic distribution of the material internal structure (crystallographic axes, grains, particles, crystals, fissuration, cracks, cavities, etc...). The segment [-1, +1] represents the ideal orthotropic distribution of the internal structure, parallel to the v_1 axis. Thus, $\underset{\sim}{M}$ is the tensor of an ideal orthotropic distribution. As real material orthotropic structures are seldom ideal, the following question arises immediately: does the representation (25), in which the involved structural tensor is only the ideal tensor $\underset{\sim}{M}$, allow an arbitrary orthotropic distribution to be taken into account ? First, we show that the representation (25) takes into account an elliptic orthotropic distribution. From (25), we can construct the structural tensor:

$$\underset{\sim}{M}_2 = \underset{\sim}{I} - \underset{\sim}{M} = \underset{\sim}{v}_2 \otimes \underset{\sim}{v}_2 \quad ; \quad M_2 = \begin{vmatrix} 0 & 0 \\ 0 & 1 \end{vmatrix}_{(\underset{\sim}{v}_1, \underset{\sim}{v}_2)} \quad (38)$$

The tensor $\underset{\sim}{M}_2$ is a very particular symmetric second order tensor, which transforms the unit vectors $\underset{\sim}{n}_\alpha$ into their projections upon the $\underset{\sim}{v}_2$ axis:

$$\underset{\sim}{n}_\alpha \left| \begin{matrix}\cos\alpha\\ \\ \sin\alpha\end{matrix}\right. \quad \overset{\underset{\sim}{M}_2}{\Rightarrow} \quad \underset{\sim}{f}_\alpha \left|\begin{matrix}0\\ \\ \sin\alpha\end{matrix}\right.$$

M_2:

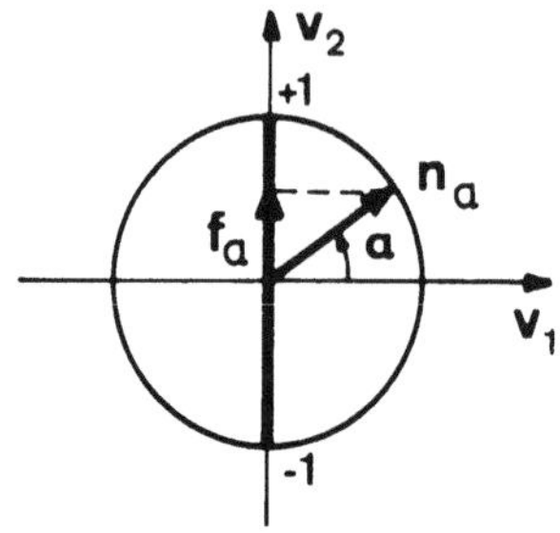

Thus, the tensor $\underset{\sim}{M}_2$ transforms the unit circle into the segment [-1, +1] lying on the $\underset{\sim}{v}_2$ axis. Consider now the structural tensor:

$$\underset{\sim}{\xi} = a\underset{\sim}{M} + b\underset{\sim}{M}_2 = b\underset{\sim}{I} + (a-b)\underset{\sim}{M} \quad ; \quad \underset{\sim}{\xi} = \begin{vmatrix} a & 0 \\ 0 & b \end{vmatrix}_{(\underset{\sim}{v}_1,\ \underset{\sim}{v}_2)} \tag{39}$$

The tensor $\underset{\sim}{\xi}$, which is involved in the representation (25), transforms the unit circle into the ellipse with principal directions $(\underset{\sim}{v}_1, \underset{\sim}{v}_2)$ and with semi-axes a and b.

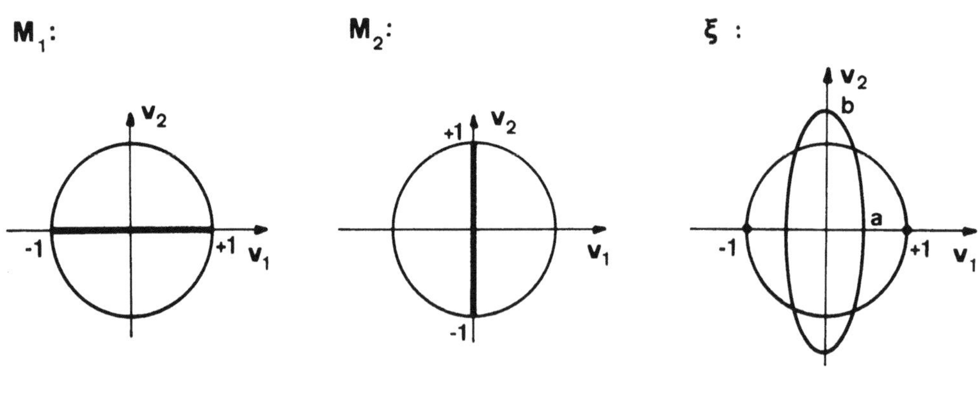

Thus, the tensor $\underset{\sim}{\xi}$ is the structural tensor for an elliptic orthotropic distribution.

In fact, representation (25) can account for an arbitrary orthotropic distribution. Indeed, the representation involves the arguments of the scalar-valued functions α_i, which are the invariants:

$$\mathrm{tr}\underset{\sim}{D} = \varepsilon_1 + \varepsilon_2 \quad ; \quad \mathrm{tr}\underset{\sim}{D}^2 = \varepsilon_1^2 + \varepsilon_2^2 \quad ; \quad \mathrm{tr}\underset{\sim}{M}\underset{\sim}{D} = \varepsilon_1 \cos^2\theta + \varepsilon_2 \sin^2\theta \tag{40}$$

where ε_1 and ε_2 are the principal values of $\underset{\sim}{D}$ and θ is the angle $(\underset{\sim}{v}_1, \underset{\sim}{E}_1)$. Let us suppose that the values of the three invariants (40) are given. The expressions of $\mathrm{tr}\underset{\sim}{D}$ and $\mathrm{tr}\underset{\sim}{D}^2$ allow the determination of the values of ε_1 and ε_2. Introducing these values in the expression of $\mathrm{tr}\underset{\sim}{M}\underset{\sim}{D}$, one obtains a single value for θ in the interval $[0, \pi/2]$, the three other solutions in the interval $[0, 2\pi]$ being obtained by the symmetries with respect to the $\underset{\sim}{v}_1$ and $\underset{\sim}{v}_2$ directions. Now, for a given arbitrary orthotropic distribution in the interval $[0, \pi/2]$, the distribution in the interval $[0, 2\pi]$ is obtained by the same symmetries. Finally, from the arbitrary scalar-valued functions α_i, arbitrary scalar-valued orthotropic functions $a(\theta)$ and $b(\theta)$ can be constructed. Thus, from representation (25), the following structural tensor can be constructed:

$$\underset{\sim}{\xi} = a(\theta)\underset{\sim}{M} + b(\theta)\underset{\sim}{M}_2 = b(\theta)\underset{\sim}{I} + [a(\theta) - b(\theta)]\underset{\sim}{M} \tag{41}$$

which is the polar representation for an arbitrary orthotropic structural tensor.

7. CONCLUSIONS

The theory of representations for tensor functions is a powerful and efficient tool for the invariant formulation of non-linear anisotropic constitutive equations. This theory specifies the invariant tensorial generators which are involved in the general form of a non-linear constitutive equation, as well as the type and the number of the independent scalar variables, which are mixt invariants of the mechanical arguments and of the structural tensors.

In this elementary introduction to the invariant formulation of anisotropic constitutive equations, we considered the very simple case of a two-dimensional equation specifying the stress tensor as a function of a single mechanical argument ; in this case, the invariant general form of the constitutive equation is given by the representation (25). In the three-dimensional case, representations of tensor functions of an arbitrary number of mechanical arguments are given in Chapter 3 for general anisotropic, orthotropic and transversely isotropic materials.

In the invariant general forms of the anisotropic non-linear constitutive equations so obtained, a certain number of arbitrary scalar-valued functions appear, the arguments of which are the well defined essential variables. It is the particular forms of these arbitrary functions which specify the invariant forms of the constitutive equations for each type of anisotropic material. The specification of the arbitrary scalar-valued functions needs coherent and well-organized experimental investigations, which really allow the observation and measurement of the evolution of the essential variables. The experimental study of the mechanical behavior of anisotropic materials involves specific difficulties, which are mentioned in Chapters 5 and 7.

REFERENCES

1. BOEHLER, J.P., Anisotropie et Comportement Rhéologique des Matériaux, Conférence Générale, *19ème Colloque du Groupe Français de Rhéologie*, Novembre 1984 (in print).

2. TRUESDELL, C. and NOLL, W., The Non-Linear Field Theory of Mechanics, *Handbuch der Physik*, III/3, Springer-Verlag, Berlin, 1965.

3. BOEHLER, J.P., Lois de Comportement Anisotrope des Milieux Continus, *Journal de Mécanique*, 17,2 (1978) : 153-190.

Chapter 3

REPRESENTATIONS FOR ISOTROPIC AND ANISOTROPIC NON-POLYNOMIAL TENSOR FUNCTIONS

J.P. Boehler
University of Grenoble, France

1. INTRODUCTION

1.1 Invariant scalar functions and form-invariant tensor functions

Material symmetries of a continuum impose definite restrictions on the form of constitutive relations. The restrictions are specified in the representations of isotropic and anisotropic tensor functions and indicate the type and the number of independent variables involved in a constitutive relation. Thus, in a properly written constitutive equation, the material symmetries are automatically verified.

In this Chapter, "tensors" and "vectors" mean second order tensors and vectors in a three-dimensional space. We restrict ourselves to constitutive functions of the arguments:

$$\underset{\sim}{A}_1,\ \underset{\sim}{A}_2,\ \ldots\underset{\sim}{A}_a,\ \underset{\sim}{W}_1,\ \underset{\sim}{W}_2,\ \ldots\underset{\sim}{W}_b,\ \underset{\sim}{v}_1,\ \underset{\sim}{v}_2,\ \ldots\underset{\sim}{v}_c \tag{1}$$

where the $\underset{\sim}{A}_i$, $\underset{\sim}{W}_j$ and $\underset{\sim}{v}_k$ are respectively an arbitrary number of symmetric tensors, skew-symmetric tensors and vectors.

Consider first a constitutive law specifying a scalar A :

$$A = f(\underset{\sim}{A}_i, \underset{\sim}{W}_j, \underset{\sim}{v}_k) \tag{2}$$

If S is the group of transformations characterizing the material symmetries of the medium, the function f is subjected to the conditions:

$$\forall \underset{\sim}{Q} \in S : f(\underset{\sim}{Q}\underset{\sim}{A}_i\underset{\sim}{Q}^t, \underset{\sim}{Q}\underset{\sim}{W}_j\underset{\sim}{Q}^t, \underset{\sim}{Q}\underset{\sim}{v}_k) = f(\underset{\sim}{A}_i, \underset{\sim}{W}_j, \underset{\sim}{v}_k) \tag{3}$$

The scalar A is called scalar invariant under the group S.

Consider now a constitutive equation specifying a symmetric second order tensor $\underset{\sim}{T}$:

$$\underset{\sim}{T} = \underset{\sim}{F}(\underset{\sim}{A}_i, \underset{\sim}{W}_j, \underset{\sim}{v}_k) \tag{4}$$

Similarly, the function $\underset{\sim}{F}$ is subjected to the conditions:

$$\forall \underset{\sim}{Q} \in S : \underset{\sim}{F}(\underset{\sim}{Q}\underset{\sim}{A}_i\underset{\sim}{Q}^t, \underset{\sim}{Q}\underset{\sim}{W}_j\underset{\sim}{Q}^t, \underset{\sim}{Q}\underset{\sim}{v}_k) = \underset{\sim}{Q}\underset{\sim}{F}(\underset{\sim}{A}_i, \underset{\sim}{W}_j, \underset{\sim}{v}_k)\underset{\sim}{Q}^t \tag{5}$$

The function $\underset{\sim}{F}$ is then called form-invariant under the group S and $\underset{\sim}{T}$ is a tensorial invariant.

1.2 Representations for polynomial scalar and tensor functions

In this Section, we suppose that the function f and the components in a reference frame of the function $\underset{\sim}{F}$ are polynomials in the components of the arguments (1). The values A and $\underset{\sim}{T}$ are then polynomial invariants under the group S.

The problem of the representation for the scalar-valued function f is to determine a basic set of polynomial scalar invariants $(I_1, I_2, \ldots I_p)$, such that an arbitrary polynomial scalar invariant of the same arguments can be expressed as a polynomial in the basic invariants. Such a set is called "integrity basis" for the considered list of arguments (1) and the group of transformations S. An integrity basis is termed irreducible if none of its subsets constitutes a complete representation by itself, i.e.

if no element of it can be expressed as a polynomial in the remainder. Even if an integrity basis is irreducible, polynomial relations may exist between its elements; but these relations do not enable any one of it to be expressed as a polynomial of the others. Such relations are called syzygies.

The problem of the representation for the tensor-valued function $\underset{\sim}{F}$ is to determine a generating set of tensors $\underset{\sim}{G}_i$, thus symmetric second order tensors which are invariant under the group S, such that the tensor $\underset{\sim}{T}$ can be expressed as the linear combination:

$$\left.\begin{aligned} &\underset{\sim}{T} = \alpha_i \underset{\sim}{G}_i \\ &\alpha_i = \alpha_i(I_1,\ I_2,\ \ldots I_p) \end{aligned}\right\} \qquad (6)$$

where the α_i are arbitrary polynomial scalar functions of the invariants of the integrity basis. The values of the functions α_i are thus polynomial scalar invariants of the arguments (1) under the group S. The representation (6) is irreducible if the integrity basis is irreducible and if none of the generators $\underset{\sim}{G}_i$ can be expressed as a linear combination of the others, with coefficients α_i.

Explicit representations for isotropic and anisotropic polynomial scalar-valued and tensor-valued functions are presented in Chapters 8 and 9 by Professor SPENCER.

1.3 Representations for non-polynomial scalar and tensor functions

In this Section, we suppose that the function f and the components of the function $\underset{\sim}{F}$ are general functions, not necessarily polynomials, of the components, in a reference frame, of the arguments (1). The values A and $\underset{\sim}{T}$ are then (general) invariants under the group S.

The representation for the scalar-valued function f is a basic set of scalar invariants $(I_1,\ I_2,\ \ldots I_n)$, such that an arbitrary scalar invariant of the arguments (1) can be expressed as a single-valued function of the basic invariants. Such a set is called "functional basis" for the considered list of arguments and the group S. A functional basis is termed irreducible if none of its elements can be expressed as a single-valued function of the others. Even if a functional basis is irreducible, functional relations may exist between its elements; but these relations do not enable any one of it to be expressed as a single-valued function of the remainder. Although they represent all functions, irreducible functional bases contain, in general, fewer terms than do the corresponding integrity bases, because here reductions are obtained with the help of general functions and not only polynomials.

A criterion characterizing functional bases is given by WINEMAN and PIPKIN [1]. We call equivalent set, under the group S, of the ordered list of arguments (1) any ordered list $(\bar{\underset{\sim}{A}}_1, \bar{\underset{\sim}{A}}_2, \ldots \bar{\underset{\sim}{A}}_a, \bar{\underset{\sim}{W}}_1, \bar{\underset{\sim}{W}}_2, \ldots \bar{\underset{\sim}{W}}_b, \bar{\underset{\sim}{v}}_1, \bar{\underset{\sim}{v}}_2, \ldots \bar{\underset{\sim}{v}}_c)$ whose elements are the transformed, by a same orthogonal transformation, of the corresponding elements of the list (1). The set of lists equivalent to a given list (1) is called the orbit of this list in the group S. A functional basis characterizes the orbits: a set of invariants constitutes a functional basis if, and only if, their values are the same for two lists of arguments lying on the same orbit and if, and only if, at least one of them, called orbit-seperator, takes different values for two lists of arguments lying on two different orbits.

The representation for the tensor-valued function $\underset{\sim}{F}$ admits the same form as in the polynomial case:

$$\left.\begin{aligned} \underset{\sim}{T} &= \alpha_i \underset{\sim}{G}_i \\ \alpha_i &= \alpha_i(I_1, I_2, \ldots I_n) \end{aligned}\right\} \qquad (7)$$

but, here, the scalar valued-functions α_i are general functions of the invariants of the functional basis of the arguments (1). The representation (7) is irreducible if the functional basis is irreducible and if none of the generators $\underset{\sim}{G}_i$ can be expressed as a linear combination of the others, formed with general functions α_i. As is the case for the bases for scalar functions, irreducible generating sets for general tensor functions contain, in general, fewer elements than do the corresponding generating sets for polynomial tensor functions, because here reductions are obtained with the help of linear combinations, the coefficients of which are general scalar-valued functions and not only polynomials.

WINEMAN and PIPKIN [1,2] showed that a representation for the polynomial case constitutes a complete representation for the non-polynomial case as well. The scalar functions α_i are then not necessarily polynomials. However, in the non-polynomial case these representations are not irreducible in general (SPENCER [3], WANG [5]). Conversely, irreducible representations for non-polynomial functions are also representations for polynomial functions, but they are not necessarily polynomials.

Finally, representations for general functions present a double advantage : on one hand, they allow restrictive assumptions for the considered functions to be avoided ; on the other hand, they contain, in general, a fewer number of basic elements and thus allow a more concise formulation for the considered constitutive equations to be obtained.

2. REPRESENTATIONS FOR ISOTROPIC SCALAR AND TENSOR FUNCTIONS

In this Section, we suppose that the functions f and $\underset{\sim}{F}$ are isotropic, i.e. the relations (3) and (5) are verified for all orthogonal transformations $\underset{\sim}{Q}$. The values A and $\underset{\sim}{T}$ are then invariant under the full orthogonal group O.

For polynomial isotropic functions of the arguments (1), the integrity bases for the scalar functions f have been established by SMITH [7] and the generating sets for the tensor functions $\underset{\sim}{F}$ by SPENCER [3]. The lists of isotropic invariants and tensor generators are too long to be reproduced here. The methods used to derive them are presented in Chapter 8 by Professor SPENCER.

The representations for general isotropic scalar-valued and tensor-valued functions of arguments (1) were studied separately by WANG [4-6] and by SMITH [8,9]. For the construction of the functional bases, WANG applies the characteristic criterion presented in Section 1.3. In order to establish the representations for the tensor functions $\underset{\sim}{F}$, WANG decomposes the list of arguments (1) into independent sublists. Irreducible generating sets associated with each sublist are obtained by geometrical reasoning. The irreducible generating set associated with the complete list (1) is then the union of the generating sets associated with all independent sublists.

The procedure used by SMITH in order to construct the functional bases for general isotropic scalar functions f constitutes an extension of the method developed by RIVLIN and ERICKSEN in [10], applied to a particular case of the arguments (1). The method consists in establishing a set of basic invariants such that once the values of these invariants are known, the components of the arguments (1) can be determined uniquely in a reference frame fully specified beforehand (orientation and sense of the coordinates are specified). Such a set of invariants constitutes a functional basis since any invariant is uniquely expressed in terms of these components. The required isotropy is assured by the preliminary choice of the reference system, which must be defined from the arguments (1) themselves. In order to establish generating sets for the general isotropic tensor functions $\underset{\sim}{F}$, SMITH employs WANG's procedure.

The representations obtained by WANG and by SMITH are not identical. Moreover, the functional bases derived by SMITH contain redundant elements. This has been analysed in [11], where after suitable amendments of the previously established bases, an irreducible functional basis is obtained. Reasonings similar to those employed in [11] allow conclusions to be drawn for the generating sets as well. The complete and irreducible representations so obtained are given in the Tables I and II below.

For example, the irreducible representation for an isotropic tensor function of two symmetric second order tensors :

$$\underset{\sim}{T} = \underset{\sim}{F}(\underset{\sim}{A}, \underset{\sim}{B}) \tag{8}$$

involves eight tensor generators :

$$\underset{\sim}{I},\ \underset{\sim}{A},\ \underset{\sim}{A}^2,\ \underset{\sim}{B},\ \underset{\sim}{B}^2,\ \underset{\sim}{A}\underset{\sim}{B}+\underset{\sim}{B}\underset{\sim}{A},\ \underset{\sim}{A}^2\underset{\sim}{B}+\underset{\sim}{B}\underset{\sim}{A}^2,\ \underset{\sim}{A}\underset{\sim}{B}^2+\underset{\sim}{B}^2\underset{\sim}{A} \tag{9}$$

and ten invariants :

$$\left.\begin{array}{l} \mathrm{tr}\underset{\sim}{A},\ \mathrm{tr}\underset{\sim}{A}^2,\ \mathrm{tr}\underset{\sim}{A}^3,\ \mathrm{tr}\underset{\sim}{B},\ \mathrm{tr}\underset{\sim}{B}^2,\ \mathrm{tr}\underset{\sim}{B}^3 \\ \mathrm{tr}\underset{\sim}{A}\underset{\sim}{B},\ \mathrm{tr}\underset{\sim}{A}^2\underset{\sim}{B},\ \mathrm{tr}\underset{\sim}{A}\underset{\sim}{B}^2,\ \mathrm{tr}\underset{\sim}{A}^2\underset{\sim}{B}^2 \end{array}\right\} \tag{10}$$

Variables	Invariants
$\underset{\sim}{A}$	$\mathrm{tr}\underset{\sim}{A},\ \mathrm{tr}\underset{\sim}{A}^2,\ \mathrm{tr}\underset{\sim}{A}^3$
$\underset{\sim}{W}$	$\mathrm{tr}\underset{\sim}{W}^2$
$\underset{\sim}{v}$	$\underset{\sim}{v}\cdot\underset{\sim}{v}$
$\underset{\sim}{A}_1,\ \underset{\sim}{A}_2$	$\mathrm{tr}\underset{\sim}{A}_1\underset{\sim}{A}_2,\ \mathrm{tr}\underset{\sim}{A}_1^2\underset{\sim}{A}_2,\ \mathrm{tr}\underset{\sim}{A}_1\underset{\sim}{A}_2^2,\ \mathrm{tr}\underset{\sim}{A}_1^2\underset{\sim}{A}_2^2$
$\underset{\sim}{A},\ \underset{\sim}{W}$	$\mathrm{tr}\underset{\sim}{A}\underset{\sim}{W}^2,\ \mathrm{tr}\underset{\sim}{A}^2\underset{\sim}{W}^2,\ \mathrm{tr}\underset{\sim}{A}^2\underset{\sim}{W}^2\underset{\sim}{A}\underset{\sim}{W}$
$\underset{\sim}{A},\ \underset{\sim}{v}$	$\underset{\sim}{v}\cdot\underset{\sim}{A}\underset{\sim}{v},\ \underset{\sim}{v}\cdot\underset{\sim}{A}^2\underset{\sim}{v}$
$\underset{\sim}{W}_1,\ \underset{\sim}{W}_2$	$\mathrm{tr}\underset{\sim}{W}_1\underset{\sim}{W}_2$
$\underset{\sim}{W},\ \underset{\sim}{v}$	$\underset{\sim}{v}\cdot\underset{\sim}{W}^2\underset{\sim}{v}$
$\underset{\sim}{v}_1,\ \underset{\sim}{v}_2$	$\underset{\sim}{v}_1\cdot\underset{\sim}{v}_2$
$\underset{\sim}{A}_1,\ \underset{\sim}{A}_2,\ \underset{\sim}{A}_3$	$\mathrm{tr}\underset{\sim}{A}_1\underset{\sim}{A}_2\underset{\sim}{A}_3$
$\underset{\sim}{A}_1,\ \underset{\sim}{A}_2,\ \underset{\sim}{W}$	$\mathrm{tr}\underset{\sim}{A}_1\underset{\sim}{A}_2\underset{\sim}{W},\ \mathrm{tr}\underset{\sim}{A}_1^2\underset{\sim}{A}_2\underset{\sim}{W},\ \mathrm{tr}\underset{\sim}{A}_1\underset{\sim}{A}_2^2\underset{\sim}{W},\ \mathrm{tr}\underset{\sim}{A}_1\underset{\sim}{W}^2\underset{\sim}{A}_2\underset{\sim}{W}$
$\underset{\sim}{A},\ \underset{\sim}{W}_1,\ \underset{\sim}{W}_2$	$\mathrm{tr}\underset{\sim}{A}\underset{\sim}{W}_1\underset{\sim}{W}_2,\ \mathrm{tr}\underset{\sim}{A}\underset{\sim}{W}_1^2\underset{\sim}{W}_2,\ \mathrm{tr}\underset{\sim}{A}\underset{\sim}{W}_1\underset{\sim}{W}_2^2$
$\underset{\sim}{W}_1,\ \underset{\sim}{W}_2,\ \underset{\sim}{W}_3$	$\mathrm{tr}\underset{\sim}{W}_1\underset{\sim}{W}_2\underset{\sim}{W}_3$
$\underset{\sim}{A}_1,\ \underset{\sim}{A}_2,\ \underset{\sim}{v}$	$\underset{\sim}{A}_1\underset{\sim}{v}\cdot\underset{\sim}{A}_2\underset{\sim}{v}$
$\underset{\sim}{A},\ \underset{\sim}{v}_1,\ \underset{\sim}{v}_2$	$\underset{\sim}{v}_1\cdot\underset{\sim}{A}\underset{\sim}{v}_2,\ \underset{\sim}{v}_1\cdot\underset{\sim}{A}^2\underset{\sim}{v}_2$
$\underset{\sim}{W}_1,\ \underset{\sim}{W}_2,\ \underset{\sim}{v}$	$\underset{\sim}{W}_1\underset{\sim}{v}\cdot\underset{\sim}{W}_2\underset{\sim}{v},\ \underset{\sim}{W}_1^2\underset{\sim}{v}\cdot\underset{\sim}{W}_2\underset{\sim}{v},\ \underset{\sim}{W}_1\underset{\sim}{v}\cdot\underset{\sim}{W}_2^2\underset{\sim}{v}$
$\underset{\sim}{W},\ \underset{\sim}{v}_1,\ \underset{\sim}{v}_2$	$\underset{\sim}{v}_1\cdot\underset{\sim}{W}\underset{\sim}{v}_2,\ \underset{\sim}{v}_1\cdot\underset{\sim}{W}^2\underset{\sim}{v}_2$
$\underset{\sim}{A},\ \underset{\sim}{W},\ \underset{\sim}{v}$	$\underset{\sim}{A}\underset{\sim}{v}\cdot\underset{\sim}{W}\underset{\sim}{v},\ \underset{\sim}{A}^2\underset{\sim}{v}\cdot\underset{\sim}{W}\underset{\sim}{v},\ \underset{\sim}{A}\underset{\sim}{W}\underset{\sim}{v}\cdot\underset{\sim}{W}^2\underset{\sim}{v}$
$\underset{\sim}{A}_1,\underset{\sim}{A}_2,\ \underset{\sim}{v}_1,\ \underset{\sim}{v}_2$	$\underset{\sim}{A}_1\underset{\sim}{v}_1\cdot\underset{\sim}{A}_2\underset{\sim}{v}_2 - \underset{\sim}{A}_1\underset{\sim}{v}_2\cdot\underset{\sim}{A}_2\underset{\sim}{v}_1$
$\underset{\sim}{A},\ \underset{\sim}{W},\ \underset{\sim}{v}_1,\ \underset{\sim}{v}_2$	$\underset{\sim}{A}\underset{\sim}{v}_1\cdot\underset{\sim}{W}\underset{\sim}{v}_2 - \underset{\sim}{A}\underset{\sim}{v}_2\cdot\underset{\sim}{W}\underset{\sim}{v}_1$
$\underset{\sim}{W}_1\ \underset{\sim}{W}_2,\ \underset{\sim}{v}_1,\ \underset{\sim}{v}_2$	$\underset{\sim}{W}_1\underset{\sim}{v}_1\cdot\underset{\sim}{W}_2\underset{\sim}{v}_2 - \underset{\sim}{W}_1\underset{\sim}{v}_2\cdot\underset{\sim}{W}_2\underset{\sim}{v}_1$

The complete list of invariants for a set of variables ($\underset{\sim}{A}_1$, ... $\underset{\sim}{A}_a$, $\underset{\sim}{W}_1$, ... $\underset{\sim}{W}_b$, $\underset{\sim}{v}_1$, ... $\underset{\sim}{v}_c$) is obtained considering all the (unordered) combinations of one, two, three and four variables given in the Table.

Table I: Irreducible isotropic functional bases

Variables	Generators
	$\mathbf{I}$
$\mathbf{A}$	$\mathbf{A}$, $\mathbf{A}^2$
$\mathbf{W}$	$\mathbf{W}^2$
$\mathbf{v}$	$\mathbf{v} \otimes \mathbf{v}$
$\mathbf{A}_1$, $\mathbf{A}_2$	$\mathbf{A}_1\mathbf{A}_2 + \mathbf{A}_2\mathbf{A}_1$, $\mathbf{A}_1^2\mathbf{A}_2 + \mathbf{A}_2\mathbf{A}_1^2$, $\mathbf{A}_1\mathbf{A}_2^2 + \mathbf{A}_2^2\mathbf{A}_1$
$\mathbf{A}$, $\mathbf{W}$	$\mathbf{AW}-\mathbf{WA}$, $\mathbf{WAW}$, $\mathbf{A}^2\mathbf{W}-\mathbf{WA}^2$, $\mathbf{WAW}^2-\mathbf{W}^2\mathbf{AW}$
$\mathbf{A}$, $\mathbf{v}$	$\mathbf{v} \otimes \mathbf{Av} + \mathbf{Av} \otimes \mathbf{v}$, $\mathbf{v} \otimes \mathbf{A}^2\mathbf{v} + \mathbf{A}^2\mathbf{v} \otimes \mathbf{v}$
$\mathbf{W}_1$, $\mathbf{W}_2$	$\mathbf{W}_1\mathbf{W}_2 + \mathbf{W}_2\mathbf{W}_1$, $\mathbf{W}_1\mathbf{W}_2^2 + \mathbf{W}_2^2\mathbf{W}_1$, $\mathbf{W}_1^2\mathbf{W}_2 + \mathbf{W}_2\mathbf{W}_1^2$
$\mathbf{W}$, $\mathbf{v}$	$\mathbf{Wv} \otimes \mathbf{Wv}$, $\mathbf{v} \otimes \mathbf{Wv} + \mathbf{Wv} \otimes \mathbf{v}$, $\mathbf{Wv} \otimes \mathbf{W}^2\mathbf{v} + \mathbf{W}^2\mathbf{v} \otimes \mathbf{Wv}$
$\mathbf{v}_1$, $\mathbf{v}_2$	$\mathbf{v}_1 \otimes \mathbf{v}_2 + \mathbf{v}_2 \otimes \mathbf{v}_1$
$\mathbf{A}$, $\mathbf{v}_1$, $\mathbf{v}_2$	$(\mathbf{v}_1 \otimes \mathbf{Av}_2 + \mathbf{Av}_2 \otimes \mathbf{v}_1) - (\mathbf{v}_2 \otimes \mathbf{Av}_1 + \mathbf{Av}_1 \otimes \mathbf{v}_2)$
$\mathbf{W}$, $\mathbf{v}_1$, $\mathbf{v}_2$	$(\mathbf{v}_1 \otimes \mathbf{Wv}_2 + \mathbf{Wv}_2 \otimes \mathbf{v}_1) - (\mathbf{v}_2 \otimes \mathbf{Wv}_1 + \mathbf{Wv}_1 \otimes \mathbf{v}_2)$

An irreducible representation for an isotropic function $\mathbf{T} = \mathbf{F}(\mathbf{A}_1, \ldots \mathbf{A}_a, \mathbf{W}_1, \ldots \mathbf{W}_b, \mathbf{v}_1, \ldots \mathbf{v}_c)$ is given by : $\mathbf{F} = \alpha_i \mathbf{G}_i$, where the α_i's are arbitrary scalar-valued functions of the invariants of the functional bases given in Table I and where the $\mathbf{G}_i$'s are the generators given in Table II, considering all (unordered) combinations of zero, one, two and three variables.

Table II: Irreducible isotropic generating sets

An irreducible functional basis is not unique. The number of its elements depends on the type of the invariants chosen for its construction. As pointed out by WANG [6], an irreducible functional basis should contain a single element. Suppose that a set of p scalar invariants, the values of which belong to $\mathbb{R}^p$, where $\mathbb{R}$ is the set of real numbers, constitutes a functional basis. A one-to-one correspondence between the set of basic invariants and $\mathbb{R}$ is a scalar invariant which constitutes by itself a new functional basis. As $\mathbb{R}$ and $\mathbb{R}^p$ have the same cardinality, such a correspondence exists. However, as these two spaces are not homeomorphic, such a correspondence is discontinuous almost everywhere and, thus, rather unwieldy in practical applications. The invariants in Table I are simple polynomial functions of the arguments (1), more precisely they are traces of products of the arguments (1) ; thus, general isotropic scalar functions are represented by polynomial invariants. Another choice of polynomial invariants can lead to irreducible functional bases containing fewer terms. The minimal number of polynomial invariants necessary to constitute a complete functional basis is at present a problem for which only partial results are known. Similar comments hold for irreducible sets of tensor generators.

As pointed out in Section 1.3, the representations for general functions contain, in general, fewer elements then the corresponding representations for polynomial functions. Consider, for example, the tensor function (8). The irreducible generating set in the polynomial case contains nine tensors :

$$\left.\begin{array}{l} \underset{\sim}{I},\ \underset{\sim}{A},\ \underset{\sim}{A}^2,\ \underset{\sim}{B},\ \underset{\sim}{B}^2,\ \underset{\sim}{A}\underset{\sim}{B}+\underset{\sim}{B}\underset{\sim}{A}, \\ \underset{\sim}{A}^2\underset{\sim}{B}+\underset{\sim}{B}\underset{\sim}{A}^2,\ \underset{\sim}{A}\underset{\sim}{B}^2+\underset{\sim}{B}^2\underset{\sim}{A},\ \underset{\sim}{A}^2\underset{\sim}{B}^2+\underset{\sim}{B}^2\underset{\sim}{A}^2 \end{array}\right\} \qquad (11)$$

whereas in the general non-polynomial case, the irreducible generating set contains the eight tensors (9), the last generator (11) being then redundant.

Consider the list of an arbitrary number of symmetric second order tensors:

$$\underset{\sim}{A}_1,\ \underset{\sim}{A}_2\ \ldots \underset{\sim}{A}_a \qquad (12)$$

The integrity basis for the list (12) contains the traces of products of up to six tensors [7], whereas the functional basis contains the traces only up to the order three (Table I). Similarly, the generating set in the polynomial case contains products of up to five tensors [3], whereas in the general case, it contains products up to only two tensors (Table II). The complete analysis for the reductions introduced in the representations when considering general functions instead of polynomial functions is presented in [12].

3. REPRESENTATIONS FOR NON-POLYNOMIAL ANISOTROPIC SCALAR AND TENSOR FUNCTIONS

3.1 Method based on the introduction of structural tensors

The proposed method is an extension of the procedure developed in Chapter 2 concerning a two-dimensional constitutive equation relating two symmetric second order tensors. Here, we consider an anisotropic medium of constitutive equation specifying a symmetric second order tensor $\underset{\sim}{T}$ which is a function of the arguments (1), i.e. of the mechanical agencies. In order to take into account the oriented internal structure of the material, we introduce additional argument tensors $\underset{\sim}{\xi}_1$, $\underset{\sim}{\xi}_2$, $\ldots\underset{\sim}{\xi}_p$, which are the structural tensors:

$$\underset{\sim}{T} = \underset{\sim}{F}(\underset{\sim}{A}_1,\ \underset{\sim}{A}_2,\ \ldots\underset{\sim}{A}_a,\ \underset{\sim}{W}_1,\ \underset{\sim}{W}_2,\ \ldots\underset{\sim}{W}_b,\ \underset{\sim}{v}_1,\ \underset{\sim}{v}_2,\ \ldots\underset{\sim}{v}_c,\ \underset{\sim}{\xi}_1,\ \underset{\sim}{\xi}_2,\ \ldots\underset{\sim}{\xi}_p) \qquad (13)$$

The relation (13) can be a constitutive equation provided it satisfies the Principle of Isotropy of Space. One of the consequences of this principle is that an arbitrary transformation $\underset{\sim}{Q}$ of the orthogonal group $\mathcal{O}$ and applied to both the medium and the agencies, produces the same orthogonal transformation of the material response:

$$\forall\underset{\sim}{Q} \in \mathcal{O} :\ \underset{\sim}{F}(\bar{\underset{\sim}{A}}_i,\ \bar{\underset{\sim}{W}}_j,\ \bar{\underset{\sim}{v}}_k,\ \bar{\underset{\sim}{\xi}}_\ell) = \bar{\underset{\sim}{F}}(\underset{\sim}{A}_i,\ \underset{\sim}{W}_j,\ \underset{\sim}{v}_k,\ \underset{\sim}{\xi}_\ell) \qquad (14)$$

where $\bar{\underset{\sim}{F}}$, $\bar{\underset{\sim}{A}}_i$, $\bar{\underset{\sim}{W}}_j$, $\bar{\underset{\sim}{v}}_k$, $\bar{\underset{\sim}{\xi}}_\ell$ are the transformed of $\underset{\sim}{F}$, $\underset{\sim}{A}_i$, $\underset{\sim}{W}_j$, $\underset{\sim}{v}_k$, $\underset{\sim}{\xi}_\ell$ by an orthogonal transformation $\underset{\sim}{Q}$. The condition (14) means that $\underset{\sim}{F}$ is an isotropic function with respect to its a + b + c + p arguments.

We introduce now the invariance group S of the structural tensors $\underset{\sim}{\xi}_\ell$:

$$\underset{\sim}{Q} \in S \Longleftrightarrow \bar{\underset{\sim}{\xi}}_\ell = \underset{\sim}{\xi}_\ell \; ; \; \ell = 1,2,\ldots p, \tag{15}$$

Taking into account the relation (15) into (14), we arrive at :

$$\forall \underset{\sim}{Q} \in S : \underset{\sim}{F}(\bar{\underset{\sim}{A}}_i, \bar{\underset{\sim}{W}}_j, \bar{\underset{\sim}{v}}_k, \underset{\sim}{\xi}_\ell) = \bar{\underset{\sim}{F}}(\underset{\sim}{A}_i, \underset{\sim}{W}_j, \underset{\sim}{v}_k, \underset{\sim}{\xi}_\ell) \tag{16}$$

For a general function $\underset{\sim}{F}$, it can be shown that (16) constitutes a converse inplication. Thus, (16) signifies that $\underset{\sim}{F}$ is an isotropic function of the agencies $\underset{\sim}{A}_i$, $\underset{\sim}{W}_j$, $\underset{\sim}{v}_k$, the type of anisotropy being characterized by the invariance group S of the structural tensors $\underset{\sim}{\xi}_\ell$. It follows that the representation of the isotropic function (14) is also a representation for $\underset{\sim}{F}$ considered as an anisotropic function of the agencies. For a given group S, the set of structural tensors is not unique. The representations for anisotropic functions so obtained are necessarily complete, but not necessarily irreducible, even if the representations employed for the isotropic functions (14) are irreducible.

Thus, two main problems arise: find the structural tensors associated with each type of anisotropy; reduce the obtained representations in applying the proposed method. This procedure is applied in Sections 3.3 to 3.5 below for the derivation of the irreducible representations for respectively general anisotropic, orthotropic and transversely isotropic tensor functions of an arbitrary number of symmetric second order tensors. The representations for tensor valued functions are composed of the functional bases and the generating sets of tensors. The functional bases constitute the representations for the corresponding scalar-valued functions.

3.2 Generalization of the Rivlin-Ericksen method

The irreducible functional bases for general orthotropic and transverse isotropic scalar-valued functions f of the complete list of arguments (1) have been established in [14], employing a generalization for anisotropic functions of the method developed by RIVLIN and ERICKSEN (cf. Section 2.) for isotropic functions. The list of invariants obtained is too long to be reproduced here.

3.3 General anisotropy

In the case of general anisotropy, the material symmetry group S is reduced to:

$$S = \{ \underset{\sim}{I}, -\underset{\sim}{I} \} \tag{17}$$

Consider the six linearly independent tensors:

$$\left.\begin{array}{l} \underset{\sim}{M}_{11} = \underset{\sim}{v}_1 \otimes \underset{\sim}{v}_1 \ , \quad \underset{\sim}{M}_{22} = \underset{\sim}{v}_2 \otimes \underset{\sim}{v}_2 \ , \quad \underset{\sim}{M}_{33} = \underset{\sim}{v}_3 \otimes \underset{\sim}{v}_3 \ , \\ \underset{\sim}{M}_{23} + \underset{\sim}{M}_{32} = \underset{\sim}{v}_2 \otimes \underset{\sim}{v}_3 + \underset{\sim}{v}_3 \otimes \underset{\sim}{v}_2 \ , \quad \underset{\sim}{M}_{31} + \underset{\sim}{M}_{13} = \underset{\sim}{v}_3 \otimes \underset{\sim}{v}_1 + \underset{\sim}{v}_1 \otimes \underset{\sim}{v}_3 \ , \\ \underset{\sim}{M}_{12} + \underset{\sim}{M}_{21} = \underset{\sim}{v}_1 \otimes \underset{\sim}{v}_2 + \underset{\sim}{v}_2 \otimes \underset{\sim}{v}_1 \ , \end{array}\right\} \tag{18}$$

where $(\underset{\sim}{v}_1, \underset{\sim}{v}_2, \underset{\sim}{v}_3)$ are three orthonormal vectors. In the $(\underset{\sim}{v}_1, \underset{\sim}{v}_2, \underset{\sim}{v}_3)$ frame, the expressions of the tensors (18) are given by:

$$\begin{array}{lll} M_{11} = \begin{vmatrix} 1 & 0 & 0 \\ 0 & 0 & 0 \\ 0 & 0 & 0 \end{vmatrix}, & M_{22} = \begin{vmatrix} 0 & 0 & 0 \\ 0 & 1 & 0 \\ 0 & 0 & 0 \end{vmatrix}, & M_{33} = \begin{vmatrix} 0 & 0 & 0 \\ 0 & 0 & 0 \\ 0 & 0 & 1 \end{vmatrix}, \\ M_{23} + M_{32} = \begin{vmatrix} 0 & 0 & 0 \\ 0 & 0 & 1 \\ 0 & 1 & 0 \end{vmatrix}, & M_{31} + M_{13} = \begin{vmatrix} 0 & 0 & 1 \\ 0 & 0 & 0 \\ 1 & 0 & 0 \end{vmatrix}, & M_{12} + M_{21} = \begin{vmatrix} 0 & 1 & 0 \\ 1 & 0 & 0 \\ 0 & 0 & 0 \end{vmatrix}. \end{array} \tag{19}$$

It is easy to show that the group (17) is the invariance group of the tensors (18), which constitute a basis in the space of symmetric second order tensors. The tensors (18) are the structural tensors for general anisotropy.

Thus, the representation for the isotropic function

$$\underset{\sim}{T} = \underset{\sim}{F}(\underset{\sim}{A}_1, \underset{\sim}{A}_2, \ldots \underset{\sim}{A}_a, \underset{\sim}{M}_{11}, \underset{\sim}{M}_{22}, \underset{\sim}{M}_{33}, \underset{\sim}{M}_{23} + \underset{\sim}{M}_{32}, \underset{\sim}{M}_{31} + \underset{\sim}{M}_{13}, \underset{\sim}{M}_{12} + \underset{\sim}{M}_{21}) \tag{20}$$

obtained from Table I and II, is a representation for $\underset{\sim}{F}$ considered as a general anisotropic function of the agencies $\underset{\sim}{A}_1, \underset{\sim}{A}_2, \ldots \underset{\sim}{A}_a$. Among the invariants of this representation the following 6a invariants appear:

$$\begin{array}{lll} tr\underset{\sim}{M}_{11}\underset{\sim}{A}_i \ , & tr\underset{\sim}{M}_{22}\underset{\sim}{A}_i \ , & tr\underset{\sim}{M}_{33}\underset{\sim}{A}_i \ , \\ tr\underset{\sim}{M}_{23}\underset{\sim}{A}_i \ , & tr\underset{\sim}{M}_{31}\underset{\sim}{A}_i \ , & tr\underset{\sim}{M}_{12}\underset{\sim}{A}_i \ . \end{array} \tag{21}$$

The analysis of the explicit expressions show that the invariants (21) are just the independent components in the $(\underset{\sim}{v}_1, \underset{\sim}{v}_2, \underset{\sim}{v}_3)$ frame of the agencies $\underset{\sim}{A}_i$: $tr\underset{\sim}{M}_{k\ell}\underset{\sim}{A}_i = A^{(i)}_{k\ell}$. Any scalar invariant of the $\underset{\sim}{A}_i$ can be expressed uniquely with respect to the invariants (21). Thus, the invariants (21) constitute a complete functional basis. As they are functionally independent, this functional basis is irreducible a fortiori.

Among the generators for the isotropic function (20) appear the six tensors (18), which constitute a complete set of tensor generators. As the tensors (18) are linearly independent, this generating set is irreducible.

Finally, an irreducible representation for the general anisotropic tensor function $\underset{\sim}{F}$ of the agencies $\underset{\sim}{A}_i$ is given by:

$$\left.\begin{array}{l} \underset{\sim}{T} = \alpha_1\underset{\sim}{M}_{11} + \alpha_2\underset{\sim}{M}_{22} + \alpha_3\underset{\sim}{M}_{33} + \alpha_4(\underset{\sim}{M}_{23} + \underset{\sim}{M}_{32}) + \\ \quad + \alpha_5(\underset{\sim}{M}_{31} + \underset{\sim}{M}_{13}) + \alpha_6(\underset{\sim}{M}_{12} + \underset{\sim}{M}_{21}) \\ \\ \alpha_i = \alpha_i(tr\underset{\sim}{M}_{11}\underset{\sim}{A}_i, \ tr\underset{\sim}{M}_{22}\underset{\sim}{A}_i, \ tr\underset{\sim}{M}_{33}\underset{\sim}{A}_i, \ tr\underset{\sim}{M}_{23}\underset{\sim}{A}_i, \ tr\underset{\sim}{M}_{31}\underset{\sim}{A}_i, \ tr\underset{\sim}{M}_{12}\underset{\sim}{A}_i) \end{array}\right\} \tag{22}$$

$$i = 1,2,\ldots a \ .$$

This result could have been established directly. Indeed, the tensor $\underset{\sim}{T}$ can always be expressed as a linear combination of the basic tensors (18), with the components T_{mn} of $\underset{\sim}{T}$ as coefficients. As the function $\underset{\sim}{F}$ does not verify any symmetry, the components T_{mn} are arbitrary functions of the 6a independent components $A^{(i)}_{k\ell}$ of the agencies $\underset{\sim}{A}_i$.

3.4 Orthotropy

In the case of orthotropy, with privileged directions $(\underset{\sim}{v}_1, \underset{\sim}{v}_2, \underset{\sim}{v}_3)$, the material symmetry group S is given by :

$$S = \left\{ \pm\underset{\sim}{I} \, , \, \underset{\sim}{S}_1, \, \underset{\sim}{S}_2, \, \underset{\sim}{S}_3 \right\} \tag{23}$$

where $\underset{\sim}{S}_1$, $\underset{\sim}{S}_2$, $\underset{\sim}{S}_3$ are the reflections with respect to the basic planes $(\underset{\sim}{v}_2, \underset{\sim}{v}_3)$, $(\underset{\sim}{v}_3, \underset{\sim}{v}_1)$, $(\underset{\sim}{v}_1, \underset{\sim}{v}_2)$ of the orthonormal privileged frame $(\underset{\sim}{v}_1, \underset{\sim}{v}_2, \underset{\sim}{v}_3)$.

Consider the three tensors:

$$\underset{\sim}{M}_1 = \underset{\sim}{v}_1 \otimes \underset{\sim}{v}_1 \quad , \quad \underset{\sim}{M}_2 = \underset{\sim}{v}_2 \otimes \underset{\sim}{v}_2 \quad , \quad \underset{\sim}{M}_3 = \underset{\sim}{v}_3 \otimes \underset{\sim}{v}_3 \, . \tag{24}$$

The expressions of the tensors (24) are given by (19). It is easy to show that the group (23) is the invariance group of the tensors (24) :

$$\underset{\sim}{Q} \in S \quad \Longleftrightarrow \quad \underset{\sim}{Q}\underset{\sim}{M}_i\underset{\sim}{Q}^t = \underset{\sim}{M}_i \; ; \; i = 1,2,3 \, . \tag{25}$$

Thus, the tensors (24) are the structural tensors for orthotropy.

The representation for the isotropic function

$$\underset{\sim}{T} = \underset{\sim}{F}(\underset{\sim}{A}_1, \, \ldots \underset{\sim}{A}_a, \, \underset{\sim}{M}_1, \, \underset{\sim}{M}_2, \, \underset{\sim}{M}_3) \tag{26}$$

obtained from Table I and II, is a complete representation for $\underset{\sim}{F}$ considered as a general orthotropic function of the agencies $\underset{\sim}{A}_1, \ldots \underset{\sim}{A}_a$. But the representation so obtained is not irreducible. The redundant elements have been analysed in [13,15]. The reductions are mainly based on the following relations, resulting from the normality and orthogonality of the vectors $(\underset{\sim}{v}_1, \underset{\sim}{v}_2, \underset{\sim}{v}_3)$:

$$\left.\begin{array}{l}
\underset{\sim}{M}_i^p = \underset{\sim}{M}_i \quad ; \quad \mathrm{tr}\underset{\sim}{M}_i = 1 \quad ; \quad i = 1,2,3 \\
\underset{\sim}{M}_i\underset{\sim}{M}_j = \underset{\sim}{0} \quad ; \; i \neq j \quad ; \quad i,j = 1,2,3 \\
(\underset{\sim}{M}_1\underset{\sim}{A} + \underset{\sim}{A}\underset{\sim}{M}_1) + (\underset{\sim}{M}_2\underset{\sim}{A} + \underset{\sim}{A}\underset{\sim}{M}_2) + (\underset{\sim}{M}_3\underset{\sim}{A} + \underset{\sim}{A}\underset{\sim}{M}_3) = 2\underset{\sim}{A} \\
\mathrm{tr}\underset{\sim}{M}_1\underset{\sim}{A} + \mathrm{tr}\underset{\sim}{M}_2\underset{\sim}{A} + \mathrm{tr}\underset{\sim}{M}_3\underset{\sim}{A} = \mathrm{tr}\underset{\sim}{A}
\end{array}\right\} \tag{27}$$

where $\underset{\sim}{A}$ is an arbitrary symmetric second order tensor.

Finally, an irreducible functional basis is composed of the invariants given in Table III, in considering all unordered combinations of one, two and three variables. An irreducible generating set is composed of the tensor generators given in Table III, in considering all unordered combinations of zero, one and two variables.

For example, the irreducible representation for $\underset{\sim}{T}$, considered as an orthotropic function of two symmetric tensors $\underset{\sim}{A}$ and $\underset{\sim}{B}$, is given by:

$$\left.\begin{aligned} \underset{\sim}{T} = {} & \alpha_1 \underset{\sim}{M}_1 + \alpha_2 \underset{\sim}{M}_2 + \alpha_3 \underset{\sim}{M}_3 + \alpha_4 (\underset{\sim}{M}_1 \underset{\sim}{A} + \underset{\sim}{A}\underset{\sim}{M}_1) + \alpha_5 (\underset{\sim}{M}_2 \underset{\sim}{A} + \underset{\sim}{A}\underset{\sim}{M}_2) + \\ & + \alpha_6 (\underset{\sim}{M}_3 \underset{\sim}{A} + \underset{\sim}{A}\underset{\sim}{M}_3) + \alpha_7 (\underset{\sim}{M}_1 \underset{\sim}{B} + \underset{\sim}{B}\underset{\sim}{M}_1) + \alpha_8 (\underset{\sim}{M}_2 \underset{\sim}{B} + \underset{\sim}{B}\underset{\sim}{M}_2) + \\ & + \alpha_9 (\underset{\sim}{M}_3 \underset{\sim}{B} + \underset{\sim}{B}\underset{\sim}{M}_3) + \alpha_{10} \underset{\sim}{A}^2 + \alpha_{11} \underset{\sim}{B}^2 + \alpha_{12} (\underset{\sim}{A}\underset{\sim}{B} + \underset{\sim}{B}\underset{\sim}{A}) \,. \\ \alpha_i = {} & \alpha_i (\mathrm{tr}\underset{\sim}{M}_1\underset{\sim}{A},\ \mathrm{tr}\underset{\sim}{M}_2\underset{\sim}{A},\ \mathrm{tr}\underset{\sim}{M}_3\underset{\sim}{A},\ \mathrm{tr}\underset{\sim}{M}_1\underset{\sim}{B},\ \mathrm{tr}\underset{\sim}{M}_2\underset{\sim}{B},\ \mathrm{tr}\underset{\sim}{M}_3\underset{\sim}{B}, \\ & \mathrm{tr}\underset{\sim}{M}_1\underset{\sim}{A}^2,\ \mathrm{tr}\underset{\sim}{M}_2\underset{\sim}{A}^2,\ \mathrm{tr}\underset{\sim}{M}_3\underset{\sim}{A}^2,\ \mathrm{tr}\underset{\sim}{M}_1\underset{\sim}{B}^2,\ \mathrm{tr}\underset{\sim}{M}_2\underset{\sim}{B}^2,\ \mathrm{tr}\underset{\sim}{M}_3\underset{\sim}{B}^2, \\ & \mathrm{tr}\underset{\sim}{M}_1\underset{\sim}{A}\underset{\sim}{B},\ \mathrm{tr}\underset{\sim}{M}_2\underset{\sim}{A}\underset{\sim}{B},\ \mathrm{tr}\underset{\sim}{M}_3\underset{\sim}{A}\underset{\sim}{B},\ \mathrm{tr}\underset{\sim}{A}^2\underset{\sim}{B},\ \mathrm{tr}\underset{\sim}{A}\underset{\sim}{B}^2,\ \mathrm{tr}\underset{\sim}{A}^3,\ \mathrm{tr}\underset{\sim}{B}^3) \end{aligned}\right\} \tag{28}$$

3.5 <u>Transverse isotropy</u>

In the case of transverse isotropy, with privileged direction $\underset{\sim}{v}_3$, the material symmetry group S is given by :

$$S = \left\{ \pm\underset{\sim}{I},\ \underset{\sim}{S}_1,\ \underset{\sim}{S}_2,\ \underset{\sim}{S}_3,\ \underset{\sim}{M}_\theta\ ,\ (0 \leqslant \theta \leqslant 2\pi) \right\} \tag{29}$$

where $\underset{\sim}{S}_1$, $\underset{\sim}{S}_2$, $\underset{\sim}{S}_3$ have the same definitions as in Section 3.4 and $\underset{\sim}{M}_\theta$ are all rotations about the $\underset{\sim}{v}_3$ axis.

Consider the tensor:

$$\underset{\sim}{M} = \underset{\sim}{v}_3 \otimes \underset{\sim}{v}_3 \tag{30}$$

It is easy to show that the invariance group of $\underset{\sim}{M}$ is the characteristic group (29):

Variables	Functional bases	Integrity bases
A	$\mathrm{tr}M_1A$, $\mathrm{tr}M_1A^2$, $\mathrm{tr}A^3$, $\mathrm{tr}M_2A$, $\mathrm{tr}M_2A^2$, $\mathrm{tr}M_3A$, $\mathrm{tr}M_3A^2$.	$\mathrm{tr}M_1A$, $\mathrm{tr}M_1A^2$, $\mathrm{tr}A^3$, $\mathrm{tr}M_1A$, $\mathrm{tr}M_2A^2$, $\mathrm{tr}M_3A$, $\mathrm{tr}M_3A^2$.
A_1, A_2	$\mathrm{tr}M_1A_1A_2$, $\mathrm{tr}A_1^2A_2$, $\mathrm{tr}A_1A_2^2$ $\mathrm{tr}M_2A_1A_2$, $\mathrm{tr}M_3A_1A_2$.	$\mathrm{tr}M_1A_1A_2$, $\mathrm{tr}M_1A_1^2A_2$, $\mathrm{tr}M_1A_1A_2^2$ $\mathrm{tr}M_2A_1A_2$, $\mathrm{tr}M_2A_1^2A_2$, $\mathrm{tr}M_2A_1A_2^2$, $\mathrm{tr}M_3A_1A_2$, $\mathrm{tr}M_3A_1^2A_2$, $\mathrm{tr}M_3A_1A_2^2$.
A_1, A_2, A_3	$\mathrm{tr}A_1A_2A_3$.	$\mathrm{tr}M_1A_1A_2A_3$, $\mathrm{tr}M_1A_2A_1A_3$, $\mathrm{tr}M_2A_1A_2A_3$, $\mathrm{tr}M_2A_2A_1A_3$, $\mathrm{tr}M_3A_1A_2A_3$, $\mathrm{tr}M_3A_2A_1A_3$.

Variables	Generators for general functions	Generators for polynomial functions
	M_1, M_2, M_3.	M_1, M_2, M_3.
A	M_1A+AM_1, A^2, M_2A+AM_2, M_3A+AM_3.	M_1A+AM_1, $M_1A^2+A^2M_1$, M_2A+AM_2, $M_2A^2+A^2M_2$, M_3A+AM_3, $M_3A^2+A^2M_3$.
A_1, A_2	$A_1A_2+A_2A_1$	$M_1A_1A_2+A_2A_1M_1$, $M_1A_2A_1+A_1A_2M_1$ $M_2A_1A_2+A_2A_1M_2$, $M_2A_2A_1+A_1A_2M_2$ $M_3A_1A_2+A_2A_1M_3$, $M_3A_2A_1+A_1A_2M_3$.

Table III : Irreducible representations for orthotropic functions

$$Q \in S \iff QMQ^t = M \tag{31}$$

Thus, M is the single structural tensor for transverse isotropy.

The representation for the isotropic function

$$T = F(A_1, \ldots A_a, M) \tag{32}$$

obtained from Table I and II, is a complete representation for F considered as a general transversely isotropic function of the agencies $A_1, \ldots A_a$. After some reductions ([13], [15]), the irreducible representations obtained are listed in Table IV.

For example, the irreducible representation of T, considered as a transversely isotropic function of two symmetric tensors A and B, is given by:

$$\left.\begin{aligned} T = {} & \alpha_1 I + \alpha_2 M + \alpha_3 A + \alpha_4 B + \alpha_5 (MA+AM) + \alpha_6 (MB+BM) + \\ & + \alpha_7 A^2 + \alpha_8 B^2 + \alpha_9 (MA^2+A^2M) + \alpha_{10} (MB^2+B^2M) + \\ & + \alpha_{11} (AB+BA) + \alpha_{12} (A^2B+BA^2) + \alpha_{13} (AB^2+B^2A) \\ \alpha_i = {} & \alpha_i (trA, trA^2, trA^3, trB, trB^2, trB^3, trAB, trA^2B, \\ & trAB^2, trMA, trMB, trMA^2, trMB^2, trMAB. \end{aligned}\right\} \tag{33}$$

3.6 Comparison with representations for polynomial anisotropic tensor functions

In the case of general anisotropy, the representations for polynomial tensor functions can be derived directly by the same arguments as those developed in Section 3.3. Thus, the irreducible representations for polynomial functions are similar to those for general functions, the scalar functions α_i in (22) being then polynomials in their arguments.

In the case of orthotropy and transverse isotropy, the integrity bases and generating sets for polynomial functions are given in Tables III and

Variables	Functional bases	Integrity bases
$\mathbf{A}$	$\mathrm{tr}\mathbf{A}$, $\mathrm{tr}\mathbf{A}^2$, $\mathrm{tr}\mathbf{A}^3$, $\mathrm{tr}\mathbf{M}\mathbf{A}$, $\mathrm{tr}\mathbf{M}\mathbf{A}^2$.	$\mathrm{tr}\mathbf{A}$, $\mathrm{tr}\mathbf{A}^2$, $\mathrm{tr}\mathbf{A}^3$, $\mathrm{tr}\mathbf{M}\mathbf{A}$, $\mathrm{tr}\mathbf{M}\mathbf{A}^2$.
$\mathbf{A}_1,\mathbf{A}_2$	$\mathrm{tr}\mathbf{A}_1\mathbf{A}_2$, $\mathrm{tr}\mathbf{M}\mathbf{A}_1\mathbf{A}_2$ $\mathrm{tr}\mathbf{A}_1^2\mathbf{A}_2$, $\mathrm{tr}\mathbf{A}_1\mathbf{A}_2^2$.	$\mathrm{tr}\mathbf{A}_1\mathbf{A}_2$, $\mathrm{tr}\mathbf{M}\mathbf{A}_1\mathbf{A}_2$, $\mathrm{tr}\mathbf{A}_1^2\mathbf{A}_2$, $\mathrm{tr}\mathbf{A}_1\mathbf{A}_2^2$, $\mathrm{tr}\mathbf{A}_1^2\mathbf{A}_2^2$, $\mathrm{tr}\mathbf{M}\mathbf{A}_1^2\mathbf{A}_2$, $\mathrm{tr}\mathbf{M}\mathbf{A}_1\mathbf{A}_2^2$.
$\mathbf{A}_1,\mathbf{A}_2,\mathbf{A}_3$	$\mathrm{tr}\mathbf{A}_1\mathbf{A}_2\mathbf{A}_3$.	$\mathrm{tr}\mathbf{M}\mathbf{A}_1\mathbf{A}_2\mathbf{A}_3$, $\mathrm{tr}\mathbf{M}\mathbf{A}_2\mathbf{A}_1\mathbf{A}_3$, $\mathrm{tr}\mathbf{M}\mathbf{A}_1\mathbf{A}_2\mathbf{A}_3^2$, $\mathrm{tr}\mathbf{M}\mathbf{A}_1\mathbf{A}_3\mathbf{A}_2^2$, $\mathrm{tr}\mathbf{M}\mathbf{A}_2\mathbf{A}_3\mathbf{A}_1^2$.
$\mathbf{A}_1,\mathbf{A}_2,\mathbf{A}_3,\mathbf{A}_4$		$\mathrm{tr}\mathbf{M}\mathbf{A}_1\mathbf{A}_2\mathbf{A}_3\mathbf{A}_4$, $\mathrm{tr}\mathbf{M}\mathbf{A}_1\mathbf{A}_4\mathbf{A}_2\mathbf{A}_3$, $\mathrm{tr}\mathbf{M}\mathbf{A}_2\mathbf{A}_1\mathbf{A}_3\mathbf{A}_4$ $\mathrm{tr}\mathbf{M}\mathbf{A}_1\mathbf{A}_4\mathbf{A}_3\mathbf{A}_2$, $\mathrm{tr}\mathbf{M}\mathbf{A}_1\mathbf{A}_2\mathbf{A}_4\mathbf{A}_3$, $\mathrm{tr}\mathbf{M}\mathbf{A}_1\mathbf{A}_3\mathbf{A}_2\mathbf{A}_4$.

Variables	Generators for general functions	Generators for polynomial functions
	$\mathbf{I}$, $\mathbf{M}$.	$\mathbf{I}$, $\mathbf{M}$.
$\mathbf{A}$	$\mathbf{A}$, $\mathbf{M}\mathbf{A}+\mathbf{A}\mathbf{M}$, $\mathbf{A}^2$, $\mathbf{M}\mathbf{A}^2+\mathbf{A}^2\mathbf{M}$.	$\mathbf{A}$, $\mathbf{M}\mathbf{A}+\mathbf{A}\mathbf{M}$, $\mathbf{A}^2$, $\mathbf{M}\mathbf{A}^2+\mathbf{A}^2\mathbf{M}$.
$\mathbf{A}_1$, $\mathbf{A}_2$	$\mathbf{A}_1\mathbf{A}_2 + \mathbf{A}_2\mathbf{A}_1$, $\mathbf{A}_1^2\mathbf{A}_2 + \mathbf{A}_2\mathbf{A}_1^2$, $\mathbf{A}_1\mathbf{A}_2^2 + \mathbf{A}_2^2\mathbf{A}_1$.	$\mathbf{M}\mathbf{A}_1\mathbf{A}_2 + \mathbf{A}_2\mathbf{A}_1\mathbf{M}$, $\mathbf{M}\mathbf{A}_2\mathbf{A}_1 + \mathbf{A}_1\mathbf{A}_2\mathbf{M}$, $\mathbf{M}\mathbf{A}_1^2\mathbf{A}_2 + \mathbf{A}_2\mathbf{A}_1^2\mathbf{M}$, $\mathbf{M}\mathbf{A}_1\mathbf{A}_2^2 + \mathbf{A}_2^2\mathbf{A}_1\mathbf{M}$.
$\mathbf{A}_1$, $\mathbf{A}_2$, $\mathbf{A}_3$		$\mathbf{M}\mathbf{A}_1\mathbf{A}_2\mathbf{A}_3 + \mathbf{A}_3\mathbf{A}_2\mathbf{A}_1\mathbf{M}$, $\mathbf{M}\mathbf{A}_1\mathbf{A}_3\mathbf{A}_2 + \mathbf{A}_2\mathbf{A}_3\mathbf{A}_1\mathbf{M}$, $\mathbf{M}\mathbf{A}_2\mathbf{A}_1\mathbf{A}_3 + \mathbf{A}_3\mathbf{A}_1\mathbf{A}_2\mathbf{M}$, $\mathbf{M}\mathbf{A}_2\mathbf{A}_3\mathbf{A}_1 + \mathbf{A}_1\mathbf{A}_3\mathbf{A}_2\mathbf{M}$, $\mathbf{M}\mathbf{A}_3\mathbf{A}_1\mathbf{A}_2 + \mathbf{A}_2\mathbf{A}_1\mathbf{A}_3\mathbf{M}$, $\mathbf{A}_1\mathbf{M}\mathbf{A}_2\mathbf{A}_3 + \mathbf{A}_3\mathbf{A}_2\mathbf{M}\mathbf{A}_1$.

Table IV: Irreducible representations for transversely isotropic functions.

IV against the corresponding functional bases and generating sets for general functions. The invariants of the integrity bases, given here in direct tensor notation, are equivalent to those given by ADKINS [16,17] in tensor component notation. We have derived the generating sets by the method indicated in the Chapters by Professor SPENCER (see also [12]). In the orthotropic case, the functional bases and the integrity bases are similar for a single tensor variable. But, from two tensor arguments, the integrity bases contain an increasingly greater number of elements. For the generating sets, this difference already appears for a single tensor argument. In the case of transverse isotropy, the two types of representations are similar for a single tensor argument. From two tensor arguments, the representations for polynomial functions contain a markedly greater number of elements then the corresponding representations for general functions.

4. REPRESENTATIONS FOR NON-POLYNOMIAL ISOTROPIC AND ORTHOTROPIC TENSOR FUNCTIONS IN A TWO-DIMENSIONAL SPACE

4.1 Isotropic functions

Consider the isotropic tensor function:

$$\underset{\sim}{T} = \underset{\sim}{F}(\underset{\sim}{A}_1, \underset{\sim}{A}_2, \ldots \underset{\sim}{A}_a) \tag{34}$$

where $\underset{\sim}{T}$ and the $\underset{\sim}{A}_i$ are symmetric second order tensors in a two-dimensional space. A complete representation for the function (34) is obtained from Table I and II for the three-dimensional case:

$$\underset{\sim}{T} = \alpha_o \underset{\sim}{I} + \alpha_i \underset{\sim}{G}_i \tag{35}$$

where $\underset{\sim}{I}$ is the two-dimensional identity tensor, the $\underset{\sim}{G}_i$'s are the generators:

$$\left.\begin{array}{l} \underset{\sim}{A}_j,\ \underset{\sim}{A}_j^2,\ \underset{\sim}{A}_k\underset{\sim}{A}_\ell + \underset{\sim}{A}_\ell\underset{\sim}{A}_k,\ \underset{\sim}{A}_k^2\underset{\sim}{A}_\ell + \underset{\sim}{A}_\ell\underset{\sim}{A}_k^2, \\ \underset{\sim}{A}_\ell^2 \underset{\sim}{A}_k + \underset{\sim}{A}_k\underset{\sim}{A}_\ell^2 \quad ; \quad k < \ell \quad ; \quad j, k, \ell = 1, 2, \ldots a. \end{array}\right\} \tag{36}$$

and α_o, α_i are arbitrary scalar-valued functions of the invariants:

$$\begin{array}{l} \mathrm{tr}\underset{\sim}{A}_j,\ \mathrm{tr}\underset{\sim}{A}_j^2,\ \mathrm{tr}\underset{\sim}{A}_j^3,\ \mathrm{tr}\underset{\sim}{A}_k\underset{\sim}{A}_\ell,\ \mathrm{tr}\underset{\sim}{A}_k^2\underset{\sim}{A}_\ell, \\ \mathrm{tr}\underset{\sim}{A}_k\underset{\sim}{A}_\ell^2,\ \mathrm{tr}\underset{\sim}{A}_k^2\underset{\sim}{A}_\ell^2,\ \mathrm{tr}\underset{\sim}{A}_m\underset{\sim}{A}_k\underset{\sim}{A}_\ell\ ;\ m < k < \ell\ ;\ j,k,\ell,m = 1,2,\ldots a \end{array} \tag{37}$$

Obviously, the representation (35) is not irreducible. It can be reduced by the application of the generalization for two-dimensional second order tensors of the Cayley-Hamilton theorem, due to RIVLIN [18].

The Cayley-Hamilton theorem for a two-dimensional second order tensor is given by:

$$\underset{\sim}{A}^2 - (\mathrm{tr}\underset{\sim}{A})\underset{\sim}{A} + \frac{1}{2}[\mathrm{tr}^2\underset{\sim}{A} - \mathrm{tr}\underset{\sim}{A}^2]\underset{\sim}{I} = \underset{\sim}{0} \tag{38}$$

If in (38) we replace $\underset{\sim}{A}$ by $\underset{\sim}{A} + \lambda\underset{\sim}{B}$, where λ is an arbitrary scalar, and equate the coefficient of λ to zero, we obtain:

$$\underset{\sim}{A}\underset{\sim}{B} + \underset{\sim}{B}\underset{\sim}{A} - (\mathrm{tr}\underset{\sim}{A})\underset{\sim}{B} - (\mathrm{tr}\underset{\sim}{B})\underset{\sim}{A} + (\mathrm{tr}\underset{\sim}{A}\ \mathrm{tr}\underset{\sim}{B} - \mathrm{tr}\underset{\sim}{A}\underset{\sim}{B})\underset{\sim}{I} = \underset{\sim}{0} \tag{39}$$

From the theorem (39), we obtain the following relations for the generators (36):

$$\begin{aligned} \underset{\sim}{A}_j^2 &= \frac{1}{2}(\mathrm{tr}\underset{\sim}{A}_j^2 - \mathrm{tr}^2\underset{\sim}{A}_j)\underset{\sim}{I} + (\mathrm{tr}\underset{\sim}{A}_j)\underset{\sim}{A}_j, \\ \underset{\sim}{A}_k\underset{\sim}{A}_\ell + \underset{\sim}{A}_\ell\underset{\sim}{A}_k &= (\mathrm{tr}\underset{\sim}{A}_\ell)\underset{\sim}{A}_k + (\mathrm{tr}\underset{\sim}{A}_k)\underset{\sim}{A}_\ell - (\mathrm{tr}\underset{\sim}{A}_k\underset{\sim}{A}_\ell - \mathrm{tr}\underset{\sim}{A}_k\mathrm{tr}\underset{\sim}{A}_\ell)\underset{\sim}{I}, \\ \underset{\sim}{A}_k^2\underset{\sim}{A}_\ell + \underset{\sim}{A}_\ell\underset{\sim}{A}_k^2 &= (\mathrm{tr}\underset{\sim}{A}_k^2)\underset{\sim}{A}_\ell + (\mathrm{tr}\underset{\sim}{A}_\ell)\underset{\sim}{A}_k^2 + \\ &\quad - \Big\{\mathrm{tr}\underset{\sim}{A}_\ell\mathrm{tr}\underset{\sim}{A}_k^2 - [\mathrm{tr}\underset{\sim}{A}_\ell\mathrm{tr}\underset{\sim}{A}_\ell\underset{\sim}{A}_k - \frac{1}{2}\mathrm{tr}\underset{\sim}{A}_k(\mathrm{tr}^2\underset{\sim}{A}_\ell - \mathrm{tr}\underset{\sim}{A}_\ell^2)]\Big\}\underset{\sim}{I}, \\ \underset{\sim}{A}_k\underset{\sim}{A}_\ell^2 + \underset{\sim}{A}_\ell^2\underset{\sim}{A}_k &= (\mathrm{tr}\underset{\sim}{A}_\ell^2)\underset{\sim}{A}_k + (\mathrm{tr}\underset{\sim}{A}_k)\underset{\sim}{A}_\ell^2 + \\ &\quad - \Big\{\mathrm{tr}\underset{\sim}{A}_k\mathrm{tr}\underset{\sim}{A}_\ell^2 - [\mathrm{tr}\underset{\sim}{A}_k\mathrm{tr}\underset{\sim}{A}_k\underset{\sim}{A}_\ell - \frac{1}{2}\mathrm{tr}\underset{\sim}{A}_\ell(\mathrm{tr}^2\underset{\sim}{A}_k - \mathrm{tr}\underset{\sim}{A}_k^2]\Big\}\underset{\sim}{I} \end{aligned} \tag{40}$$

and the following relations for the invariants (37):

$$\mathrm{tr}\,\underset{\sim}{A}_j^3 = \frac{3}{2}\,\mathrm{tr}\underset{\sim}{A}_j\;\mathrm{tr}\underset{\sim}{A}_j^2 - \frac{1}{2}\,\mathrm{tr}^3\underset{\sim}{A}_j,$$

$$\mathrm{tr}\underset{\sim}{A}_k^2\,\underset{\sim}{A}_\ell = \mathrm{tr}\underset{\sim}{A}_k\;\mathrm{tr}\underset{\sim}{A}_k\underset{\sim}{A}_\ell - \frac{1}{2}\,\mathrm{tr}\underset{\sim}{A}_\ell(\mathrm{tr}^2\underset{\sim}{A}_k - \mathrm{tr}\underset{\sim}{A}_k^2),$$

$$\mathrm{tr}\underset{\sim}{A}_k\underset{\sim}{A}_\ell^2 = \mathrm{tr}\underset{\sim}{A}_\ell\;\mathrm{tr}\underset{\sim}{A}_k\underset{\sim}{A}_\ell - \frac{1}{2}\,\mathrm{tr}\underset{\sim}{A}_k(\mathrm{tr}^2\underset{\sim}{A}_\ell - \mathrm{tr}\underset{\sim}{A}_\ell^2), \tag{41}$$

$$\mathrm{tr}\underset{\sim}{A}_k^2\underset{\sim}{A}_\ell^2 = \mathrm{tr}\underset{\sim}{A}_k\;\mathrm{tr}\underset{\sim}{A}_\ell\;\mathrm{tr}\underset{\sim}{A}_k\underset{\sim}{A}_\ell + \frac{1}{2}(\mathrm{tr}\underset{\sim}{A}_k^2\;\mathrm{tr}\underset{\sim}{A}_\ell^2 - \mathrm{tr}^2\underset{\sim}{A}_k\;\mathrm{tr}^2\underset{\sim}{A}_\ell),$$

$$\mathrm{tr}\underset{\sim}{A}_m\underset{\sim}{A}_k\underset{\sim}{A}_\ell = \frac{1}{2}[\mathrm{tr}\underset{\sim}{A}_m\;\mathrm{tr}\underset{\sim}{A}_k\underset{\sim}{A}_\ell + \mathrm{tr}\underset{\sim}{A}_k\;\mathrm{tr}\underset{\sim}{A}_\ell\underset{\sim}{A}_m + \mathrm{tr}\underset{\sim}{A}_\ell\;\mathrm{tr}\underset{\sim}{A}_m\underset{\sim}{A}_k - \mathrm{tr}\underset{\sim}{A}_m\;\mathrm{tr}\underset{\sim}{A}_k\;\mathrm{tr}\underset{\sim}{A}_\ell].$$

Taking into account the relations (40) and (41), we obtain from (36) and (37) the following irreducible representation for the isotropic function (34):

$$\left.\begin{aligned} &\underset{\sim}{T} = \phi_1\underset{\sim}{I} + \phi_2\underset{\sim}{A}_1 + \phi_3\underset{\sim}{A}_2 + \ldots + \phi_{a+1}\underset{\sim}{A}_a \\ &\phi_i = \phi_i(\mathrm{tr}\underset{\sim}{A}_j,\ \mathrm{tr}\underset{\sim}{A}_j^2,\ \mathrm{tr}\underset{\sim}{A}_k\underset{\sim}{A}_\ell), \\ &\qquad k < \ell \quad ; \quad j,k,\ell = 1,2,\ldots a. \end{aligned}\right\} \tag{42}$$

It is worthwhile to point out that the representation (42) obtained is quasi-linear.

4.2 Orthotropic tensor functions

In the case of two-dimensional orthotropy, with privileged directions $(\underset{\sim}{v}_2,\ \underset{\sim}{v}_3)$, the material symmetry group of transformations is given by:

$$S = \left\{\pm\,\underset{\sim}{I},\ \underset{\sim}{S}_2,\ \underset{\sim}{S}_3\right\} \tag{43}$$

where $\underset{\sim}{S}_2$ and $\underset{\sim}{S}_3$ are the reflections with respect to the $\underset{\sim}{v}_2$ and $\underset{\sim}{v}_3$ axes. It is easy to show that S is the invariance group of the structural tensor:

$$\underset{\sim}{M} = \underset{\sim}{v}_3 \otimes \underset{\sim}{v}_3 \tag{44}$$

Thus, the representations of the isotropic function:

$$\underset{\sim}{T} = \underset{\sim}{F}(\underset{\sim}{A}_1, \underset{\sim}{A}_2, \ldots \underset{\sim}{A}_a, \underset{\sim}{M}) \tag{45}$$

is a representation for $\underset{\sim}{F}$, considered as an orthotropic function of the agencies $\underset{\sim}{A}_i$.

Taking into account the relations:

$$\mathrm{tr}\underset{\sim}{M} = \mathrm{tr}\underset{\sim}{M}^2 = 1 \tag{46}$$

we obtain directly from (42) the following irreducible representation:

$$\left.\begin{aligned} &\underset{\sim}{T} = \phi_1\underset{\sim}{I} + \phi_2\underset{\sim}{M} + \phi_3\underset{\sim}{A}_1 + \phi_4\underset{\sim}{A}_2 + \ldots + \phi_{a+2}\underset{\sim}{A}_a \\ &\phi_i = \phi_i(\mathrm{tr}\underset{\sim}{A}_j, \mathrm{tr}\underset{\sim}{A}_j^2, \mathrm{tr}\underset{\sim}{M}\underset{\sim}{A}_j, \mathrm{tr}\underset{\sim}{A}_k\underset{\sim}{A}_\ell), \\ &\qquad k < \ell \quad ; \quad j,k,\ell = 1,2, \ldots a \end{aligned}\right\} \tag{47}$$

which is a quasi-linear form.

REFERENCES

1. WINEMAN, A.S. and PIPKIN, A.C., Material Symmetry Restrictions on Constitutive Equations, *Arch. Rat. Mech. An.*, 17 (1964) : 184-214.

2. PIPKIN, A.C. and WINEMAN, A.S., Material Symmetry Restrictions on Non-Polynomial Constitutive Equations, *Arch. Rat. Mech. An.*, 12 (1963) : 420-426.

3. SPENCER, A.J.M., Theory of Invariants, in *Continuum Physics*, ed. by C. Eringen, Academic Press, (1971) : 239-353.

4. WANG, C.C., On Representations for Isotropic Functions, Part I and II, *Arch. Rat. Mech. An.*, 33 (1969) : 249-287.

5. WANG, C.C., A New Representation Theorem for Isotropic Functions, Part I and II, *Arch. Rat. Mech. An.*, 36 (1970) : 166-223.

6. WANG, C.C., Corrigendum, *Arch. Rat. Mech. An.*, 43 (1971) : 392-395.

7. SMITH, G.F., On Isotropic Integrity Bases, *Arch. Rat. Mech. An.*, 18 (1965) : 282-292.

8. SMITH, G.F., On a Fundamental Error in Two Papers of C.C. WANG, *Arch. Rat. Mech. An.*, 36 (1970) : 161-165.

9. SMITH, G.F., On Isotropic Functions of Symmetric Tensors, Skew-Symmetric Tensors and Vectors, *Int.J.Engng. Sci.*, 19 (1971) : 899-916.

10.RIVLIN, R.S. and ERICKSEN, J.L., Stress-Deformation Relations for Isotropic Materials, *J. Rat. Mech. An.*, 4 (1955) : 323-425.

11.BOEHLER, J.P., On Irreducible Representations for Isotropic Scalar Functions, *ZAMM*, 57 (1977) : 323-327.

12.BOEHLER, J.P.,Lois de Comportement Anisotrope des Milieux Continus, *Journal de Mécanique*, 17,2 (1978) : 153-190.

13.BOEHLER, J.P., A Simple Derivation of Representations for Non-Polynomial Constitutive Equations in Some Cases of Anisotropy, *ZAMM*, 59 (1979) : 157-167.

14.BOEHLER, J.P.,Functional Bases for Orthotropic and Transverse Isotropic Non-Polynomial Functions, (in preparation).

15.BOEHLER, J.P. and RACLIN, J.,Representations Irréductibles des Fonctions Tensorielles Anisotropes Non-Polynomiales de Deux Tenseurs Symétriques, *Arch. Mech. Stos.*, 19,3 (1977) : 431-444.

16.ADKINS, J.E., Further Symmetry Relations for Transversely Isotropic Materials, *Arch. Rat. Mech. An.*, 5 (1960) : 263-274.

17.ADKINS, J.E., Symmetry Relations for Orthotropic and Transversely Isotropic Materials, *Arch. Rat. Mech. An.*, 4 (1960) : 193-213.

18.RIVLIN, R.S., Further Remarks on the Stress-Deformation Relations for Isotropic Materials, *J. Rat. Mech. An.*, 4 (1955) : 681-702.

Chapter 4

ANISOTROPIC LINEAR ELASTICITY

J.P. Boehler
University of Grenoble

1. INTRODUCTION

Consider a symmetric second order tensor $\underset{\sim}{T}$ which is a function $\underset{\sim}{F}$ of a symmetric second order tensor $\underset{\sim}{D}$. If $\underset{\sim}{F}$ is a transversely isotropic function of $\underset{\sim}{D}$, its irreducible representation is obtained from Table IV of Chapter 3:

$$\left.\begin{aligned} \underset{\sim}{T} &= \alpha_0 \underset{\sim}{I} + \alpha_1 \underset{\sim}{M} + \alpha_2 \underset{\sim}{D} + \alpha_3 (\underset{\sim}{M}\underset{\sim}{D}+\underset{\sim}{D}\underset{\sim}{M}) + \alpha_4 \underset{\sim}{D}^2 + \alpha_5 (\underset{\sim}{M}\underset{\sim}{D}^2+\underset{\sim}{D}^2\underset{\sim}{M}) \\ \alpha_i &= \alpha_i (\mathrm{tr}\underset{\sim}{D},\ \mathrm{tr}\underset{\sim}{D}^2,\ \mathrm{tr}\underset{\sim}{D}^3,\ \mathrm{tr}\underset{\sim}{M}\underset{\sim}{D},\ \mathrm{tr}\underset{\sim}{M}\underset{\sim}{D}^2) \end{aligned}\right\} \quad (1)$$

where $\underset{\sim}{M}$ is the structural tensor:

$$\underset{\sim}{M} = \underset{\sim}{v}_3 \otimes \underset{\sim}{v}_3 \quad , \quad (2)$$

$\underset{\sim}{v}_3$ being the privileged axis of transverse isotropy.

If $\underset{\sim}{F}$ is an orthotropic function, its irreducible representation is obtained from Table III of Chapter 3:

$$\left.\begin{aligned} \underset{\sim}{T} &= \alpha_1 \underset{\sim}{M}_1 + \alpha_2 \underset{\sim}{M}_2 + \alpha_3 \underset{\sim}{M}_3 + \alpha_4(\underset{\sim}{M}_1\underset{\sim}{D}+\underset{\sim}{D}\underset{\sim}{M}_1) + \alpha_5(\underset{\sim}{M}_2\underset{\sim}{D}+\underset{\sim}{D}\underset{\sim}{M}_2) + \\ &\quad + \alpha_6(\underset{\sim}{M}_3\underset{\sim}{D}+\underset{\sim}{D}\underset{\sim}{M}_3) + \alpha_7\underset{\sim}{D}^2. \\ \alpha_i &= \alpha_i(\mathrm{tr}\underset{\sim}{M}_1\underset{\sim}{D},\ \mathrm{tr}\underset{\sim}{M}_2\underset{\sim}{D},\ \mathrm{tr}\underset{\sim}{M}_3\underset{\sim}{D},\ \mathrm{tr}\underset{\sim}{M}_1\underset{\sim}{D}^2,\ \mathrm{tr}\underset{\sim}{M}_2\underset{\sim}{D}^2,\ \mathrm{tr}\underset{\sim}{M}_3\underset{\sim}{D}^2,\ \mathrm{tr}\underset{\sim}{D}^3) \end{aligned}\right\} \qquad (3)$$

where $\underset{\sim}{M}_1$, $\underset{\sim}{M}_2$, $\underset{\sim}{M}_3$ are the structural tensors:

$$\underset{\sim}{M}_1 = \underset{\sim}{v}_1 \otimes \underset{\sim}{v}_1 \quad ; \quad \underset{\sim}{M}_2 = \underset{\sim}{v}_2 \otimes \underset{\sim}{v}_2 \quad ; \quad \underset{\sim}{M}_3 = \underset{\sim}{v}_3 \otimes \underset{\sim}{v}_3 \ , \qquad (4)$$

$(\underset{\sim}{v}_1, \underset{\sim}{v}_2, \underset{\sim}{v}_3)$ being the privileged orthonormal frame of orthotropy.

Equations (1) and (3) can describe any material response which is expressible by an explicit relation between two symmetric second order tensors. Let $\underset{\sim}{T}$ be the stress tensor and $\underset{\sim}{D}$ the deformation tensor. The representations (1) and (3) are then the most general invariant forms of non-linear elasticity for the corresponding types of anisotropy.

The aim of this Chapter is to present the application of the theory of representations for tensor functions to the invariant formulation of anisotropic linear constitutive equations relating two symmetric second order tensors; for this class of mechanical behaviors, a potential function for the stress tensor may or may not exist. Thus, in the case of elasticity, without entering into the question concerning the existence of a strain-energy function for all elastic materials, we first analyze elastic materials (of Cauchy-elastic materials), for which the existence of a strain-energy function is not assumed, and then hyperelastic materials (or Green-elastic materials), for which a strain-energy function exists. The obtained invariant formulations involve only second order tensors in linear combination. For hyperelastic materials, the transitions from the invariant formulations to the classical formulations will be given explicitly in the cases of transverse isotropy and orthotropy (cf. [1], [2]); the case of general anisotropy is analyzed in [3]. Kinematic constraints result in further restrictions on the constitutive equations for anisotropic linear elasticity; such restrictions are discussed in Chapter 10, Section 2, by Professor SPENCER.

In the Sections below, we use the following derivatives:

$$\left.\begin{aligned} &\frac{\partial \mathrm{tr}\underset{\sim}{D}}{\partial \underset{\sim}{D}} = \underset{\sim}{I} \;; \quad \frac{\partial \mathrm{tr}\underset{\sim}{D}^2}{\partial \underset{\sim}{D}} = 2\underset{\sim}{D} \;; \quad \frac{\partial \mathrm{tr}\underset{\sim}{D}^3}{\partial \underset{\sim}{D}} = 3\underset{\sim}{D}^2 \;; \\ &\frac{\partial \mathrm{tr}\underset{\sim}{M}_i\underset{\sim}{D}}{\partial \underset{\sim}{D}} = \underset{\sim}{M}_i \;; \quad \frac{\partial \mathrm{tr}\underset{\sim}{M}_i\underset{\sim}{D}^2}{\partial \underset{\sim}{D}} = \underset{\sim}{M}_i\underset{\sim}{D}+\underset{\sim}{D}\underset{\sim}{M}_i \;; \quad i = 1,2,3 \;. \end{aligned}\right\} \tag{5}$$

2. TRANSVERSE ISOTROPY

2.1 Invariant formulation

Consider a linear elastic solid, which is transversely isotropic whith respect to an undistorted state, which will be taken as the reference configuration. The linear restriction of the representation (1) is given by:

$$\left.\begin{aligned} &\underset{\sim}{T} = \alpha_o\underset{\sim}{I} + \alpha_1\underset{\sim}{M} + \alpha_2\underset{\sim}{D} + \alpha_3(\underset{\sim}{M}\underset{\sim}{D}+\underset{\sim}{D}\underset{\sim}{M}) \;; \\ &\alpha_o = a_o + b_o\mathrm{tr}\underset{\sim}{D} + c_o\mathrm{tr}\underset{\sim}{M}\underset{\sim}{D} \;; \\ &\alpha_1 = a_1 + b_1\mathrm{tr}\underset{\sim}{D} + c_1\mathrm{tr}\underset{\sim}{M}\underset{\sim}{D} \;; \\ &\alpha_2 = a_2 \;; \quad \alpha_3 = a_3 \;. \end{aligned}\right\} \tag{6}$$

Thus, the linear elastic behavior of a transversely isotropic solid depends on eight material constants. In an undistorted state, the elastic solid is subjected to the following transversely isotropic stress:

$$\underset{\sim}{T}_u = a_o\underset{\sim}{I} + a_1\underset{\sim}{M} \tag{7}$$

Suppose that the undistorted state is a natural state, i.e. the stress $\underset{\sim}{T}_u$ is zero:

$$\underset{\sim}{D} \equiv \underset{\sim}{0} \quad => \quad \underset{\sim}{T}_u = a_o\underset{\sim}{I} + a_1\underset{\sim}{M} = \underset{\sim}{0} \tag{8}$$

From relation (8), we obtain:

$$\left.\begin{aligned} tr\underset{\sim}{T}_u &= 3a_o+a_1 = 0 \\ tr\underset{\sim}{M}\underset{\sim}{T}_u &= a_o+a_1 = 0 \end{aligned}\right\} => a_o = a_1 = 0 \tag{9}$$

Thus, when the solid admits a natural state, the transversely isotropic linear elastic behavior depends on six material constants.

Suppose now that the elastic solid is a hyperelastic material, i.e. its mechanical response $\underset{\sim}{T}$ is the derivative of a scalar-valued strain-energy function W:

$$\underset{\sim}{T} = \frac{\partial W}{\partial \underset{\sim}{D}} \tag{10}$$

The representation of the strain-energy function is obtained from Table IV of Chapter 3:

$$W = W(tr\underset{\sim}{D},\ tr\underset{\sim}{M}\underset{\sim}{D},\ tr\underset{\sim}{D}^2,\ tr\underset{\sim}{M}\underset{\sim}{D}^2,\ tr\underset{\sim}{D}^3) \tag{11}$$

Introducing (11) in (10) and using the expressions (5), we obtain :

$$\begin{aligned} \underset{\sim}{T} = {} & \frac{\partial W}{\partial tr\underset{\sim}{D}}\underset{\sim}{I} + \frac{\partial W}{\partial tr\underset{\sim}{M}\underset{\sim}{D}}\underset{\sim}{M} + 2\frac{\partial W}{\partial tr\underset{\sim}{D}^2}\underset{\sim}{D} + \\ & + \frac{\partial W}{\partial tr\underset{\sim}{M}\underset{\sim}{D}^2}(\underset{\sim}{M}\underset{\sim}{D}+\underset{\sim}{D}\underset{\sim}{M}) + 3\frac{\partial W}{\partial tr\underset{\sim}{D}^3}\underset{\sim}{D}^2 . \end{aligned} \tag{12}$$

Identifying (12) with the linear form (6) of the representation (1), we obtain:

$$\left.\begin{aligned} & \frac{\partial W}{\partial tr\underset{\sim}{D}} = b_o tr\underset{\sim}{D} + c_o tr\underset{\sim}{M}\underset{\sim}{D} \ ; \quad \frac{\partial W}{\partial tr\underset{\sim}{D}^2} = \frac{1}{2}a_2 \ ; \quad \frac{\partial W}{\partial tr\underset{\sim}{D}^3} = 0 \ ; \\ & \frac{\partial W}{\partial tr\underset{\sim}{M}\underset{\sim}{D}} = b_1 tr\underset{\sim}{D} + c_1 tr\underset{\sim}{M}\underset{\sim}{D} \ ; \quad \frac{\partial W}{\partial tr\underset{\sim}{M}\underset{\sim}{D}^2} = a_3 \ . \end{aligned}\right\} \tag{13}$$

The derivatives (13) are subjected to the following integrability condition:

$$\frac{\partial}{\partial tr\underset{\sim}{D}}\left(\frac{\partial W}{\partial tr\underset{\sim}{M}\underset{\sim}{D}}\right) = \frac{\partial}{\partial tr\underset{\sim}{M}\underset{\sim}{D}}\left(\frac{\partial W}{\partial tr\underset{\sim}{D}}\right) \quad \Rightarrow \quad c_o = b_1 . \tag{14}$$

Thus, the linear hyperelastic behavior of a transversely isotropic solid depends on five independent material constants.

Finally, the invariant form of transverse isotropic linear hyperelasticity is given by [2]:

$$\underset{\sim}{T} = (b_o tr\underset{\sim}{D}+b_1 tr\underset{\sim}{M}\underset{\sim}{D})\underset{\sim}{I} + (b_1 tr\underset{\sim}{D}+c_1 tr\underset{\sim}{M}\underset{\sim}{D})\underset{\sim}{M} + a_2\underset{\sim}{D} + a_3\,(\underset{\sim}{M}\underset{\sim}{D}+\underset{\sim}{D}\underset{\sim}{M}) \tag{15}$$

Integrating the partial derivatives (13) and taking into account relation (14), we obtain the following invariant form for the strain-energy function:

$$W = \frac{1}{2}\, b_o tr^2\underset{\sim}{D} + \frac{1}{2}\, c_1 tr^2\underset{\sim}{M}\underset{\sim}{D} + b_1 tr\underset{\sim}{M}\underset{\sim}{D}tr\underset{\sim}{D} + \frac{1}{2}\, a_2 tr\underset{\sim}{D}^2 + a_3 tr\underset{\sim}{M}\underset{\sim}{D}^2 \tag{16}$$

It is worthwhile to point out that the expression (15) furnishes a formulation of transverse isotropic hyperelasticity which involves only second order tensors in linear combination. The transition to non-linear cases (1) is direct and does not necessitate expansions of the strain-energy function or introduction of higher order tensors, as it is the case in the classical formulation ([4], [5]).

2.2 Classical formulation

In the classical formulation, the law of transverse isotropic linear hyperlasticity is written as follows:

$$\underset{\sim}{T} = \underset{\approx}{A}\,\underset{\sim}{D} \tag{17}$$

where $\underset{\approx}{A}$ is a fourth order tensor, specifically the tensor of transversely isotropic constants in the linear case. The expression of $\underset{\approx}{A}$ in any

$(\bullet, \bullet, \underset{\sim}{v}_3)$ privileged orthonormal frame is given by:

$$A = \begin{vmatrix} a & f & d & 0 & 0 & 0 \\ f & a & d & 0 & 0 & 0 \\ d & d & c & 0 & 0 & 0 \\ 0 & 0 & 0 & g & 0 & 0 \\ 0 & 0 & 0 & 0 & g & 0 \\ 0 & 0 & 0 & 0 & 0 & \frac{1}{2}(a-f) \end{vmatrix} \qquad (18)$$

where a, c, d, f, g, are five independent material constants.

The transition from the invariant formulation (15) to the classical formulation (17)-(18) has been established in [2]:

$$\left.\begin{aligned} & b_0 = f \;;\; b_1 = d-f \;;\; c_1 = c+a-2d-4g \;; \\ & a_2 = a-f \;;\; a_3 = 2g-a+f \;. \end{aligned}\right\} \qquad (19)$$

Thus, an alternative invariant form of transverse isotropic linear hyper-elasticity, employing the material constants of the matrix (18), is given by:

$$\begin{aligned} \underset{\sim}{T} = {} & [f\,\mathrm{tr}\underset{\sim}{D}+(d-f)\mathrm{tr}\underset{\sim}{M}\underset{\sim}{D}]\underset{\sim}{I} + [(d-f)\mathrm{tr}\underset{\sim}{D}+(c+a-2d-4g)\mathrm{tr}\underset{\sim}{M}\underset{\sim}{D}]\underset{\sim}{M} + \\ & + (a-f)\underset{\sim}{D} + (2g-a+f)(\underset{\sim}{M}\underset{\sim}{D}+\underset{\sim}{D}\underset{\sim}{M}) \end{aligned} \qquad (20)$$

3. ORTHOTROPY

3.1 Invariant formulation

Consider a linear elastic solid, which is orthotropic with respect to an undistorted state. The linear restriction of the representation (3) is given by:

$$\left.\begin{aligned} & \underset{\sim}{T} = \alpha_1\underset{\sim}{M}_1 + \alpha_2\underset{\sim}{M}_2 + \alpha_3\underset{\sim}{M}_3 + \alpha_4(\underset{\sim}{M}_1\underset{\sim}{D}+\underset{\sim}{D}\underset{\sim}{M}_1) + \alpha_5(\underset{\sim}{M}_2\underset{\sim}{D}+\underset{\sim}{D}\underset{\sim}{M}_2) + \alpha_6(\underset{\sim}{M}_3\underset{\sim}{D}+\underset{\sim}{D}\underset{\sim}{M}_3) \;; \\ & \alpha_1 = a_1 + b_1\mathrm{tr}\underset{\sim}{M}_1\underset{\sim}{D} + c_1\mathrm{tr}\underset{\sim}{M}_2\underset{\sim}{D} + d_1\mathrm{tr}\underset{\sim}{M}_3\underset{\sim}{D} \;; \\ & \alpha_2 = a_2 + b_2\mathrm{tr}\underset{\sim}{M}_1\underset{\sim}{D} + c_2\mathrm{tr}\underset{\sim}{M}_2\underset{\sim}{D} + d_2\mathrm{tr}\underset{\sim}{M}_3\underset{\sim}{D} \;; \\ & \alpha_3 = a_3 + b_3\mathrm{tr}\underset{\sim}{M}_1\underset{\sim}{D} + c_3\mathrm{tr}\underset{\sim}{M}_2\underset{\sim}{D} + d_3\mathrm{tr}\underset{\sim}{M}_3\underset{\sim}{D} \;; \\ & \alpha_4 = a_4 \;;\; \alpha_5 = a_5 \;;\; \alpha_6 = a_6 \;. \end{aligned}\right\} \qquad (21)$$

Thus, the linear elastic behavior of an orthotropic solid depends on 15 material constants. In an undistorted state, the elastic solid is subjected to the following orthotropic stress:

$$\mathbf{T}_u = a_1\mathbf{M}_1 + a_2\mathbf{M}_2 + a_3\mathbf{M}_3 \tag{22}$$

If the undistroted state is a natural state, we obtain:

$$\mathbf{D} \equiv \mathbf{0} \Rightarrow \quad \mathbf{T}_u = a_1\mathbf{M}_1 + a_2\mathbf{M}_2 + a_3\mathbf{M}_3 = \mathbf{0} \tag{23}$$

The relation (23) leads to:

$$\mathrm{tr}\mathbf{M}_1\mathbf{T}_u = a_1 = 0 \;\;;\;\; \mathrm{tr}\mathbf{M}_2\mathbf{T}_u = a_2 = 0 \;\;;\;\; \mathrm{tr}\mathbf{M}_3\mathbf{T}_u = a_3 = 0\,. \tag{24}$$

Thus, when the solid admits a natural state, the orthotropic linear elastic behavior depends on 12 material constants.

In the case of an orthotropic hyperelastic material, the representation of the scalar-valued strain-energy function is obtained from Table III of Chapter 3:

$$W = W(\mathrm{tr}\mathbf{M}_1\mathbf{D},\ \mathrm{tr}\mathbf{M}_2\mathbf{D},\ \mathrm{tr}\mathbf{M}_3\mathbf{D},\ \mathrm{tr}\mathbf{M}_1\mathbf{D}^2,\ \mathrm{tr}\mathbf{M}_2\mathbf{D}^2,\ \mathrm{tr}\mathbf{M}_3\mathbf{D}^2,\ \mathrm{tr}\mathbf{D}^3) \tag{25}$$

Introducing (25) in (10) and employing the expressions (5), we obtain:

$$\begin{aligned}\mathbf{T} = {} & \frac{\partial W}{\partial \mathrm{tr}\mathbf{M}_1\mathbf{D}}\mathbf{M}_1 + \frac{\partial W}{\partial \mathrm{tr}\mathbf{M}_2\mathbf{D}}\mathbf{M}_2 + \frac{\partial W}{\partial \mathrm{tr}\mathbf{M}_3\mathbf{D}}\mathbf{M}_3 + \frac{\partial W}{\partial \mathrm{tr}\mathbf{M}_1\mathbf{D}^2}(\mathbf{M}_1\mathbf{D}+\mathbf{D}\mathbf{M}_1) + \\ & + \frac{\partial W}{\partial \mathrm{tr}\mathbf{M}_2\mathbf{D}^2}(\mathbf{M}_2\mathbf{D}+\mathbf{D}\mathbf{M}_2) + \frac{\partial W}{\partial \mathrm{tr}\mathbf{M}_3\mathbf{D}^2}(\mathbf{M}_3\mathbf{D}+\mathbf{D}\mathbf{M}_3) + 3\,\frac{\partial W}{\partial \mathrm{tr}\mathbf{D}^3}\mathbf{D}^2\,.\end{aligned} \tag{26}$$

Identifying (26) with the linear form (21) of the representation (3), we obtain:

$$\frac{\partial W}{\partial \mathrm{tr} M_1 D} = b_1 \mathrm{tr} M_1 D + c_1 \mathrm{tr} M_2 D + d_1 \mathrm{tr} M_3 D \quad ; \quad \frac{\partial W}{\partial \mathrm{tr} M_1 D^2} = a_4 \quad ;$$

$$\frac{\partial W}{\partial \mathrm{tr} M_2 D} = b_2 \mathrm{tr} M_1 D + c_2 \mathrm{tr} M_2 D + d_2 \mathrm{tr} M_3 D \quad ; \quad \frac{\partial W}{\partial \mathrm{tr} M_2 D^2} = a_5 \quad ; \tag{27}$$

$$\frac{\partial W}{\partial \mathrm{tr} M_3 D} = b_3 \mathrm{tr} M_1 D + c_3 \mathrm{tr} M_2 D + d_3 \mathrm{tr} M_3 D \quad ; \quad \frac{\partial W}{\partial \mathrm{tr} M_3 D^2} = a_6 \quad ; \quad \frac{\partial W}{\partial \mathrm{tr} D^3} = 0 .$$

The derivatives (27) are subjected to the following three integrability conditions:

$$\left.\begin{aligned}
\frac{\partial W}{\partial \mathrm{tr} M_1 D}\left(\frac{\partial W}{\partial \mathrm{tr} M_2 D}\right) &= \frac{\partial W}{\partial \mathrm{tr} M_2 D}\left(\frac{\partial W}{\partial \mathrm{tr} M_1 D}\right) \quad \Rightarrow \quad c_1 = b_2 \quad ; \\
\frac{\partial W}{\partial \mathrm{tr} M_1 D}\left(\frac{\partial W}{\partial \mathrm{tr} M_3 D}\right) &= \frac{\partial W}{\partial \mathrm{tr} M_3 D}\left(\frac{\partial W}{\partial \mathrm{tr} M_1 D}\right) \quad \Rightarrow \quad b_3 = d_1 \quad ; \\
\frac{\partial W}{\partial \mathrm{tr} M_2 D}\left(\frac{\partial W}{\partial \mathrm{tr} M_3 D}\right) &= \frac{\partial W}{\partial \mathrm{tr} M_3 D}\left(\frac{\partial W}{\partial \mathrm{tr} M_2 D}\right) \quad \Rightarrow \quad d_2 = c_3 \quad .
\end{aligned}\right\} \tag{28}$$

Thus, the linear hyperelastic behavior of an orthotropic solid depends on nine independent material constants.

Finally, the invariant formulation of orthotropic linear hyperelasticity is given by [1]):

$$\begin{aligned}
T = {} & (b_1 \mathrm{tr} M_1 D + b_2 \mathrm{tr} M_2 D + d_1 \mathrm{tr} M_3 D) M_1 + (b_2 \mathrm{tr} M_1 D + c_2 \mathrm{tr} M_2 D + c_3 \mathrm{tr} M_3 D) M_2 + \\
& + (d_1 \mathrm{tr} M_1 D + c_3 \mathrm{tr} M_2 D + d_3 \mathrm{tr} M_3 D) M_3 + a_4 (M_1 D + D M_1) + \\
& + a_5 (M_2 D + D M_2) + a_6 (M_3 D + D M_3)
\end{aligned} \tag{29}$$

The integration of the partial derivatives (27) results in :

$$\begin{aligned}
W = {} & \frac{1}{2} b_1 \mathrm{tr}^2 M_1 D + \frac{1}{2} c_2 \mathrm{tr}^2 M_2 D + \frac{1}{2} d_3 \mathrm{tr}^2 M_3 D + c_3 \mathrm{tr} M_2 D \, \mathrm{tr} M_3 D + d_1 \mathrm{tr} M_3 D \, \mathrm{tr} M_1 D + \\
& + b_2 \mathrm{tr} M_1 D \, \mathrm{tr} M_2 D + a_4 \mathrm{tr} M_1 D^2 + a_5 \mathrm{tr} M_2 D^2 + a_6 \mathrm{tr} M_3 D^2
\end{aligned} \tag{30}$$

Comments on the invariant forms (29) and (30) are similar to those given in Section 2.1.

3.2 Classical formulation

In the classical formulation, the expression of the hyperelastic fourth order tensor $\underset{\approx}{A}$ in the privileged orthonormal frame $(\underset{\sim}{v}_1, \underset{\sim}{v}_2, \underset{\sim}{v}_3)$ is given by:

$$\underset{\sim}{T} = \underset{\approx}{A}\underset{\sim}{D} \quad ; \quad A = \begin{vmatrix} a & f & e & 0 & 0 & 0 \\ f & b & d & 0 & 0 & 0 \\ e & d & c & 0 & 0 & 0 \\ 0 & 0 & 0 & g & 0 & 0 \\ 0 & 0 & 0 & 0 & h & 0 \\ 0 & 0 & 0 & 0 & 0 & j \end{vmatrix} \tag{31}$$

where a, b, c, d, e, f, g, h, j, are nine independent material constants.

The transition from the invariant formulation (29) to the classical formulation (31) has been established in [1], [3]:

$$\left.\begin{array}{l} b_1 = a+2g-2h-2j \;;\; c_2 = b+2h-2j-2g \;; \\ d_3 = c+2j-2g-2h \;;\; a_4 = h+j-g \;;\; a_5 = j+g-h \;; \\ a_6 = g+h-j \;;\; c_3 = d \;;\; d_1 = e \;;\; b_2 = f \;. \end{array}\right\} \tag{32}$$

Thus, an alternative invariant form of orthotropic linear hyperelasticity is given by:

$$\begin{aligned} \underset{\sim}{T} = & [(a+2g-2h-2j)\mathrm{tr}\underset{\sim}{M}_1\underset{\sim}{D} + f\ \mathrm{tr}\underset{\sim}{M}_2\underset{\sim}{D} + e\ \mathrm{tr}\underset{\sim}{M}_3\underset{\sim}{D}]\underset{\sim}{M}_1 \quad + \\ & + [f\ \mathrm{tr}\underset{\sim}{M}_1\underset{\sim}{D} + (b+2h-2j-2g)\mathrm{tr}\underset{\sim}{M}_2\underset{\sim}{D} + d\ \mathrm{tr}\underset{\sim}{M}_3\underset{\sim}{D}]\underset{\sim}{M}_2 \quad + \\ & + [e\ \mathrm{tr}\underset{\sim}{M}_1\underset{\sim}{D} + d\ \mathrm{tr}\underset{\sim}{M}_2\underset{\sim}{D} + (c+2j-2g-2h)\mathrm{tr}\underset{\sim}{M}_3\underset{\sim}{D}]\underset{\sim}{M}_3 \quad + \\ & + (h+j-g)(\underset{\sim}{M}_1\underset{\sim}{D} + \underset{\sim}{D}\underset{\sim}{M}_1) + (j+g-h)(\underset{\sim}{M}_2\underset{\sim}{D} + \underset{\sim}{D}\underset{\sim}{M}_2) + (g+h-j)(\underset{\sim}{M}_3\underset{\sim}{D} + \underset{\sim}{D}\underset{\sim}{M}_3) . \end{aligned} \tag{33}$$

4. ISOTROPY

Consider an elastic solid, which is isotropic with respect to an undistorted state. The most general invariant form of non-linear isotropic elasticity is given by the following irreducible representation obtained from Table I of Chapter 3:

$$\left.\begin{aligned} \underset{\sim}{T} &= \alpha_o \underset{\sim}{I} + \alpha_1 \underset{\sim}{D} + \alpha_2 \underset{\sim}{D}^2 \\ \alpha_i &= \alpha_i(\mathrm{tr}\underset{\sim}{D},\ \mathrm{tr}\underset{\sim}{D}^2,\ \mathrm{tr}\underset{\sim}{D}^3) \end{aligned}\right\} \tag{34}$$

The linear restriction of the representation (34) is given by:

$$\left.\begin{aligned} \underset{\sim}{T} &= \alpha_o \underset{\sim}{I} + \alpha_1 \underset{\sim}{D} \\ \alpha_o &= a_o + b_o \mathrm{tr}\underset{\sim}{D} \quad ; \\ \alpha_1 &= a_1 . \end{aligned}\right\} \tag{35}$$

Thus, isotropic linear elasticity depends on three material constants. In an undistorted state, the elastic solid is subjected to the following isotropic stress:

$$\underset{\sim}{T}_u = a_o \underset{\sim}{I} \tag{36}$$

If the undistorted state is a natural state, we obtain:

$$\underset{\sim}{D} \equiv \underset{\sim}{0} \quad \Rightarrow \quad \underset{\sim}{T}_u = a_o \underset{\sim}{I} = \underset{\sim}{0} \quad \Rightarrow \quad a_o = 0 \tag{37}$$

Thus, if the solid admits a natural state, the isotropic linear elasticity depends on two material constants:

$$\underset{\sim}{T} = (b_o \mathrm{tr}\underset{\sim}{D})\underset{\sim}{I} + a_1 \underset{\sim}{D} \tag{38}$$

The invariant form (38) is equivalent to the classical form:

$$\underset{\sim}{T} = (\lambda \mathrm{tr}\underset{\sim}{D})\underset{\sim}{I} + 2\mu \underset{\sim}{D} \tag{39}$$

where λ and μ are Lamé's coefficients.

If the solid is a hyperelastic material, the strain-energy function is given by:

$$W = \frac{1}{2} b_o \mathrm{tr}^2 \underset{\sim}{D} + \frac{1}{2} a_1 \mathrm{tr} \underset{\sim}{D}^2 = \frac{1}{2} \lambda \mathrm{tr}^2 \underset{\sim}{D} + \mu \mathrm{tr} \underset{\sim}{D}^2 \tag{40}$$

In the isotropic case, the assumption of the existence of a strain-energy function introduces no integrability conditions.

REFERENCES

1. BOEHLER, J.P., A Simple Derivation of Representations for Non-Polynomial Constitutive Equations in Some Cases of Anisotropy, *ZAMM*, 59 (1979): 157-167.
2. BOEHLER, J.P., Sur les Formes Invariantes dans le Sous-Groupe Orthotrope de Révolution..., *ZAMM*, 55 (1975): 609-611.
3. BOEHLER, J.P., Contributions Théoriques et Expérimentales à l'Etude des Milieux Plastiques Anisotropes, *Thèse de Doctorat ès Sciences*, Grenoble, 1975.
4. GOLDENBLAT, I.I., *Some Problems of the Mechanics of Deformable Media*, Noordhoff, Groningen, 1962.
5. LEHMANN, Th., Anisotrope Plastische Formänderungen, *Rheol. Acta*, 3 (1964): 281-285.

Chapter 5

YIELDING AND FAILURE OF TRANSVERSELY ISOTROPIC SOLIDS

J.P. Boehler
University of Grenoble, France

1. INTRODUCTION

A suitable framework allowing to describe yielding and failure of anisotropic materials with the required generality and pertinence is furnished by the theory of representations for tensor functions. In this Chapter, invariant formulations of the plastic behavior and failure of transversely isotropic solids are developed. For other material symmetries, the corresponding constitutive relations can be derived in a similar manner.

Employing the results of Chapter 3, a general theory of the rate independent flow of transversely isotropic materials is developed. General invariant forms of both the flow law and the yield criterion are obtained. The type and the number of form-invariant tensor generators and of scalar invariants entering the flow law, as well as the rules for constructing failure criteria as scalar-valued functions of invariants of the stress and structural tensors are established. The restrictions imposed by the material symmetries are then automatically satisfied.

The essential features of the anisotropic flow law and their consequences in experimental investigations of the mechanical behavior of anisotropic solids are analyzed in Section 3. Specific forms of failure criteria are proposed for glass-fiber reinforced composites and compared with experimental data (Section 4).

A simplified theory employing a fourth order tensor of plastic anisotropy and a tensorially linear description of the material response is discussed in Section 5, in order to arrive at a workable form of the constitutive relation and at simple forms of the failure criteria for transversely isotropic solids.

2. GENERAL THEORY

Consider a transversely isotropic solid with privileged direction $\underset{\sim}{v}_3$. For further investigations of the plastic behavior, we restrict our attention to the general invariant form of the constitutive relation for a rate type material. Let $\underset{\sim}{T}$ denote the stress tensor and $\underset{\sim}{D}$ the rate of deformation tensor. In the case of transverse isotropy, the irreducible representation of $\underset{\sim}{T}$ considered as a tensor function of $\underset{\sim}{D}$ is obtained from Table IV of Chapter 3:

$$\left.\begin{aligned} &\underset{\sim}{T} = \alpha_o \underset{\sim}{I} + \alpha_1 \underset{\sim}{M} + \alpha_2 \underset{\sim}{D} + \alpha_3 (\underset{\sim}{M}\underset{\sim}{D}+\underset{\sim}{D}\underset{\sim}{M}) + \alpha_4 \underset{\sim}{D}^2 + \alpha_5 (\underset{\sim}{M}\underset{\sim}{D}^2+\underset{\sim}{D}^2\underset{\sim}{M}) \\ &\alpha_i = \alpha_i \,(\mathrm{tr}\underset{\sim}{D},\ \mathrm{tr}\underset{\sim}{D}^2,\ \mathrm{tr}\underset{\sim}{D}^3,\ \mathrm{tr}\underset{\sim}{M}\underset{\sim}{D},\ \mathrm{tr}\underset{\sim}{M}\underset{\sim}{D}^2) \end{aligned}\right\} \quad (1)$$

where $\underset{\sim}{M}$ is the structural tensor for transverse isotropy:

$$\underset{\sim}{M} = \underset{\sim}{v}_3 \otimes \underset{\sim}{v}_3 \quad (2)$$

Representation (1) can describe perfectly plastic behavior when it satisfies the condition of homogeneity of order zero with respect to time ([1], [2], [3]). The only time involving variable being the rate of deformation tensor $\underset{\sim}{D}$, the constitutive equation (1) is necessarily homogeneous of order zero with respect to $\underset{\sim}{D}$:

$$\frac{\partial \underset{\sim}{T}}{\partial \underset{\sim}{D}} \underset{\sim}{D} = \underset{\sim}{0} \quad \text{if} \quad \frac{\partial \underset{\sim}{T}}{\partial \underset{\sim}{D}} \neq \underset{\sim}{0} \tag{3}$$

The homogeneity condition (3) results in the following equations of Euler for the scalar functions α_i in (1):

$$\left.\begin{aligned} &\frac{\partial \alpha_i}{\partial E_j} E_i + n\alpha_i = 0 \\ \text{where:} \quad & n = \begin{cases} 0 \\ 1 \\ 2 \end{cases} \quad \text{when } i = \begin{cases} 0,\ 1 \\ 2,\ 3 \\ 4,\ 5 \end{cases} \; ; \; j = 1,2,3,4,5. \end{aligned}\right\} \tag{4}$$

In (4), the following notation is used for the arguments of the functions α_i:

$$\left.\begin{aligned} &E_1 = \mathrm{tr}\underset{\sim}{D} \; ; \; E_2 = \mathrm{tr}^{1/2} \underset{\sim}{D}^2 \; ; \; E_3 = \mathrm{tr}^{1/3} \underset{\sim}{D}^3, \\ &E_4 = \mathrm{tr}\underset{\sim}{M}\underset{\sim}{D} \; ; \; E_5 = \mathrm{tr}^{1/2} \underset{\sim}{M}\underset{\sim}{D}^2 . \end{aligned}\right\} \tag{5}$$

The solutions of equations (4) are:

$$\left.\begin{aligned} &\alpha_o = P(p,q,r,s) \quad ; \; \alpha_1 = Q(p,q,r,s) \quad ; \; \alpha_2 = \frac{1}{E_2} R(p,q,r,s,) \; ; \\ &\alpha_3 = \frac{1}{E_2} S(p,q,r,s); \; \alpha_4 = \frac{1}{E_2^2} T(p,q,r,s,); \alpha_5 = \frac{1}{E_2^2} U(p,q,r,s,) \quad , \end{aligned}\right\} \tag{6}$$

where P, Q, R, S, T and U are arbitrary scalar-valued functions of the following four dimensionless kinematic variables:

$$p = \frac{E_1}{E_2} \; ; \; q = \frac{E_3}{E_2} \; ; \; r = \frac{E_4}{E_2} \; ; \; s = \frac{E_5}{E_2} . \tag{7}$$

Introducing the solutions (6) into the representation (1), we obtain the general invariant form for the plastic constitutive law of a transversely isotropic solid:

$$\underset{\sim}{T} = \beta_o \underset{\sim}{I} + \beta_1 \underset{\sim}{M} + \frac{\beta_2}{E_2} \underset{\sim}{D} + \frac{\beta_3}{E_2} (\underset{\sim}{M}\underset{\sim}{D}+\underset{\sim}{D}\underset{\sim}{M}) + \frac{\beta_4}{E_2^2} \underset{\sim}{D}^2 + \frac{\beta_5}{E_2^2}(\underset{\sim}{M}\underset{\sim}{D}^2+\underset{\sim}{D}^2\underset{\sim}{M}) \; ; \; \beta_i = \beta_i(p,q,r,s) \tag{8}$$

The tensor form (8) is equivalent to the system of five scalar relations which specify the following five invariants of the functional basis of $\underset{\sim}{T}$ in the transversely isotropic case:

$$\mathrm{tr}\underset{\sim}{T} \quad , \quad \mathrm{tr}\underset{\sim}{T}^2 \quad , \quad \mathrm{tr}\underset{\sim}{T}^3 \quad , \quad \mathrm{tr}\underset{\sim}{M}\underset{\sim}{T} \quad , \quad \mathrm{tr}\underset{\sim}{M}\underset{\sim}{T}^2 \tag{9}$$

These relations are established in calculating the tensors $\underset{\sim}{T}^2$, $\underset{\sim}{T}^3$, $\underset{\sim}{M}\underset{\sim}{T}+\underset{\sim}{T}\underset{\sim}{M}$ and $\underset{\sim}{M}\underset{\sim}{T}^2+\underset{\sim}{T}^2\underset{\sim}{M}$ from the tensor relation (8) and then taking the traces of the obtained expressions. Some reductions are obtained by the application of the generalization for 3 three-dimensional second order tensors of the Cayley-Hamilton theorem, due to RIVLIN [4].

The Cayley-Hamilton theorem for a three-dimensional second order tensor is given by:

$$\underset{\sim}{A}^3 - (\mathrm{tr}\underset{\sim}{A})\underset{\sim}{A}^2 + \frac{1}{2}(\mathrm{tr}^2\underset{\sim}{A}-\mathrm{tr}\underset{\sim}{A}^2)\underset{\sim}{A} - \frac{1}{3}(\mathrm{tr}\underset{\sim}{A}^3-\frac{3}{2}\mathrm{tr}^2\underset{\sim}{A}\ \mathrm{tr}\underset{\sim}{A} + \frac{1}{2}\mathrm{tr}^3\underset{\sim}{A})\underset{\sim}{I} = \underset{\sim}{0} \tag{10}$$

If in (10) we replace $\underset{\sim}{A}$ by $\underset{\sim}{A}+\lambda\underset{\sim}{B}+\mu\underset{\sim}{C}$, wher λ and μ are arbitrary scalars, and equate to zero the coefficient of $\lambda\mu$, we obtain the folloqing generalization of (10):

$$\begin{aligned}&\underset{\sim}{A}\underset{\sim}{B}\underset{\sim}{C} + \underset{\sim}{A}\underset{\sim}{C}\underset{\sim}{B} + \underset{\sim}{C}\underset{\sim}{A}\underset{\sim}{B} + \underset{\sim}{B}\underset{\sim}{A}\underset{\sim}{C} + \underset{\sim}{B}\underset{\sim}{C}\underset{\sim}{A} + \underset{\sim}{C}\underset{\sim}{B}\underset{\sim}{A} +\\ &\quad - (\mathrm{tr}\underset{\sim}{B}\underset{\sim}{C}-\mathrm{tr}\underset{\sim}{B}\mathrm{tr}\underset{\sim}{C})\underset{\sim}{A} - (\mathrm{tr}\underset{\sim}{C}\underset{\sim}{A}-\mathrm{tr}\underset{\sim}{C}\mathrm{tr}\underset{\sim}{A})\underset{\sim}{B} - (\mathrm{tr}\underset{\sim}{A}\underset{\sim}{B}-\mathrm{tr}\underset{\sim}{A}\mathrm{tr}\underset{\sim}{B})\underset{\sim}{C} +\\ &\quad - \mathrm{tr}\underset{\sim}{A}(\underset{\sim}{B}\underset{\sim}{C}+\underset{\sim}{C}\underset{\sim}{B}) - \mathrm{tr}\underset{\sim}{B}(\underset{\sim}{C}\underset{\sim}{A}+\underset{\sim}{A}\underset{\sim}{C}) - \mathrm{tr}\underset{\sim}{C}(\underset{\sim}{A}\underset{\sim}{B}+\underset{\sim}{B}\underset{\sim}{A}) +\\ &\quad - (\mathrm{tr}\underset{\sim}{A}\mathrm{tr}\underset{\sim}{B}\mathrm{tr}\underset{\sim}{C}-\mathrm{tr}\underset{\sim}{A}\mathrm{tr}\underset{\sim}{B}\underset{\sim}{C}-\mathrm{tr}\underset{\sim}{C}\mathrm{tr}\underset{\sim}{A}\underset{\sim}{B}-\mathrm{tr}\underset{\sim}{B}\mathrm{tr}\underset{\sim}{A}\underset{\sim}{C}+\mathrm{tr}\underset{\sim}{A}\underset{\sim}{B}\underset{\sim}{C}+\mathrm{tr}\underset{\sim}{C}\underset{\sim}{B}\underset{\sim}{A})\underset{\sim}{I} \equiv \underset{\sim}{0}\end{aligned} \tag{11}$$

The method allowing the expressions of the invariants (9) to be reduced by application of the identity (11) is presented in [1]. Finally, after reduction, we obtain the following system of scalar relations:

$$\left.\begin{aligned}\mathrm{tr}\underset{\sim}{T} &= g_1(p,q,r,s\ )\\ \mathrm{tr}\underset{\sim}{T}^2 &= g_2(p,q,r,s\ )\\ \mathrm{tr}\underset{\sim}{T}^3 &= g_3(p,q,r,s\ )\\ \mathrm{tr}\underset{\sim}{M}\underset{\sim}{T} &= g_4(p,q,r,s\ )\\ \mathrm{tr}\underset{\sim}{M}\underset{\sim}{T}^2 &= g_5(p,q,r,s\ )\end{aligned}\right\} \tag{12}$$

When the four dimensionless kinematic variables (7) are independent, their elimination from the five relations (12) results in the most general invariant form of the yield criterion for a transversely isotropic solid:

$$f(\mathrm{tr}\underset{\sim}{T},\ \mathrm{tr}\underset{\sim}{T}^2,\ \mathrm{tr}\underset{\sim}{T}^3,\ \mathrm{tr}\underset{\sim}{M}\underset{\sim}{T},\ \mathrm{tr}\underset{\sim}{M}\underset{\sim}{T}^2) = 0 \qquad (13)$$

Singular cases of yield loci correspond to the case where the dimensionless kinematic variables (7) are not independent, i.e. when kinematic constraints are imposed on the motion of the solid.

In the general form (13) of the yield condition, the independent variables involve, in addition to the isotropic invariants $\mathrm{tr}\underset{\sim}{T}$, $\mathrm{tr}\underset{\sim}{T}^2$, $\mathrm{tr}\underset{\sim}{T}^3$, two mixed invariants $\mathrm{tr}\underset{\sim}{M}\underset{\sim}{T}$, $\mathrm{tr}\underset{\sim}{M}\underset{\sim}{T}^2$, which account for the anisotropic character of the material behavior. They specify the orientation of the principal directions of the stress tensor with respect to the privileged axis of the transversely isotropic material. In an arbitrary privileged orthonormal frame $(\cdot, \cdot, \underset{\sim}{v}_3)$, the mixed invariants are expressed by:

$$\mathrm{tr}\underset{\sim}{M}\underset{\sim}{T} = T_{33} \quad ; \quad \mathrm{tr}\underset{\sim}{M}\underset{\sim}{T}^2 = T_{23}^{\ 2} + T_{31}^{\ 2} + T_{33}^{\ 2} \qquad (14)$$

The general invariant form (13) is of importance in the sense that it specifies the minimal number and the type of independent variables to be observed in experiments.

In order to derive the general invariant form of the flow law, we invert the plastic constitutive law (8). The method consists in calculating from (8) the expressions of the tensors:

$$\underset{\sim}{T},\ \underset{\sim}{T}^2,\ \underset{\sim}{M}\underset{\sim}{T}+\underset{\sim}{T}\underset{\sim}{M},\ \underset{\sim}{M}\underset{\sim}{T}^2+\underset{\sim}{T}^2\underset{\sim}{M} \qquad (15)$$

The above-mentioned generalization (11) of the Cayley-Hamilton theorem allows the obtained expressions to be reduced; the employed method is presented in [1]. After reduction, the tensors (15) are expressed with respect to the tensors $\underset{\sim}{I}$, $\underset{\sim}{M}$, $\underset{\sim}{D}$, $\underset{\sim}{D}^2$, $\underset{\sim}{M}\underset{\sim}{D}+\underset{\sim}{D}\underset{\sim}{M}$ and $\underset{\sim}{M}\underset{\sim}{D}^2+\underset{\sim}{D}^2\underset{\sim}{M}$, which are the tensor generators of the constitutive equations (8). The elimination of $\underset{\sim}{D}^2$, $\underset{\sim}{M}\underset{\sim}{D}+\underset{\sim}{D}\underset{\sim}{M}$ and $\underset{\sim}{M}\underset{\sim}{D}^2+\underset{\sim}{D}^2\underset{\sim}{M}$ between the reduced expressions of the tensors (15) results in the constitutive equations (8) solved with respect to $\underset{\sim}{D}$:

$$\left.\begin{aligned} \frac{\underset{\sim}{D}}{\sqrt{tr\underset{\sim}{D}^2}} &= \Psi_0\underset{\sim}{I} + \Psi_1\underset{\sim}{M} + \Psi_2\underset{\sim}{T} + \Psi_3(\underset{\sim}{M}\underset{\sim}{T}+\underset{\sim}{T}\underset{\sim}{M}) + \Psi_4\underset{\sim}{T}^2 + \Psi_5(\underset{\sim}{M}\underset{\sim}{T}^2+\underset{\sim}{T}^2\underset{\sim}{M}) \\ \Psi_i &= \Psi_i(p,q,r,s) \end{aligned}\right\} \quad (16)$$

In (16), the scalar-valued functions Ψ_i can also be expressed as functions of the basic stress invariants (9), since the dimensionless kinematic variables p,q,r, and s are expressed with respect to the invariants (9), when eliminating them from relations (12). However, the invariants (9) are no longer independent variables, as they must verify the yield criterion (13), which is the condition for plastic motion. Thus, in (16) the scalar-valued functions Ψ_i can be expressed as:

$$\left.\begin{aligned} &\Psi_i = \Psi_i(tr\underset{\sim}{T},\ tr\underset{\sim}{T}^2,\ tr\underset{\sim}{T}^3,\ tr\underset{\sim}{M}\underset{\sim}{T},\ tr\underset{\sim}{M}\underset{\sim}{T}^2) \\ &\text{with the restriction (13):} \\ &f(tr\underset{\sim}{T},\ tr\underset{\sim}{T}^2,\ tr\underset{\sim}{T}^3,\ tr\underset{\sim}{M}\underset{\sim}{T},\ tr\underset{\sim}{M}\underset{\sim}{T}^2) = 0 \end{aligned}\right\} \quad (17)$$

Relation (16) constitutes the most general invariant form of the plastic flow law for a transversely isotropic solid. By (16), the rate of deformation tensor is specified only to within a scalar multiplier. The presence in (16) of tensor generators involving the structural tensor $\underset{\sim}{M}$ results, in general, in non-coincidence of the principal directions of $\underset{\sim}{T}$ and $\underset{\sim}{D}$, except when the principal directions of $\underset{\sim}{D}$ coincide with that of $\underset{\sim}{M}$, i.e. with $(\cdot, \cdot, \underset{\sim}{v}_3)$. This property is further analyzed in Section 3.

It is worthwhile to point out that once the scalar-valued functions β_i of the invariant form (8) for the plastic constitutive law are known, both the general form (13) of the yield condition and the general form (16) of the flow law are specified. Thus, the flow law and the yield condition are interrelated. However, in this general approach by the theory of representations for tensor functions to the formulation of the plastic behavior of solids, the flow law is not necessarily associated with the yield condition by the rule of plastic potential. Theories employing the plastic potential result in restricted forms of the flow law, whatever general flow potential is assumed. This point is further discussed in Chapter 14 by Professor BETTEN.

Further details, as well as the analysis of incompressibility and plane strain, are given in [1-3].

A quasi-linear form of the representation (1) is obtained by neglecting the generators involving $\underset{\sim}{D}^2$, hence:

$$\left.\begin{aligned} \underset{\sim}{T} &= \alpha_0 \underset{\sim}{I} + \alpha_1 \underset{\sim}{M} + \alpha_2 \underset{\sim}{D} + \alpha_3 (\underset{\sim}{M}\underset{\sim}{D}+\underset{\sim}{D}\underset{\sim}{M}) \\ \alpha_i &= \alpha_i(\mathrm{tr}\underset{\sim}{D},\ \mathrm{tr}\underset{\sim}{D}^2,\ \mathrm{tr}\underset{\sim}{D}^3,\ \mathrm{tr}\underset{\sim}{M}\underset{\sim}{D},\ \mathrm{tr}\underset{\sim}{M}\underset{\sim}{D}^2) \end{aligned}\right\} \qquad (18)$$

A similar procedure leads to the same general invariant form (13) for the yield criterion and to the following invariant quasi-linear form for the flow law:

$$\left.\begin{aligned} &\frac{\underset{\sim}{D}}{\sqrt{\mathrm{tr}\underset{\sim}{D}^2}} = \Psi_0 \underset{\sim}{I} + \Psi_1 \underset{\sim}{M} + \Psi_2 \underset{\sim}{T} + \Psi_3 (\underset{\sim}{M}\underset{\sim}{T}+\underset{\sim}{T}\underset{\sim}{M}) \\ &\Psi_i = \Psi_i(p,q,r,s) \\ &\text{or} \begin{cases} \Psi_i = \Psi_i(\mathrm{tr}\underset{\sim}{T},\ \mathrm{tr}\underset{\sim}{T}^2,\ \mathrm{tr}\underset{\sim}{T}^3,\ \mathrm{tr}\underset{\sim}{M}\underset{\sim}{T},\ \mathrm{tr}\underset{\sim}{M}\underset{\sim}{T}^2) \\ f(\mathrm{tr}\underset{\sim}{T},\ \mathrm{tr}\underset{\sim}{T}^2,\ \mathrm{tr}\underset{\sim}{T}^3,\ \mathrm{tr}\underset{\sim}{M}\underset{\sim}{T},\ \mathrm{tr}\underset{\sim}{M}\underset{\sim}{T}^2) = 0 \end{cases} \end{aligned}\right\} \qquad (19)$$

In this case, the explicit calculation necessary to derive (13) and (19) from the quasi-linear form (1) and the homogeneity condition (3) is given in [3].

3. PLASTIC DEFORMATIONS IN UNIAXIAL AND TRIAXIAL TESTS

In this Section, we analyze the essential feature of the anisotropic flow law (16), i.e. the non-coincidence of the principal directions of the stress and rate of deformation tensors in "off-axis" specimens, as well as the consequences on the plastic deformations produced in uniaxial and triaxial tests.

We consider first triaxial tests consisting of axial compressions or tractions σ_n on oriented specimens that are subjected to confining pressures p. We denote by θ the angle between the axis $(\underset{\sim}{e}_3)$ of the specimen and the privileged direction $\underset{\sim}{v}_3$ of the material (Fig.1).

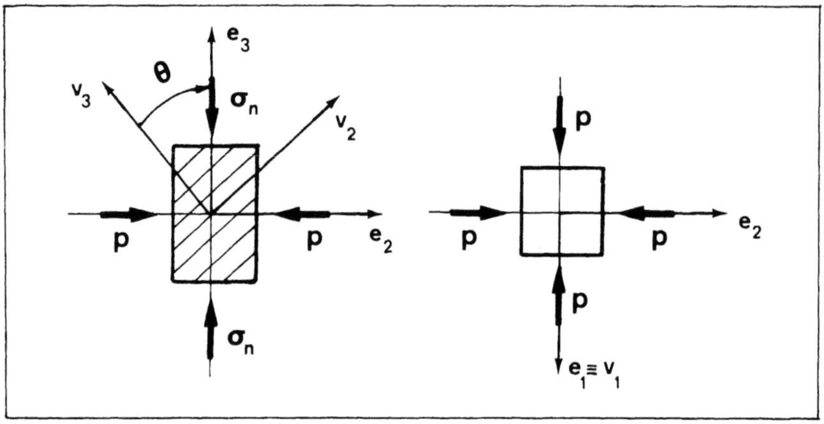

Fig.1 - Triaxial tests on transversely isotropic materials

If we assume that stresses are prescribed during the test, the stress tensor $\underset{\sim}{T}$ is defined in the axes of the specimens by:

$$T = \begin{vmatrix} p & 0 & 0 \\ 0 & p & 0 \\ 0 & 0 & \sigma_n \end{vmatrix}_{(\underset{\sim}{v}_1,\underset{\sim}{e}_2,\underset{\sim}{e}_3)} \tag{20}$$

For the stress (20), we obtain the following components for the rate of deformation tensor, as obtained from the law (16) and expressed in the principal frame $(\underset{\sim}{v}_1,\underset{\sim}{e}_2,\underset{\sim}{e}_3)$ of the stress tensor:

$$\left.\begin{aligned}
\frac{D_{11}}{\sqrt{\mathrm{tr}\underset{\sim}{D}^2}} &= \Psi o + p\Psi_2 + p^2\Psi_4 \\
\frac{D_{22}}{\sqrt{\mathrm{tr}\underset{\sim}{D}^2}} &= \Psi_o + \Psi_1\sin^2\theta + (\Psi_2+2\Psi_3\sin^2\theta)p + (\Psi_4+2\Psi_5\sin^2\theta)p^2 \\
\frac{D_{33}}{\sqrt{\mathrm{tr}\underset{\sim}{D}^2}} &= \Psi_o + \Psi_1\cos^2\theta + (\Psi_2+2\Psi_3\cos^2\theta)\sigma_n + (\Psi_4+2\Psi_5\cos^2\theta)\sigma_n^2 \\
\frac{D_{23}}{\sqrt{\mathrm{tr}\underset{\sim}{D}^2}} &= [\Psi_1 + \Psi_3(\sigma_n+p) + \Psi_5(\sigma_n^2+p^2)]\sin\theta\cos\theta \quad ; \quad D_{31} = D_{12} = 0.
\end{aligned}\right\} \tag{21}$$

One can conclude that the principal directions of the stress tensor $\underset{\sim}{T}$ and the rate of deformation tensor $\underset{\sim}{D}$ coincide if, and only if:

$$[\Psi_1 + \Psi_3(\sigma_n+p) + \Psi_5(\sigma_n^2+p^2)]\sin\theta\cos\theta = 0 \qquad (22)$$

Thus, for the so-called "in-axis" tests, i.e. when the privileged direction $\underset{\sim}{v}_3$ of the material coincide with one of the principal directions of the stress tensor ($\theta = 0$ or $\pi/2$), the principal directions of $\underset{\sim}{T}$ and $\underset{\sim}{D}$ are the same. For "off-axis" tests ($\theta \neq 0$ and $\pi/2$), the principal directions of $\underset{\sim}{T}$ and $\underset{\sim}{D}$ cannot coincide in general.

In fact, in a standard compression test under confining pressure, mixed boundary conditions are prescribed, since D_{33} rather than $T_{33} = \sigma_n$ is enforced in the axial direction $\underset{\sim}{e}_3$ of the specimens. Thus, on the ends of the specimens, the principal directions of the imposed rate of deformation tensor are the axes of the specimens. On the lateral sides, the principal stresses $T_{11} = T_{22} = p$ are prescribed, without shear stress. Thus, on the lateral sides of the specimens, the principal directions of the imposed stress tensor are also the axes of the specimens. Now, if the developed stress and kinematic fields were homogeneous, the axes ($\underset{\sim}{v}_1$, $\underset{\sim}{e}_2$, $\underset{\sim}{e}_3$) of the specimens would be the principal directions of both the stress tensor and the rate of deformation tensor. As this is impossible in general for off-axis tests, the developed stress and kinematic fields in specimens of transversely isotropic material for $0 < \theta < \pi/2$ are heterogeneous; in fact, a shear stress $T_{23} \neq 0$ appears on the contact of the specimens with the platens and the specimens tend to assume a S-shape.

Experimental evidence of the S-shaped deformation of off-axis specimens of transversely isotropic materials subjected to standard compressions under confining pressure is given in ([2], [5], [6]) for a stratified rock (diatomite). Fig.2 shows an undeformed specimen, which was drilled out at the orientation $\theta = 60°$; Fig.2b shows the deformed shape of the specimen after axial compression (σ_n = 13.2 MPa) under a confining pressure p = 6 MPa.

The same reservations apply to other tests and particularly to the standard uniaxial tensile tests, where the ends of the specimens are

a) b)

Fig.2 - Nonhomogeneity of deformation in compression tests under confining pressure on a stratified rock (after [2]); a) specimen before loading ; b) specimen after unloading.

rigidly clamped and subjected to a translation parallel to the axis of the specimens, without allowing rotation. The standard uniaxial test on anisotropic materials is analyzed in ([7], [8]). The S-shaped deformation of off-axis specimens is studied numerically in ([9], [10]) for composite materials and analyzed experimentally in ([8], [9]) for rolled sheet-steel.

As standard tests on anisotropic materials are generally interpreted on the assumption of homogeneity for the developed stress and kinematic fields, the experimental data obtained for "off-axis" specimens are erroneous. This indicates a need to imagine new experimental techniques, suitable for testing the mechanical behavior of anisotropic solids.

In order to avoid the S-shape effect in standard uniaxial tensile tests on off-axis specimens, a new testing procedure has been proposed in [8], [10], [11]. The numerical analysis for composite materials ([9], [10]) and the experimental results obtained for rolled sheet-steel ([8], [9], [10]) indicate that the new testing device allows homogeneous kinematic fields to be obtained in off-axis specimens and thus leads to reliable measurement of the mechanical properties of anisotropic solids (see also

Chapter 7, Section 4.1).

Consider now a hydrostatic pressure test:

$$\underset{\sim}{T} = p\underset{\sim}{I} \tag{23}$$

The non-vanishing components of the rate of deformation tensor, as obtained from the law (16) and expressed in a privileged frame $(\cdot, \cdot, \underset{\sim}{v}_3)$ of the material, are the following:

$$\left.\begin{aligned}
&\frac{D_{11}}{\sqrt{\mathrm{tr}\underset{\sim}{D}^2}} = \frac{D_{22}}{\sqrt{\mathrm{tr}\underset{\sim}{D}^2}} = \Psi_0 + \Psi_2 p + \Psi_4 p^2, \\
&\frac{D_{33}}{\sqrt{\mathrm{tr}\underset{\sim}{D}^2}} = \Psi_0 + \Psi_1 + (\Psi_2+2\Psi_3)p + (\Psi_4+2\Psi_5)p^2 \\
&D_{23} = D_{31} = D_{12} = 0
\end{aligned}\right\} \tag{24}$$

The rate of volume change is given by:

$$\frac{\mathrm{tr}\underset{\sim}{D}}{\sqrt{\mathrm{tr}\underset{\sim}{D}^2}} = 3\Psi_0 + \Psi_1 + (3\Psi_2+2\Psi_3)p + (3\Psi_4+2\Psi_5)p^2 \tag{25}$$

One can conclude that the principal directions of the rate of deformation tensor $\underset{\sim}{D}$ coincide with the privileged directions $(\cdot, \cdot, \underset{\sim}{v}_3)$ of the transversely isotropic material. Since $D_{33} \neq D_{11} = D_{22}$, the deviatoric part of $\underset{\sim}{D}$ is different from zero; this constitutes an essential difference from the behavior under hydrostatic pressure of isotropic materials.

An experimental study has been developed in [12] in order to show that yielding and hardening of a soft rock occur under hydrostatic pressure and that the directional properties of an anisotropic material result in a non-isotropic deformation under isotropic pressures. The material used was diatomite, a soft stratified rock. Cylindrical specimens were drilled out from blocks in seven different directions, oriented at $\theta = 0°$, $15°$, $30°$, $45°$, $60°$, $75°$ and $90°$ with respect to the privileged axis $\underset{\sim}{v}_3$ of the material, which is the normal to the strata.

In a first series of experiments, virgin specimens were subjected to monotically increasing isotropic stress, up to the required intensity and then unloaded. In Fig.3, the irreversible deformations $\Delta L_i/L_i$ in each $\underset{\sim}{v}_i$ direction are plotted against the hydrostatic pressure. Marked permanent deformation occurs between 10 and 15 MPa; in the isotropic plane $(\underset{\sim}{v}_1, \underset{\sim}{v}_2)$, permanent deformations are much smaller. For higher pressures, significant hardening occurs and the pressure-volume change relation tends toward a locking behavior when no volume changes occur.

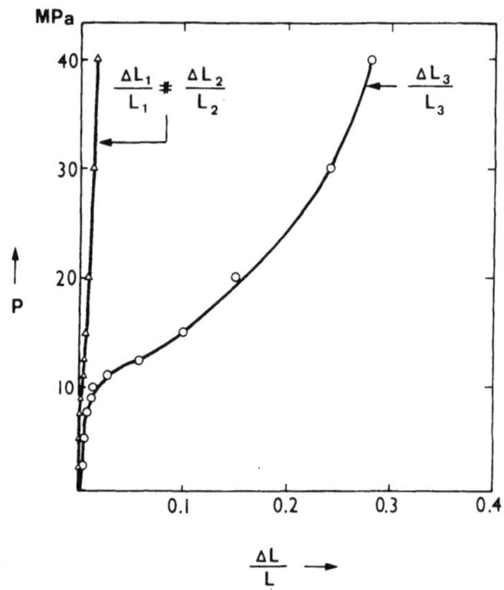

Fig.3 - Deformations of diatomite under isotropic pressure (after [12]).

The directional character of deformation under isotropic pressures was studied in the second series of tests for the seven values of the orientation angle θ. The pressure went up to 100 MPa. A general view of deformed specimens is given in Fig.4. The off-axis cylindrical specimens $(0 < \theta < \pi/2)$ were deformed into elliptic inclined cylinders; this confirms the previsions of the theory as to the orientation of the strain principal directions $(\cdot, \cdot, \underset{\sim}{v}_3)$ with respect to the oriented cylinders, as well as the regularity of the material response. In Fig.5, the deformed and undeformed cross-sections are superimposed. Fig.5a shows the permanent deformation in the plane of isotropy (orientation θ = 0); Fig.5b (orientation θ = 90°) furnishes information as to the strain ellipsoid for a transversely isotropic material subjected to an isotropic pressure.

Fig.4 - View of oriented cylinders of diatomite deformed by isotropic stress (after [12]).

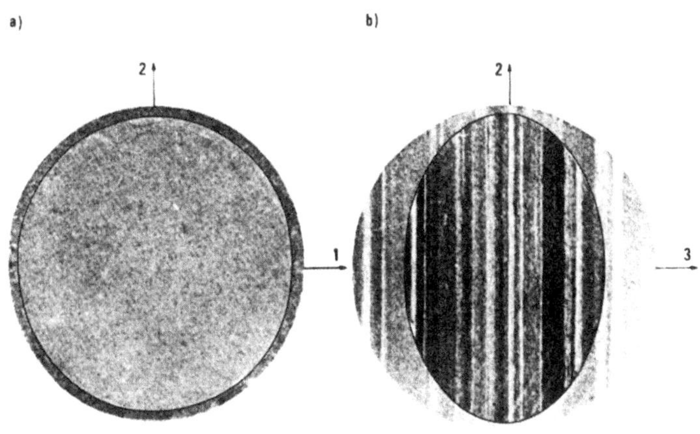

Fig.5 - Permanent deformation of cylindrical specimens of diatomite under isotropic pressure (after [12]). a): θ = 0°; b): θ = 90°.

4. FAILURE CRITERIA FOR GLASS/EPOXY COMPOSITES UNDER CONFINING PRESSURE

4.1 Introduction

The general invariant form (13) of the yield criterion for transversely isotropic materials requires specifications for each class of materials. Such specifications can be obtained from experimental investigations. In this Section, we consider failure of glass-fiber reinforced composites subjected to simple tensile tests and compression tests under different confining pressures. An experimental study on glass-fiber mat laminates discloses complex phenomena in the failure range of such materials: pronounced anisotropy of the discontinuous type associated with the appearance of different failure modes, marked stress-sign sensitivity and hydrostatic stress effects. We propose an elaborate failure condition, which accounts for the different observed phenomena and furnishes correct predictions for both directional strengths and corresponding failure modes. Much of this Section is taken from [13].

4.2 General form of the yield criterion for triaxial tests

An equivalent form of the general invariant form (13) of the yield criterion for transversely isotropic materials is given by:

$$g(\mathrm{tr}\underset{\sim}{T},\ \mathrm{tr}\underset{\sim}{S}^2,\ \mathrm{tr}\underset{\sim}{S}^3,\ \mathrm{tr}\underset{\sim}{M}\underset{\sim}{S},\ \mathrm{tr}\underset{\sim}{M}\underset{\sim}{S}^2) = 0 \tag{26}$$

where $\underset{\sim}{S}$ is the stress deviator:

$$\underset{\sim}{S} = \underset{\sim}{T} - (\tfrac{1}{3}\,\mathrm{tr}\underset{\sim}{T})\underset{\sim}{I} \tag{27}$$

We consider triaxial tests consisting of axial compressions or tractions σ_n on specimens oriented at $\theta°$ to the privileged axis $\underset{\sim}{v}_3$ of the material and subjected to confining pressures p (cf. Section 3 and Fig.1). For such stress conditions, the arguments of the criterion (26) can be expressed in terms of the three variables σ_n, p and θ:

$$\left.\begin{aligned} &tr\underset{\sim}{T} = 2p+\sigma_n \quad ; \quad tr\underset{\sim}{S}^2 = \frac{2}{3}(\sigma_n-p)^2 \quad ; \quad tr\underset{\sim}{S}^3 = \frac{2}{9}(\sigma_n-p)^3 \; ; \\ &tr\underset{\sim}{M}\underset{\sim}{S} = \frac{1}{3}(\sigma_n-p)(3\cos^2\theta-1) \quad ; \quad tr\underset{\sim}{M}\underset{\sim}{S}^2 = \frac{1}{9}(\sigma_n-p)(3\cos^2\theta+1). \end{aligned}\right\} \tag{28}$$

Thus, if restricted to tests under confining pressure, the five arguments of the general form (26) are no longer independent; they are related by:

$$\left\{\begin{aligned} &tr\underset{\sim}{M}\underset{\sim}{S}^2 = \frac{1}{3}tr\underset{\sim}{S}^2 + \frac{\varepsilon}{\sqrt{6}}\, tr\underset{\sim}{M}\underset{\sim}{S}\; tr^{1/2}\underset{\sim}{S}^2 \\ &tr\underset{\sim}{S}^3 = \varepsilon\,\frac{2}{3\sqrt{6}}\, tr^{3/2}\underset{\sim}{S}^2 \end{aligned}\right. \quad \text{where} \quad \left\{\begin{aligned} &\varepsilon = 1 \text{ if } \sigma_n-p > 0 \\ &\varepsilon = -1 \text{ if } \sigma_n-p < 0 \end{aligned}\right. \tag{29}$$

Only three invariants remain independent and the general form of the yield condition for triaxial tests is reduced to:

$$h(tr\underset{\sim}{T},\ tr\underset{\sim}{S}^2,\ tr\underset{\sim}{M}\underset{\sim}{S}) = 0 \tag{30}$$

Further details, as well as the visualization of the yield criterion (30) in the space of the basic invariants, are given in [14].

4.3 Failure modes and directional strengths

We performed a number of tests on a chopped strand mat laminate (glass fiber/epoxy resin). Such materials are transversely isotropic, the privileged axis $\underset{\sim}{v}_3$ being the normal to the mats' plane. Specimens with a parallelepipedic shape were cut in seven different orientations θ to the $\underset{\sim}{v}_3$ axis (θ = 0, 15, 30, 45, 60, 75 and 90°) and subjected to axial compressions under four different confining pressures: p = 0, 25, 50 and 75 MPa. Simple tensile tests were performed on specimens with a special shape ([13]).

The compression tests under confining pressures disclose two different modes of failure. For the specimens inclined at θ = 45, 60 and 75°, whatever the confining pressure, failure occurs by sliding with friction on a plane parallel to the mats; this well-defined failure mode will be called "parallel mode". For the other orientations (θ = 0, 15, 30 and

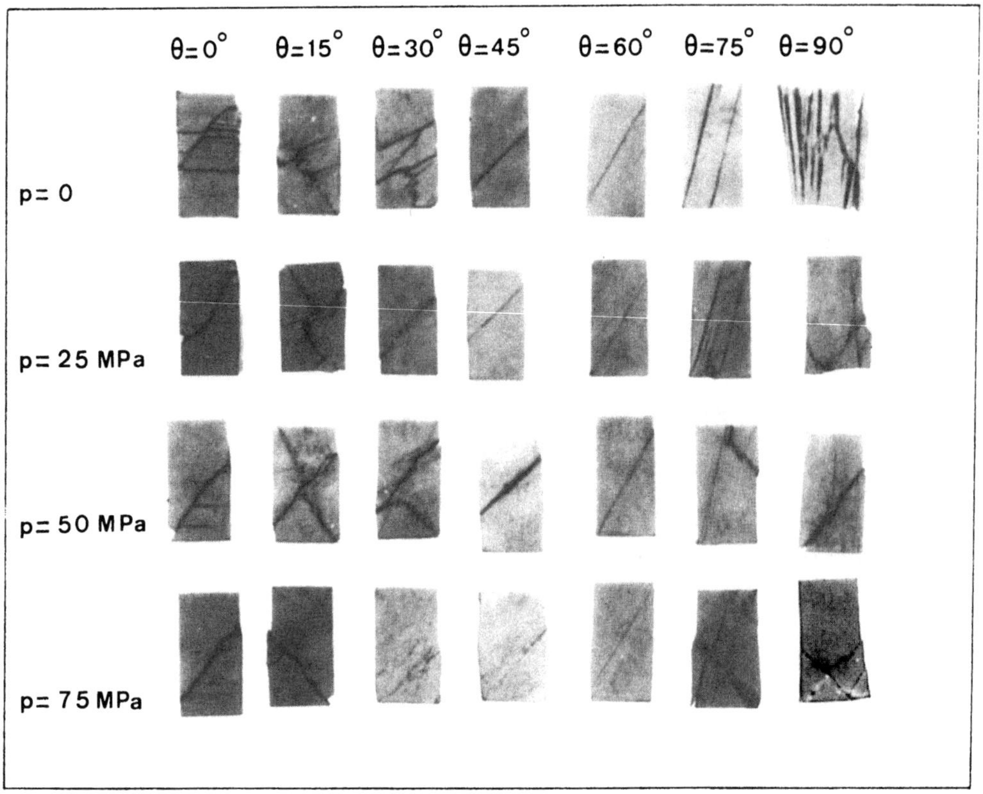

Fig.6 - Failure modes in compression under confining pressures (after [13]).

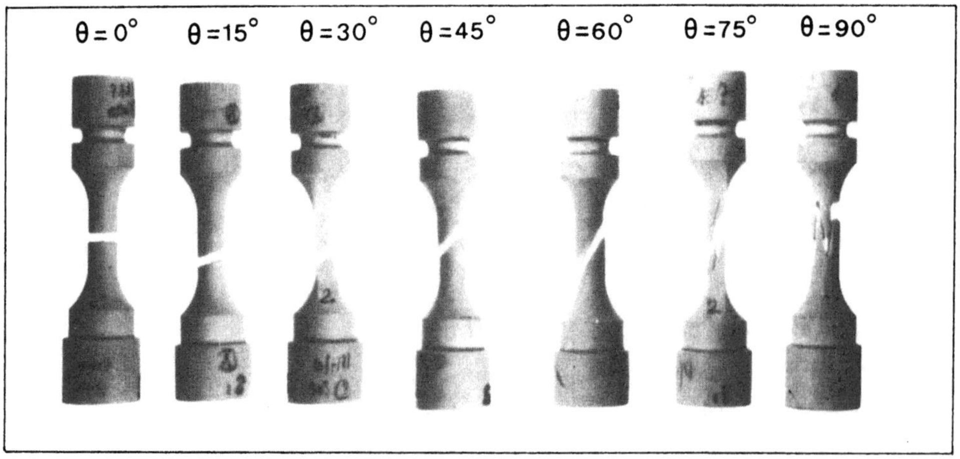

Fig.7 - Failure modes in simple tension (after[13]).

and 90°), failure occurs in planes crossing the mats; this mode of failure will be called "across mode". Simple compression parallel to the mats (θ = 90°, p = 0) is a particular case, where failure is produced by an extension parallel to the mats; when the confining pressure increases, this particular failure mode is replaced by the "across mode". These phenomena are visible in Fig.6, where one deformed specimen, after failure, is presented for each orientation and each confining pressure; the visible sides are those with inclined mat planes.

Failure modes in simple tensile tests are quite different from that observed in compression (Fig.7). For the orientations θ = 0 to 60°, failure occurs by an extension parallel to the mats, which is progressively accompanied by a shear stress parallel to the strata when θ increases from 15 to 60°. For θ = 90°, failure occurs in several planes crossing the mats' planes. At θ = 75°, mixed phenomena are observed.

For the compression tests under confining pressures, the obtained values for the limit over-stresses σ_n-p are presented in Table I. Whatever the confining pressure, a marked drop in the directional strength is observed at the transitions from θ = 30° to θ = 45° and from θ= 90° to θ = 75°, which correspond to the transitions from the "across mode" to the "parallel mode" of failure. Thus, the anisotropy of the triaxial strengths is of the discontinuous type [1]. In simple tensile tests, the

p(MPa)	θ = 0	θ = 15°	θ = 30°	θ = 45°	θ = 60°	θ = 75°	θ = 90°	
0	17	21	20	38	50	84	230	traction
0	480	410	340	160	155	208	325	compression
25	505	458	375	180	165	225	383	"
50	525	476	405	220	180	268	414	"
75	550	484	415	318	198	290	449	"

Table I - Directional limit over-stresses σ_n-p (MPa) in compression under confining pressures and in simple tension (after [13]).

directional strengths increase strongly for θ > 45°.

The ratios r of the directional strengths in simple compression with respect to that in simple tension (Table II) decrease when the orientation angle θ increases. The values of the ratios r are very high for θ = 0° to θ = 30° and a marked drop is observed between the orientations θ = 30° and θ = 45°, corresponding to the transition from the "across mode" to the "parallel mode" of failure in compression. These results disclose the strong sensitivity of the material to the sign of the normal stress, as well as the importance of the effects due to the orientation of the applied stress tensor.

θ	0°	15°	30°	45°	60°	75°	90°
r	27,6	19,3	16,9	4,5	3,1	2,5	1,4

Table II - Ratios r of the simple compressive strengths with respect to the simple tensile strengths (after [13]).

4.4 Proposed failure criteria

The criterion of Tsai and Wu [15], which is widely used in engineering, takes into account sensitivity of composite materials to the sign of normal stresses, as well as hydrostatic pressure effects. Nevertheless, the comparison between the experimental values of the directional strengths and the predictions of the Tsai and Wu criterion shows that this criterion does not adequately predict the actual behavior of mat laminates in compression under confining pressure [13]; in fact, this criterion underestimates strongly directional strengths for the orientations θ = 15, 30 and 45°. Thus, it is necessary to develop a more elaborate yield condition.

The analysis of the directional strengths in compression under confining pressures shows that for the tested composite the anisotropy of the strength is of the discontinuous type and that failure occurs with two different well-defined modes, according to the orientation of the specimen. As a reliable criterion must account for this phenomenon, we propose a failure condition composed of two continuously derivable branches, each branch characterizing a failure mode. A similar approach has been proposed by Hashin [16] in his study on failure criteria for unidirectional fiber-reinforced composites under tensile stress.

For compressive stress states under confining pressures and the "parallel mode" of failure, a generalization of Coulomb's law of friction is proposed:

$$|\sigma_{23}| = (1+a_0p+a_1p^2)\sigma_{33}\ \tan\phi_0 + (1+a_2p+a_3p^2)c_0 \qquad (31)$$

where σ_{23} is the stress tangential to the failure plane, σ_{33} is the stress normal to this plane and p is the confining pressure (Fig.8). The constants ϕ_0 and c_0 are the friction angle for p = 0 and the shear strength in the mat's plane respectively. The material constants a_i account for the influence of the confining pressure.

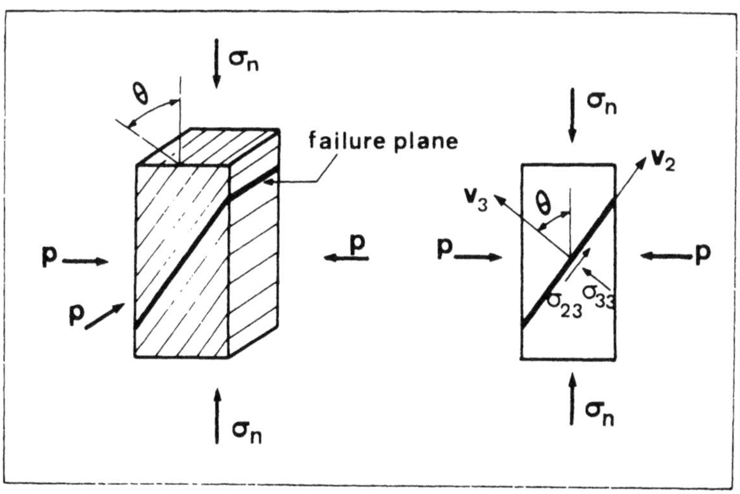

Fig.8 - Failure in "parallel mode" (after [13]).

Expressing the σ_{33} and σ_{23} stresses in terms of the axial over-stress σ_n-p , the confining pressure p and the orientation θ, and employing relations (28), we obtain:

$$\left.\begin{aligned} |\sigma_{23}| &= |\sigma_n - p|\cos\theta\sin\theta = \sqrt{\mathrm{tr}\underset{\sim}{M}\underset{\sim}{S}^2 - \mathrm{tr}^2\underset{\sim}{M}\underset{\sim}{S}} \\ \sigma_{33} &= (\sigma_n - p)\cos^2\theta + p = \mathrm{tr}\underset{\sim}{M}\underset{\sim}{S} + \frac{1}{3}\mathrm{tr}\underset{\sim}{T} \\ 3p &= \mathrm{tr}\underset{\sim}{T} - \sqrt{\frac{3}{2}\mathrm{tr}\underset{\sim}{S}^2} \end{aligned}\right\} \qquad (32)$$

From relations (32), one can conclude that the proposed criterion (31) is a particular form of the general criterion (30). Finally, the proposed criterion (31) admits the following form:

$$\begin{aligned} (\sigma_n - p)\cos\theta\sin\theta = (1+a_o p+a_1 p^2)(\sigma_n\cos^2\theta + p\sin^2\theta)\tan\phi_o + \\ + (1+a_2 p + a_3 p^2)c_o \end{aligned} \qquad (33)$$

From relation (33), we notice that oriented simple compression tests allow the values of the constants ϕ_o and c_o to be determined. The four other constants a_i specify the evolution of the directional strenghts with the confining pressure.

For compressive stress states under confinig pressure and the "across mode" of failure, a criterion is proposed that relates the axial overstress $\sigma_n - p$, the confining pressure p and the orientation θ by the following expression:

$$\begin{aligned} (1+a_1 p+a_2 p^2)\,(\sigma_n - p)^2 + (b_o + b_1 p + b_2 p^2)\,(\sigma_n - p)^2\cos^2\theta + \\ + (c_o + c_1 p + c_2 p^2)\,(\sigma_n - p)^2\cos^4\theta = k^2 \end{aligned} \qquad (34)$$

where a_i, b_i and c_i are material constants and k is the constant of the criterion. Employing relations (28), we obtain:

$$\left.\begin{aligned} & 0 < \sigma_n - p = \sqrt{\frac{3}{2}\mathrm{tr}\underset{\sim}{S}^2} \;;\; \cos^2\theta = \frac{1}{3} + \frac{\mathrm{tr}\underset{\sim}{M}\underset{\sim}{S}}{\sqrt{\frac{3}{2}\mathrm{tr}\underset{\sim}{S}^2}} \;; \\ & 3p = \mathrm{tr}\underset{\sim}{T} - \sqrt{\frac{3}{2}\mathrm{tr}\underset{\sim}{S}^2}. \end{aligned}\right\} \qquad (35)$$

From relation (35), one can conclude that the proposed criterion (34) is a particular form of the general criterion (30). Oriented simple compression tests allow the values of the material constants b_o and c_o and the constant k of the criterion to be determined. The six other material constants account for the influence of the confining pressure.

When applied to simple tensile tests, the criterion (34) admits the reduced form:

$$\sigma_n = \frac{k^2}{1+b_o\cos^2\theta+c_o\cos^4\theta} \qquad (36)$$

This form is more general than the corresponding reduced form of the criterion (33) and will be used for the prediction of the directional simple tensile strengths. The three constants k, b_o and c_o must be recalculated, because the failure modes in simple tensile tests are different from the "across mode".

4.5 Comparison with experimental results

The values of the material constants involved in the proposed criteria (33), (34) and (36) were calculated so as to obtain the best theoretical representation for the experimental data; the values obtained are given in [13].

In Figure 9, experimental data, as well as the theoretical curves obtained in application of the proposed criteria are presented. Whatever the confining pressure p and the orientation θ, the theoretical predictions fit the experimental results well.

Observe also that the criteria (33) and (34) proposed for failure under compression with confining pressures are compatible. Indeed, whatever the confining pressure, for the orientations corresponding to failure in "across mode", the criterion (33) specifying the failure strengths for the "parallel mode" predicts limit-stresses which are higher than those predicted by the criterion (34) specific for failure in "across mode"; conversely, whatever the confining pressure, for the orientations corresponding to failure in "parallel mode", the criterion (34) speci-

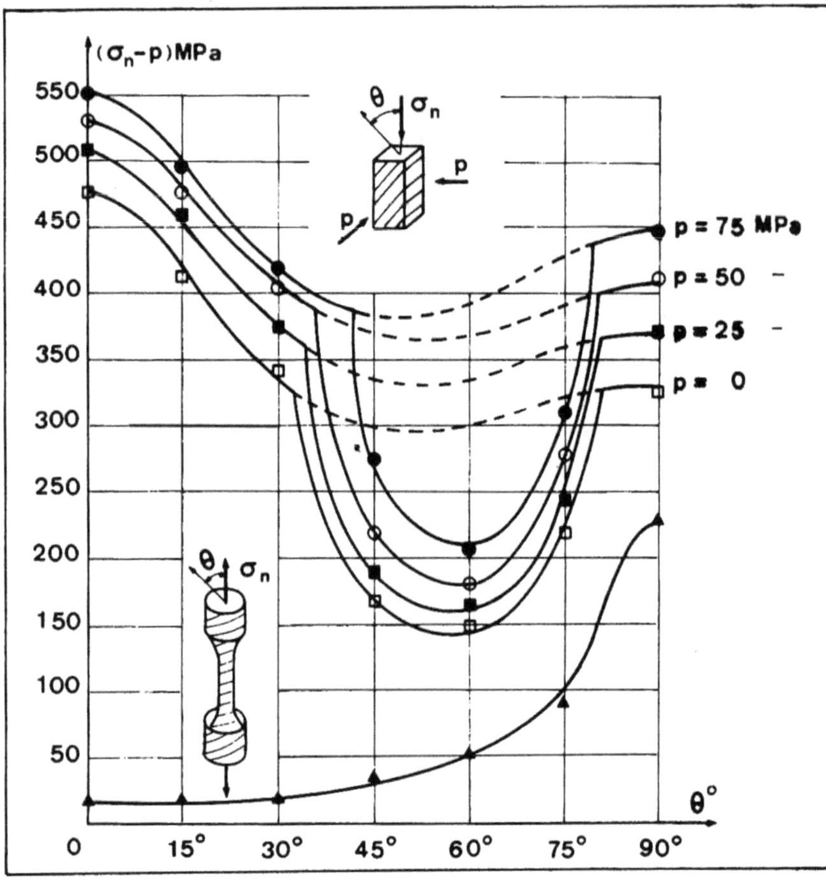

Fig.9 - Comparison between experimental data and the proposed failure criteria (after [13]).

fying the failure strengths for the "across mode" predicts limit-stresses which are higher than those predicted by the criterion (33) specific for failure in "parallel mode". Finally, the lower envelope of the two proposed criteria furnishes correct predictions for the values of failure strengths, as well as for the orientations corresponding to a given failure mode.

5. SIMPLILFIED THEORY

5.1 Plastic behavior

In order to describe the plastic behavior of an anisotropic solid in a simple manner, as well as to be able to employ the methods already developed for isotropic solids, a simplified theory was proposed in [17], [18]. The underlying idea is to replace the anisotropic material by an equivalent isotropic material by considering, instead of the stress tensor $\underset{\sim}{T}$, the following transformed tensor involving anisotropic effects:

$$\bar{\underset{\sim}{T}} = \underset{\approx}{A}\,\underset{\sim}{T} \qquad (37)$$

where $\underset{\approx}{A}$ is a fourth order tensor, specifically the tensor of plastic anisotropy coefficients. In the case of transverse isotropy, the expression of $\underset{\approx}{A}$ in any $(\bullet, \bullet, \underset{\sim}{v}_3)$ privileged orthonormal frame is given by:

$$A = \mathrm{diag.}(\gamma, \gamma, \alpha, \beta/2, \beta/2, \gamma/2) \qquad (38)$$

where α, β, γ are three independent material constants.

In the simplified theory, a quasi-linear isotropic relation between the transformed stress tensor $\bar{\underset{\sim}{T}}$ and the rate of deformation tensor $\underset{\sim}{D}$ is used. The corresponding irreducible representation is given by:

$$\left.\begin{aligned} &\bar{\underset{\sim}{T}} = \phi_o \underset{\sim}{I} + \phi_1 \underset{\sim}{D} \quad \text{or} \quad \underset{\sim}{T} = \phi_o \underset{\approx}{A}^{-1} \underset{\sim}{I} + \phi_1 \underset{\approx}{A}^{-1} \underset{\sim}{D} \\ &\phi_i = \phi_i(\mathrm{tr}\underset{\sim}{D}, \mathrm{tr}\underset{\sim}{D}^2, \mathrm{tr}\underset{\sim}{D}^3) \end{aligned}\right\} \qquad (39)$$

In this case, the homogeneity condition (3) results in the following equations of Euler:

$$\frac{\partial \phi_i}{\partial E_j} E_j + n\phi_i = 0 \qquad (40)$$

where: $n = \begin{cases} 0 \\ 1 \end{cases}$ when $i = \begin{cases} 0 \\ 1 \end{cases}$; $j = 1,2,3$;

$$E_1 = \mathrm{tr}\underset{\sim}{D} \; ; \; E_2 = \mathrm{tr}^{1/2}\underset{\sim}{D}^2 \; ; \; E_3 \mathrm{tr}^{1/3}\underset{\sim}{D}^3$$

The solutions of equations (40) are:

$$\phi_o = A(p,q) \quad ; \quad \phi_1 = \frac{1}{E_2} B(p,q) \tag{41}$$

where p and q are the following two dimensionless kinematic variables:

$$p = \frac{E_1}{E_2} \quad ; \quad q = \frac{E_3}{E_2} \tag{42}$$

Introducing the solutions (42) into the representation (39), the invariant form for the plastic constitutive law of a transversely isotropic solid is obtained:

$$\left.\begin{aligned} &\underset{\sim}{\bar{T}} = A\underset{\sim}{I} + \frac{B}{E_2}\underset{\sim}{D} \\ &A = A(p,q) \quad ; \quad B = B(p,q) \end{aligned}\right\} \tag{43}$$

The tensor form (43) is equivalent to the system of scalar relations between the independent isotropic invariants of $\underset{\sim}{\bar{T}}$ and $\underset{\sim}{D}$:

$$\left.\begin{aligned} \mathrm{tr}\underset{\sim}{\bar{T}} &= 3A + pB \\ \mathrm{tr}\underset{\sim}{\bar{T}}^2 &= 3A^2 + 2pAB + B^2 \\ \mathrm{tr}\underset{\sim}{\bar{T}}^3 &= 3A^3 + 3pA^2B + 3AB^2 + qB^3 \end{aligned}\right\} \tag{44}$$

When the two dimensionless kinematic variables (42) are independent, their elimination from (44) results in the following expression of the yield criterion:

$$f(\mathrm{tr}\underset{\sim}{\bar{T}},\ \mathrm{tr}\underset{\sim}{\bar{T}}^2,\ \mathrm{tr}\underset{\sim}{\bar{T}}^3) = 0 \tag{45}$$

which is a particular form of the criterion (13) of the general theory, since, [19]:

$$\bar{\underset{\sim}{T}} = \underset{\approx}{A}\,\underset{\sim}{T} = [(\alpha+\gamma-2\beta)\mathrm{tr}\underset{\sim}{M}\underset{\sim}{T}]\underset{\sim}{M} + \gamma\underset{\sim}{T} + (\beta-\gamma)(\underset{\sim}{M}\underset{\sim}{T}+\underset{\sim}{T}\underset{\sim}{M}) \tag{46}$$

where $\underset{\sim}{M} = \underset{\sim}{v}_3 \otimes \underset{\sim}{v}_3$.

The relations (39) of the simplified theory can be written as a particular case of the quasi-linear form (18), [19]:

$$\begin{aligned}\underset{\sim}{T} = \frac{1}{\gamma}\,\phi_o\underset{\sim}{I} + [(\frac{1}{\alpha} - \frac{1}{\gamma})\phi_o + (\frac{1}{\alpha} + \frac{1}{\gamma} - \frac{2}{\beta})\phi_1\mathrm{tr}\underset{\sim}{M}\underset{\sim}{D}]\underset{\sim}{M} \; + \\ + \frac{1}{\gamma}\,\phi_1\underset{\sim}{D} + (\frac{1}{\beta} - \frac{1}{\gamma})\phi_1(\underset{\sim}{M}\underset{\sim}{D}+\underset{\sim}{D}\underset{\sim}{M}),\end{aligned} \tag{47}$$

where ϕ_o and ϕ_1 are specified in (41). The flow law is obtained by inversion of (47); this is equivalent to the inversion of (43) and the introduction of (46):

$$\frac{\underset{\sim}{D}}{\mathrm{tr}^{1/2}\underset{\sim}{D}^2} = -\frac{A}{B}\underset{\sim}{I} + \frac{(\alpha+\gamma-2\beta)\mathrm{tr}\underset{\sim}{M}\underset{\sim}{T}}{B}\underset{\sim}{M} + \frac{\gamma}{B}\underset{\sim}{T} + \frac{\beta-\gamma}{B}(\underset{\sim}{M}\underset{\sim}{T}+\underset{\sim}{T}\underset{\sim}{M}). \tag{48}$$

The flow law (48) of the simplified theory complies with the general theory in its quasi-linear form (18). Finally, the simplified theory (39) is a particular case of (1).

In any $(\bullet, \bullet, \underset{\sim}{v}_3)$ privileged orthonormal frame, the component form of the flow law (48) is given by:

$$\frac{B}{\mathrm{tr}^{1/2}\underset{\sim}{D}^2}\underset{\sim}{D} = \begin{vmatrix} -A+\gamma T_{11} & \gamma T_{12} & \beta T_{13} \\ \gamma T_{12} & -A+\gamma T_{22} & \beta T_{23} \\ \beta T_{13} & \beta T_{23} & -A+\alpha T_{33} \end{vmatrix} \tag{49}$$

It is easy to see that the flow law (49) of the simplified theory possesses the same essential features as the flow law (16) of the general theory, i.e.:

- non-coincidence of the principal directions of the stress and rate of deformation tensors in "off-axis" tests.
- non-isotropic deformation under hydrostatic pressure.
- non-association with the yield criterion (45) by the rule of plastic potential, which appears as a special case.

If the flow is incompressible, the following restriction results from (48):

$$-3A + (\alpha-\gamma)\mathrm{tr}\underset{\sim}{M}\underset{\sim}{T} + \gamma\mathrm{tr}\underset{\sim}{T} = 0 \tag{50}$$

The components of $\underset{\sim}{D}$ in the privileged $(\bullet, \bullet, \underset{\sim}{v}_3)$ frame become:

$$\left.\begin{aligned}
D_{11} &= \frac{\lambda}{3}(2\gamma T_{11} - \gamma T_{22} - \alpha T_{33}) \quad ; \quad & D_{23} &= \lambda\beta T_{23} \quad ; \\
D_{22} &= \frac{\lambda}{3}(-\gamma T_{11} + 2\gamma T_{22} - \alpha T_{33}) \quad ; \quad & D_{31} &= \lambda\beta T_{31} \quad ; \\
D_{33} &= \frac{\lambda}{3}(-\gamma T_{11} - \gamma T_{22} + 2\alpha T_{33}) \quad ; \quad & D_{12} &= \lambda\alpha T_{12} \quad .
\end{aligned}\right\} \tag{51}$$

where: $\lambda = \dfrac{\mathrm{tr}^{1/2}\underset{\sim}{D}^2}{B}$.

When further restrictions of plane flow is made so that :

$$D_{11} = D_{31} = D_{12} = 0 \tag{52}$$

the reactive stresses express as follows:

$$T_{11} = \frac{1}{2}(T_{22} + \frac{\alpha}{\gamma}T_{33}) \quad ; \quad T_{31} = T_{12} = 0 \tag{53}$$

and the components (51) of the rate of deformation tensor $\underset{\sim}{D}$ become:

$$D_{22} = \frac{\lambda}{2}(\gamma T_{22} - \alpha T_{33}) \quad ; \quad D_{33} = \frac{\lambda}{2}(-\gamma T_{22} + \alpha T_{33}) \quad ; \quad D_{12} = \lambda\beta T_{12}. \tag{54}$$

Relations (53) will be used in further developments concerning the analysis of specific yield criteria in the case of plane flow.

5.2 Yield criteria

In the simplified theory, isotropic yield criteria are generalized for transversely isotropic materials in a straightforward manner by the substitution of the stress tensor $\underset{\sim}{T}$ for the transformed stress tensor $\underset{\sim}{\bar{T}}$. Thus, for cohesive materials, the generalization of the Von Mises criterion for transversely isotropic solids takes the form, [17], [18]:

$$\mathrm{tr}\underset{\sim}{\bar{S}}^2 = 2k^2 \tag{55}$$

where $\underset{\sim}{\bar{S}}$ is the deviator of the transformed stress tensor $\underset{\sim}{\bar{T}}$:

$$\underset{\sim}{\bar{S}} = \underset{\sim}{\bar{T}} - (\frac{1}{3}\,\mathrm{tr}\underset{\sim}{\bar{T}})\underset{\sim}{I} \tag{56}$$

From (56) and (46), we obtain:

$$\begin{aligned}\mathrm{tr}\underset{\sim}{\bar{S}}^2 = & -\frac{1}{3}\gamma^2\mathrm{tr}^2\underset{\sim}{T} + \gamma^2\mathrm{tr}\underset{\sim}{T}^2 + \frac{2}{3}(\alpha^2+\gamma^2+\alpha\gamma-3\beta^2)\mathrm{tr}^2\underset{\sim}{M}\underset{\sim}{T} + \\ & + 2(\beta^2-\gamma^2)\mathrm{tr}\underset{\sim}{M}\underset{\sim}{T}^2 - \frac{2}{3}\,\gamma(\alpha-\gamma)\mathrm{tr}\underset{\sim}{M}\underset{\sim}{T}\mathrm{tr}\underset{\sim}{T}\,.\end{aligned} \tag{57}$$

Thus, the criterion (55) of the simplified theory appears as a particular case of the yield condition (13) of the general theory.

We shall first check whether the form (55) is versatile enough to allow an interpretation of experimental results concerning uniaxial traction and compression of transversely isotropic materials. In Figure 10, ($\underset{\sim}{e}_2$, $\underset{\sim}{e}_3$) are the principal directions of the applied stress tensor, $\underset{\sim}{v}_3$ is the privileged direction of the transversely isotropic solid and θ is the angle ($\underset{\sim}{v}_3$, $\underset{\sim}{e}_3$). The principal values of the applied stress tensor are denoted by σ_n and σ_t ; in the case of simple traction or compression, σ_t is equal to zero. If R_θ denotes the yield point in simple traction or compression in the direction $\underset{\sim}{e}_3$, the following expression is obtained from (55):

$$R_\theta{}^2 = \frac{3k^2}{(\alpha\cos^2\theta + \gamma\sin^2\theta)^2 + 3(\beta^2 - \alpha\gamma)\sin^2\theta\cos^2\theta} \tag{58}$$

The variation of R_θ passes through an extremum for:

$$\tan^2\theta = \frac{2\alpha^2 + \alpha\gamma - 3\beta^2}{2\gamma^2 + \alpha\gamma - 3\beta^2} \tag{59}$$

In the case where $\gamma > \alpha$, depending upon the values of the material plastic anisotropy constants α, β and γ, the strength variation is of the type shown in Figure 11. The strength variation:

a) admits a maximum for $\theta_{max} < \frac{\pi}{4}$ if $\beta^2 < \frac{1}{3}(2\alpha^2 + \alpha\gamma)$

b) admits a minimum for $\theta_{min} > \frac{\pi}{4}$ if $\beta^2 > \frac{1}{3}(2\gamma^2 + \alpha\gamma)$

c) decreases monotically if $\frac{1}{3}(2\alpha^2 + \alpha\gamma) < \beta^2 < \frac{1}{3}(2\gamma^2 + \alpha\gamma)$.

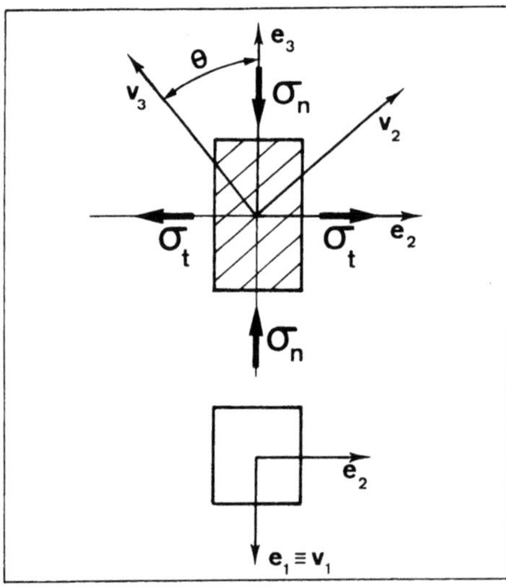

Fig.10 - Transversely isotropic material in strength tests.

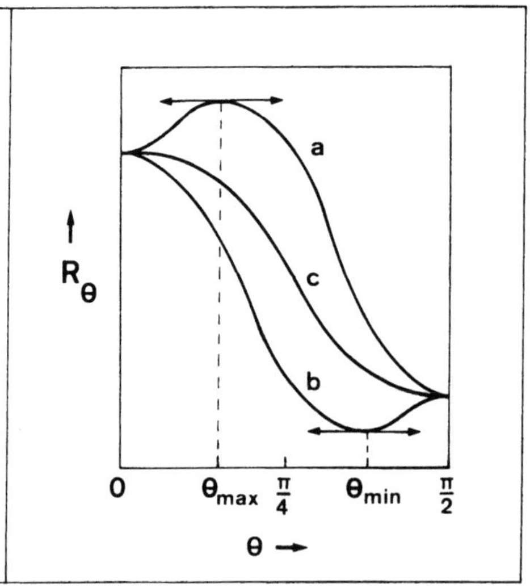

Fig.11 - Theoretical directional strength variation in simple compression or traction of cohesive transversely isotropic materials (after [2]).

We remark that the curves of Figure 11 allow for an appropriate interpretation of the experimental facts presented in Chapter 1 for clays and rolled sheet-steel.

We consider now simple shear tests, for which $\sigma_n = -\sigma_t$ (cf. Figure 10). If C_θ denotes the yield point in simple shear, the following expression is obtained from (55) for the variation of C_θ:

$$C_\theta^2 = \frac{3k^2}{(\alpha^2+\alpha\gamma+\gamma^2)(\cos^2\theta-\sin^2\theta)^2 + 6\beta^2\sin^2\theta\cos^2\theta} \tag{60}$$

which is symmetric with respect to the axis $\theta = \pi/4$.

If we let $\theta = 0$ and $\theta = \frac{\pi}{2}$ in (58) and $\theta = \frac{\pi}{4}$ in (60), we obtain the following meaning for the three material plastic anisotropy constants:

$$\alpha = \frac{k\sqrt{3}}{R_o} \quad ; \quad \beta = \frac{k\sqrt{2}}{C_{\pi/4}} \quad ; \quad \gamma = \frac{k\sqrt{3}}{R_{\pi/4}} \; . \tag{61}$$

In the case of plane flow, taking into account relations (53), the yield condition (55) results in:

$$\sigma_n^2 [(\alpha\cos^2\theta-\gamma\sin^2\theta)^2+4\beta^2\sin^2\theta\cos^2\theta]+\sigma_t^2[(\alpha\sin^2\theta-\gamma\cos^2\theta)^2+4\beta^2\sin^2\theta\cos^2\theta] +$$

$$+ 2\sigma_n\sigma_t[(\alpha+\gamma)^2\sin^2\theta\cos^2\theta-\alpha\gamma-4\beta^2\sin^2\theta\cos^2\theta] = 4 k^2 \tag{62}$$

where σ_n and σ_t are the principal stresses in the plane of deformation. The analysis of the curves (62), as well as the analysis of the criterion (55) in plane strain and axisymmetric three-dimensional stresses is given in [1], [20].

For materials with internal friction, a simple criterion can be developed starting with an invariant form of the Drucker-Prager criterion for isotropic solids. In the simplified theory, the generalization for transversely isotropic materials of the Drucker-Prager criterion takes the form, [17], [18]:

$$\mathrm{tr}\bar{\underset{\sim}{T}} + a\,\mathrm{tr}^{1/2}\bar{\underset{\sim}{S}}^2 = c \tag{63}$$

where $\mathrm{tr}\bar{S}^2$ is given by (57) and $\mathrm{tr}\bar{T}$ expresses as follows:

$$\mathrm{tr}\bar{T} = (\alpha-\gamma)\mathrm{tr}MT + \gamma\,\mathrm{tr}T \tag{64}$$

In (63), the material constant a is associated with internal friction. When applied to uniaxial compression or traction, the criterion (63) results in the following variation of R_θ:

$$R_\theta = \frac{c}{\alpha\cos^2\theta+\gamma\sin^2\theta\pm a\sqrt{\frac{2}{3}}\,[(\alpha\cos^2\theta+\gamma\sin^2\theta)^2+3(\beta^2-\alpha\gamma)\cos^2\theta\sin^2\theta]^{1/2}} \tag{65}$$

where the + sign before the constant a holds for the strength in simple compression and the - sign holds for the strength in simple traction. Similar curves to those of Figure 11 are obtained in this case, except that their form is now influenced by the constant a. The respective discussion, as well as the analysis of the criterion (63) in plane flow, plane stress and axisymmetric three-dimensional stresses are given in [1], [20]. Comparisons with experiments are discussed in [2].

REFERENCES

1. BOEHLER, J.P., Contributions Théoriques et Expérimentales à l'Etude des Milieux Plastiques Anisotropes, *Thèse de Doctorat ès Sciences*, Grenoble, 1975.
2. BOEHLER, J.P. and SAWCZUK, A., On Yielding of Oriented Solids, *Acta Mechanica*, 27 (1977): 185-206.
3. BOEHLER, J.P., Lois de Comportement Anisotrope des Milieux Continus, *Journal de Mécanique*, 17, 2 (1978): 153-190.
4. RIVLIN, R.S., Further Remarks on the Stress-Deformation Relations for Isotropic Materials, *J. Rat. Mech. An.*, 4 (1955): 681-702.
5. ALLIROT, D. and BOEHLER, J.P., Evolution des Propriétés Mécaniques d'une Roche Stratifiée sous Pression de Confinement, *Proc. 4th Int. Congress on Rock Mechanics*, A.A. Balkema, Rotterdam, Vol.1, 1979, 15-22.

6. ALLIROT, D., BOEHLER, J.P. and SAWCZUK, A., Pressure-Induced Evolution of Anisotropies in Stratified Rock, *Studia Geotechnica et Mechanica*, III, 2-4 (1981): 59-73.

7. BOEHLER, J.P., Anisotropie et Comportement Rhéologique des Matériaux, Conférence Générale, *19ème Colloque du Groupe Français de Rhéologie* (in print).

8. BOEHLER, J.P., EL AOUFI, L. and RACLIN, J., On Experimental Testing Methods for Anisotropic Materials, *Res Mechanica* (in print).

9. EL AOUFI, L., Contributions à l'Etude Expérimentale et Numérique du Comportement Mécanique des Matériaux Anisotropes, *Thèse de Doctorat de Spécialité*, Grenoble, 1984.

10. BOEHLER, J.P. and EL AOUFI, L., Hétérogénéité des Champs de Contraintes et de Déformations dans les Matériaux Anisotropes, *19ème Colloque du Groupe Français de Rhéologie* (in print).

11. RACLIN, J., Contributions Théoriques et Expérimentales à l'Etude de la Plasticité, de l'Ecrouissage et de la Rupture des Solides Anisotropes, *Thèse de Doctorat ès Sciences*, Grenoble, 1984.

12. ALLIROT, D., BOEHLER, J.P. and SAWCZUK, A., Irreversible Deformations of an Anisotropic Rock under Hydrostatic Pressure, *Int. J. Rock Mech. Min. Sci. & Geomech. Abstr.*, 14 (1977): 77-83.

13. BOEHLER, J.P. and RACLIN, J., Failure Criteria for Glass-Fiber Reinforced Composites under Confining Pressure, *J. Struct. Mech.*, 13, 3 & 4 (1985): 371-393.

14. BOEHLER, J.P. and DELAFIN, M., Failure Criteria for Unidirectional Fiber-Reinforced Composites, in: *Mechanical Behavior of Anisotropic Solids, Proc. Euromech Coll. 115 - Colloque International 295 du CNRS*, edited by J.P. Boehler, Editions du CNRS (Paris) - Martinus Nijhoff (The Hague), 1982, 449-470.

15. TSAI, S.W. and WU, E.M., A General Theory of Strength for Anisotropic Materials, *J. Composites Materials*, 5 (1971): 58-80.

16. HASHIN, Z., Failure Criteria for Unidirectional Fiber Composites, *J. Appl. Mech.*, 102,2 (1980): 329-334.

17. BOEHLER, J.P., Contribution à l'Etude de l'Equilibre Limite des Sols Anisotropes, *Thèse de Doctorat de Spécialité*, Grenoble 1968.

18. BOEHLER, J.P. and SAWCZUK, A., Equilibre Limite des Sols Anisotropes, *J. de Mécanique*, 9, 1 (1970): 5-33.

19. BOEHLER, J.P., Sur les Formes Invariantes dans le Sous-Groupe Orthotrope de Révolution des Transformations Orthogonales..., *ZAMM*, 55 (1975): 609-611.

20. BOEHLER, J.P. and SAWCZUK, A., Analyse Géométrique des Critères d'Ecoulement Plastique Anisotrope, in: *Problèmes de Rhéologie*, PWN, Varsovie, 1977, 69-75.

Chapter 6

ON A RATIONAL FORMULATION OF ISOTROPIC AND ANISOTROPIC HARDENING

J.P. Boehler
University of Grenoble, France

1. INTRODUCTION

The aim of this work is to develop a unified and rational formulation of isotropic and anisotropic hardening within the framework of the tensor functions representation theory. We consider the evolution of the yield criteria during plastic deformations and we assume that this evolution depends on the strain history only through the present value of the plastic strain. For more complicated situations, the proposed concepts can be developed in a straightforward manner.

The classical concepts of isotropic, kinematic and anisotropic hardening are analysed in detail. It is first shown that kinematic hardening can be split into specific isotropic and anisotropic hardening, the contribution of the anisotropic part being of a very restrictive orthotropic type. Secondly, it is shown that certain materials, when subjected to isotropic plastic strain, undergo an isotropic evolution of their internal structure, which results in a modification of the shape of the initial yield condition, without appearance of anisotropic properties. Examples are given for this phenomenon, which, obviously, cannot be taken into account by the classical formulation of isotropic or kinematic hardening. Finally, the classical concepts of isotropic, kinematic and anisotropic hardening are not disjointed and do not cover the full range of hardening phenomena. Consequently, there appears a need for a more general concept and a rational formulation of isotropic and anisotropic hardening.

In our analysis of the classical formulations, we show that the proposed rules are equivalent to very restrictive transformations of the stress tensor. We define the generalized concepts of isotropic and anisotropic hardening by, respectively, general isotropic and anisotropic transformations of the stress tensor. Physically, such a formulation is based on the assumption that the modifications of the materials' internal structure during plastic deformations result in an evolution of the mechanical properties, which can be taken into account by specific transformations of the stress tensor. By application of the therorems of representations for tensor functions, it is shown that the proposed procedure enables the transformation of an arbitrary initial yield criterion into an arbitrary isotropic or anisotropic subsequent criterion.

Within the proposed concept, general isotropic and anisotropic hardening rules are developed, in considering first-order polynomial isotropic or anisotropic transformations of the stress tensor and applying the procedure to materials obeying the von Mises and linear criterion. Although the considered stress transformation and yield criteria are very simple, the proposed general rules constitute a framework, which is already wide enough to be able to account for the observed mechanical phenomena and to include the classical concepts of isotropic, kinematic and anisotropic hardening as special cases.

The essential results of the proposed rational formulation are presented in this Chapter, which was originally published under the same title in: *Plasticity Today*, edited by A. Sawczuk and G. Bianchi, Elsevier Applied Science Publishers, London, 1985: 483-502.

Further details and developments on the subject will be included in forthcoming papers [6], [7].

2. CLASSICAL FORMULATION OF ISOTROPIC, KINEMATIC AND ANISOTROPIC HARDENING

2.1 Isotropic hardening

Consider first an isotropic material. The general invariant form of the initial yield criterion is given by:

$$f(\mathrm{tr}\underset{\sim}{T},\ \sqrt{\mathrm{tr}\underset{\sim}{S}^2},\ \sqrt[3]{\mathrm{tr}\underset{\sim}{S}^3}) = 0 \quad ; \quad \underset{\sim}{S} = \underset{\sim}{T} - (\tfrac{1}{3}\,\mathrm{tr}\underset{\sim}{T})\underset{\sim}{I} \tag{1}$$

In the classical concept of isotropic hardening [3], it is assumed that the yield surface (1) undergoes a homothetic transformation of the ratio $\lambda(\underset{\sim}{P})$ in the stress space, where $\underset{\sim}{P}$ is the plastic strain tensor. This transformation is equivalent to a homothetic transformation of the stress tensor $\underset{\sim}{T}$, with the ratio $1/\lambda(\underset{\sim}{P})$. Thus, the classical isotropic hardening rule can be written in the form:

$$\left.\begin{array}{l} f(\mathrm{tr}\underset{\sim}{T},\ \sqrt{\mathrm{tr}\underset{\sim}{S}^2},\ \sqrt[3]{\mathrm{tr}\underset{\sim}{S}^3}) = 0 \Rightarrow f(\frac{1}{\lambda}\mathrm{tr}\underset{\sim}{T},\ \frac{1}{\lambda}\sqrt{\mathrm{tr}\underset{\sim}{S}^2},\ \frac{1}{\lambda}\sqrt[3]{\mathrm{tr}\underset{\sim}{S}^3}) = 0 \\ \text{where} \qquad \lambda(\underset{\sim}{P}) > 0 \quad ; \begin{cases} \lambda(\underset{\sim}{P}) > 1\text{: hardening} \\ \lambda(\underset{\sim}{P}) < 1\text{: softening} \end{cases} \end{array}\right\} \tag{2}$$

If the material is anisotropic, the general form of the initial criterion is given by:

$$f(\underset{\sim}{T},\ \underset{\sim}{\xi}_1,\ \underset{\sim}{\xi}_2,\ \ldots,\ \underset{\sim}{\xi}_p) = 0 \tag{3}$$

where the $\underset{\sim}{\xi}_i$ are structural tensors, taking into account the anisotropic material behavior [4], [5]. In order to derive the invariant form of (3), the theory of representation of anisotropic tensor functions [4], [21] is very useful, for it indicates the type and the minimal number of basic invariants involved. By application of this theory, we obtain the general invariant form of the initial criterion:

$$f(\mathrm{tr}\,\underset{\sim}{T},\ \sqrt{\mathrm{tr}\,\underset{\sim}{S}^2},\ \sqrt[3]{\mathrm{tr}\,\underset{\sim}{S}^3},\ J_1,\ J_2,\ \ldots,\ J_n) = 0 \tag{4}$$

where the J_k are mixed invariants of the stress deviator $\underset{\sim}{S}$ and the structural tensors $\underset{\sim}{\xi}_i$ and are homogeneous of order 1 with respect to $\underset{\sim}{S}$. The type of anisotropy of the material is specified by the invariance group of the J_k [4].

For example, the initial yield criterion of an orthotropic material is given by:

$$f(\mathrm{tr}\,\underset{\sim}{T},\ \sqrt{\mathrm{tr}\,\underset{\sim}{S}^2},\ \sqrt[3]{\mathrm{tr}\,\underset{\sim}{S}^3},\ \mathrm{tr}\,\underset{\sim}{M}_1\underset{\sim}{S},\ \mathrm{tr}\,\underset{\sim}{M}_2\underset{\sim}{S},\ \sqrt{\mathrm{tr}\,\underset{\sim}{M}_1\underset{\sim}{S}^2},\ \sqrt{\mathrm{tr}\,\underset{\sim}{M}_2\underset{\sim}{S}^2}) = 0$$

$$\underset{\sim}{M}_1 = \underset{\sim}{v}_1 \otimes \underset{\sim}{v}_1 \quad ; \quad \underset{\sim}{M}_2 = \underset{\sim}{v}_2 \otimes \underset{\sim}{v}_2 \tag{5}$$

where $(\underset{\sim}{v}_1,\ \underset{\sim}{v}_2)$ are two of the three privileged orthogonal directions of the medium.

Applying the classical concept of isotropic hardening, the subsequent criterion of an initially anisotropic material is given by:

$$f(\mathrm{tr}\,\underset{\sim}{T},\ \sqrt{\mathrm{tr}\,\underset{\sim}{S}^2},\ \sqrt[3]{\mathrm{tr}\,\underset{\sim}{S}^3},\ J_k) = 0 \Rightarrow f(\frac{1}{\lambda}\mathrm{tr}\,\underset{\sim}{T},\ \frac{1}{\lambda}\sqrt{\mathrm{tr}\,\underset{\sim}{S}^2},\ \frac{1}{\lambda}\sqrt[3]{\mathrm{tr}\,\underset{\sim}{S}^3},\ \frac{1}{\lambda}J_k) = 0 \tag{6}$$

Such a transformation does not modify the invariance group of the J_k. Thus, the initial type of anisotropy is preserved in the subsequent criterion.

Finally, for initially isotropic or anisotropic materials, the classical concept of isotropic hardening is equivalent to a homothetic transformation of the stress tensor:

$$\underset{\sim}{T} \Rightarrow \bar{\underset{\sim}{T}} = \frac{1}{\lambda(\underset{\sim}{P})}\underset{\sim}{T} \tag{7}$$

For this type of isotropic hardening rule, we suggest the name "homothetic isotropic hardening" (see also Section 2.5 below).

2.2 Kinematic Hardening

In order to account for the Bauschinger effect, different kinematic hardening rules were proposed [18], [25]. The classical concept of kinematic hardening assumes that the yield surface undergoes a rigid translation $\underset{\sim}{C}(\underset{\sim}{P})$ in the stress space, where $\underset{\sim}{C}(\underset{\sim}{P})$ is a symmetric second order tensor. This transformation is equivalent to the translation $-\underset{\sim}{C}(\underset{\sim}{P})$ of the stress tensor:

$$\underset{\sim}{T} \Rightarrow \underset{\sim}{\bar{T}} = \underset{\sim}{T} - \underset{\sim}{C}(\underset{\sim}{P}) \tag{8}$$

We consider first an isotropic material, with initial criterion (1). The translation tensor $\underset{\sim}{C}(\underset{\sim}{P})$ can be split into isotropic and deviatoric parts:

$$\underset{\sim}{C}(\underset{\sim}{P}) = \frac{1}{3}\bar{c}(\underset{\sim}{P})\underset{\sim}{I} + \underset{\sim}{C}'(\underset{\sim}{P}) \quad ; \quad \mathrm{tr}\underset{\sim}{C}'(\underset{\sim}{P}) = 0 \tag{9}$$

By the isotropic part $\frac{1}{3}\bar{c}(\underset{\sim}{P})\underset{\sim}{I}$, which induces a translation of $\underset{\sim}{T}$ parallel to the isotropic axis of the stress space, the initial criterion (1) is transformed into:

$$f(\mathrm{tr}\underset{\sim}{T} - \bar{c}(\underset{\sim}{P}), \sqrt{\mathrm{tr}\underset{\sim}{S}^2}, \sqrt[3]{\mathrm{tr}\underset{\sim}{S}^3}) = 0 \tag{10}$$

Thus, the subsequent criterion remains isotropic.

By the deviatoric part $\underset{\sim}{C}'(\underset{\sim}{P})$, which induces a translation orthogonal to the isotropic subspace of the full six-dimensional stress space, the initial isotropic criterion (1) is transformed into:

$$f(\mathrm{tr}\underset{\sim}{T}, \sqrt{\mathrm{tr}\underset{\sim}{\bar{S}}^2}, \sqrt[3]{\mathrm{tr}\underset{\sim}{\bar{S}}^3}) = 0 \; ; \; \underset{\sim}{\bar{S}} = \underset{\sim}{\bar{T}} - (\tfrac{1}{3}\mathrm{tr}\underset{\sim}{\bar{T}})\underset{\sim}{I} \tag{11}$$

where the isotropic invariants of the transformed stress tensor $\underset{\sim}{\bar{T}}$ are related to that of $\underset{\sim}{T}$ by:

$$\left.\begin{aligned}
\mathrm{tr}\underset{\sim}{\bar{T}} &= \mathrm{tr}\underset{\sim}{T} \\
\mathrm{tr}\underset{\sim}{\bar{S}}^2 &= \mathrm{tr}\underset{\sim}{S}^2 - 2(2c_1'+c_2')\mathrm{tr}\underset{\sim}{M}_1\underset{\sim}{S} - 2(c_1'+2c_2')\mathrm{tr}\underset{\sim}{M}_2\underset{\sim}{S} + 2(c_1'^2+c_2'^2+c_1'c_2') \\
\mathrm{tr}\underset{\sim}{\bar{S}}^3 &= \mathrm{tr}\underset{\sim}{S}^3 + 3(c_1'+c_2')\mathrm{tr}\underset{\sim}{S}^2 - 3(2c_1'+c_2')\mathrm{tr}\underset{\sim}{M}_1\underset{\sim}{S}^2 - 3(c_1'+2c_2')\mathrm{tr}\underset{\sim}{M}_2\underset{\sim}{S}^2 \\
&\quad - 3c_2'(2c_1'+c_2')\mathrm{tr}\underset{\sim}{M}_1\underset{\sim}{S} - 3c_1'(c_1'+2c_2')\mathrm{tr}\underset{\sim}{M}_2\underset{\sim}{S} + 3c_1'c_2'(c_1'+c_2')
\end{aligned}\right\} \tag{12}$$

where c'_1, c'_2, c'_3 are the principal values of $\underset{\sim}{C}'(\underset{\sim}{P})$ and:

$$\underset{\sim}{M}_1 = \underset{\sim}{e}_1 \otimes \underset{\sim}{e}_1 \; ; \quad \underset{\sim}{M}_2 = \underset{\sim}{e}_2 \otimes \underset{\sim}{e}_2 \tag{13}$$

($\underset{\sim}{e}_1$, $\underset{\sim}{e}_2$) being the two first principal directions of $\underset{\sim}{C}'(\underset{\sim}{P})$.

Comparing (11) and (12) with (5), we can conclude that the subsequent criterion is that of an orthotropic medium, with privileged directions $\underset{\sim}{e}_1$, $\underset{\sim}{e}_2$, $\underset{\sim}{e}_3$. It is worthwhile to point out that the induced orthotropy is of a very restrictive type: for a general orthotropic medium, seven independent invariants are involved in the yield criterion (5), whereas in (11) only the three combinations (12) are involved.

Finally, kinematic hardening can be split into isotropic and orthotropic parts. The isotropic part transforms an initial isotropic criterion into a subsequent isotropic criterion by a translation of the stress tensor parallel to the isotropic axis of the stress space. For this type of hardening rule, we suggest the name "translational isotropic hardening" The orthotropic part transforms an initial isotropic criterion into a subsequent orthotropic criterion of a very restrictive form. For this type of anisotropic hardening rule, we suggest the name "translational orthotropic hardening".

If the material is initially anisotropic, the translational isotropic hardening rule leaves the type of anisotropy unchanged, whereas the translational orthotropic hardening rule modifies the type of the initial anisotropy. Further details will be given in ref. 6.

2.3 Anisotropic hardening

The analysis of Section 2.2 shows that the anisotropic part of kinematic hardening is a very restrictive type of orthotropic hardening. In fact, kinematic hardening cannot take into account most observed phenomena of anisotropic hardening [9]. Different, more elaborate anisotropic hardening rules have been proposed [1,2,22,24], in order to account for change in shape and size and for rotation and translation of the initial yield criterion. The aforementioned rules are based on anisotropic linear transformations of the stress tensor. A more general approach, based on the theory of representation for tensor functions, has been developed in ref. 17 for initially isotropic and in ref. 8 and 9 for initially anisotropic material. More information will be given in ref. 6.

2.4 Hardening phenomena which cannot be described by the classical formulations

Several important hardening phenomena cannot be taken into account by the classical approach of isotropic, kinematic and anisotropic hardening. We shall present here two examples for two different initally isotropic materials subjected to isotropic plastic strains.

The first example concerns granular materials without cohesion, obeying the isotropic linear criterion:

$$\mathrm{tr}\underset{\sim}{T} - a\sqrt{\mathrm{tr}\underset{\sim}{S}^2} = 0 \tag{14}$$

If such materials are subjected to isotropic compressions $\underset{\sim}{P} = \varepsilon\underset{\sim}{I}$, they remain isotropic, but the density increases with ε. The increase of density induces an increase in the internal friction, without appearance of cohesion. Such phenomena have been observed experimentally on sands, for example [23]. Thus, the subsequent criterion is given by:

$$\mathrm{tr}\underset{\sim}{T} - \bar{a}(\varepsilon)\sqrt{\mathrm{tr}\underset{\sim}{S}^2} = 0, \text{ with } \bar{a}(0) = a \; ; \; \frac{d\bar{a}}{d\varepsilon} < 0 \tag{15}$$

By the homothetic isotropic hardening rule, the initial criterion is unchanged; by translational isotropic hardening, the initial friction is unchanged and there appears non-zero cohesion (Fig.1). Thus, the subsequent criterion (15) cannot be obtained by the classical concepts of isotropic hardening.

The second example concerns rolled sheet-steel obeying the von Mises criterion:

$$\sqrt{\mathrm{tr}\underset{\sim}{S}^2} = \sqrt{2}k \tag{16}$$

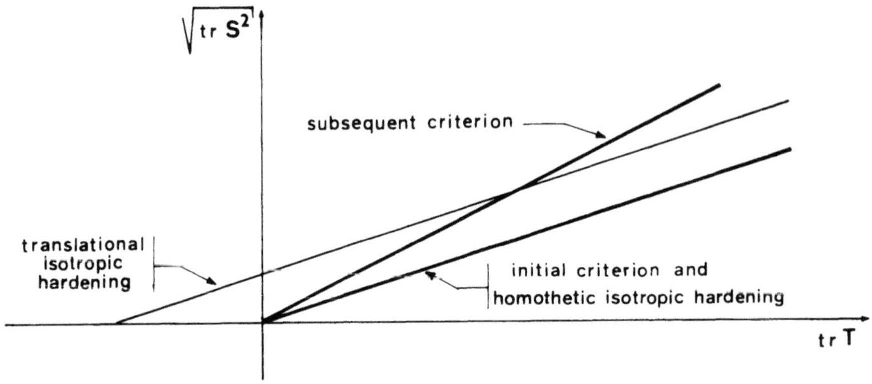

Fig.1 - Granular materials subjected to isotropic plastic strain

For orthotropic rolled sheet-steel, a generalization of the von Mises criterion has been proposed, taking into account the initial anisotropy and the anisotropy induced by the plastic strain [8]; the theoretical predictions are well in agreement with experimental results [9]. For an initially isotropic steel subjected to a plane isotropic plastic strain $\underset{\sim}{P} = \varepsilon \underset{\sim}{I}$, the subsequent criterion, developed with respect to the principal plane stresses, is given by:

$$(\sigma_1^2 + \sigma_2^2)(\frac{2}{3} + a\varepsilon) + 2\,\sigma_1\sigma_2(-\frac{1}{3} + b\varepsilon) = 2k^2(1 + c\varepsilon) \tag{17}$$

wher a, b and c are material constants. In the general case ($a+2b \neq 0$), equation (17) represents in the (σ_1, σ_2) plane an ellipse with the same principal axes as the von Mises ellipse, but with a different size and shape.

By the homothetic isotropic hardening rule, the initial von Mises ellipse is transformed into a homothetic ellipse without change of shape; by the translational isotropic hardening rule, the initial criterion (16) is unchanged (Fig.2). Thus, the subsequent criterion (17) cannot be obtained by the classical concepts of isotropic hardening.

In these two examples, an initial isotropic criterion is transformed into a subsequent isotropic criterion with change in size or (and) change in shape. For such hardening phenomena, we suggest the name "isotropic

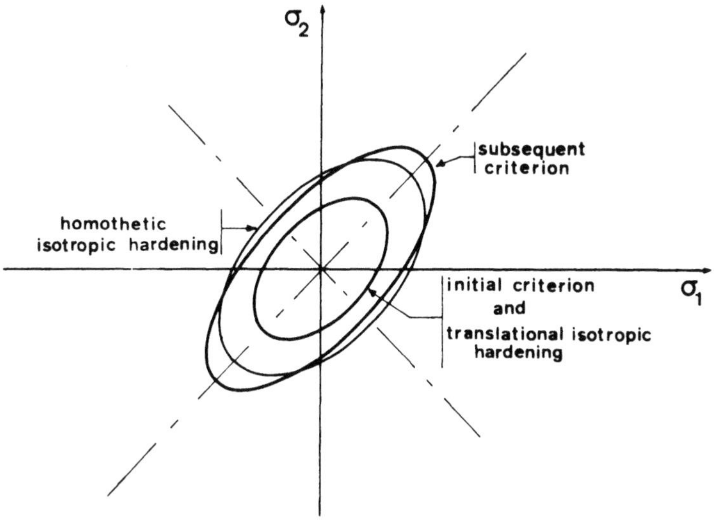

Fig.2 - Von Mises materials subjected to isotropic plastic plane strain

hardening" in a more general sense that the classical one.
That is why we suggest the name "homothetic isotropic hardening" for the classical concept of isotropic hardening and "translational isotropic hardening" for the isotropic part of the classical concept of kinematic hardening, in order to underline that the classical concepts are special cases of a more general formulation of isotropic hardening.

2.5 Conclusions

The classical concepts of isotropic, kinematic and anisotropic hardening:

(1) are not disjointed. Indeed, kinematic hardening is a combination of a special isotropic hardening (translational isotropic hardening) and a special anisotropic hardening (translational orthotropic hardening).

(2) give restrictive forms for the subsequent criteria.

(3) cannot take into account several important hardening phenomena.

Thus, there appears a need for a more general concept and a rational formulation of isotropic and anisotropic hardening.

3. GENERAL FORMULATION OF ISOTROPIC AND ANISOTROPIC HARDENING

3.1 Proposed general concept

In our analysis of the classical formulation of isotropic, kinematic and anisotropic hardening, we have noticed that the proposed rules can be interpreted by very restrictive transformations of the stress tensor.

A general concept of isotropic and anisotropic hardening can be based on the assumption that the evolution of mechanical properties, due to the modifications of the material internal structure, can be taken into account by specific transformations of the stress tensor involved in the constitutive law. A similar process, in a more restrictive manner, has been proposed for the modelling of damage growth [11, 14, 15, 16].

Within the proposed scheme, the generalized concepts of isotropic and anisotropic hardening are defined by, respectively, general isotropic and anisotropic transformations of the stress tensor. The invariant forms of these general transformations will be specified by application of the theory of representation of tensor functions. The main object of Section 3 is to set up the mathematical tool involved in the proposed general concept of isotropic and anisotropic hardening. How to derive specific transformation rules of the stress tensor from microscopic observations or phenomenological considerations will not be treated here. Proposed general isotropic and anisotropic hardening rules will be presented in Section 4.

3.2 Influence of the plastic strain on the hardening rule

The influence of the plastic strain $\underset{\sim}{P}$ has been mathematically analysed for initially isotropic [17] and anisotropic materials [8], by the application of the theorems of representation for tensor functions. The results of these analyses are: if the initial isotropic or anisotropic criterion undergoes an isotropic transformation, the involved plastic strain is necessarily isotropic; a non-zero plastic strain deviator induces an anisotropic transformation of the yield criterion.

Physically, these theoretical results can be interpreted in the following manner:

(a) A non-isotropic plastic strain induces the appearance of a preferred orientation of the material internal structure, which results in anisotropic hardening on the macroscale. For example, an initially isotropic criterion is transformed into a subsequent anisotropic criterion.

(b) Possibly there may exist materials for which a non-isotropic plastic strain induces only isotropic modifications of the internal structure. But, in this case, only the isotropic part of $\underset{\sim}{P}$ is involved in the law of transformation.

Finally, if the hardening is isotropic, either the plastic strain is isotropic, or only its isotropic part is involved in the hardening rule. If the hardening is anisotropic, the deviatoric part of the plastic strain is necessarily different from zero. Further details will be developed in ref. 6.

3.3 Isotropic hardening

3.3.1 Initially isotropic materials

Consider first an initially isotropic material. The invariant form of the initial criterion is given by (1). The general concept of isotropic hardening is introduced by the general isotropic transformation of the stress tensor $\underset{\sim}{T}$, involving the plastic strain $\varepsilon\underset{\sim}{I}$:

$$\underset{\sim}{T} \Rightarrow \underset{\sim}{\bar{T}} = \underset{\sim}{\bar{T}}(\underset{\sim}{T}, \varepsilon\underset{\sim}{I}) \qquad (18)$$

The invariant form of (18) is given by the theory of representation for tensor functions:

$$\left.\begin{aligned} \bar{\underset{\sim}{T}} &= \alpha_o \underset{\sim}{I} + \alpha_1 \underset{\sim}{S} + \alpha_2 \underset{\sim}{S}^2 \\ \alpha_i &= \alpha_i(\mathrm{tr}\underset{\sim}{T},\ \sqrt{\mathrm{tr}\underset{\sim}{S}^2},\ \sqrt[3]{\mathrm{tr}\underset{\sim}{S}^3},\ \varepsilon) \end{aligned}\right\} \qquad (19)$$

Employing the Cayley-Hamilton theorem, the tensorial relation (19) can be written in the form of relations between the basic isotropic invariants of $\underset{\sim}{T}$ and those of the transformed tensor $\bar{\underset{\sim}{T}}$:

$$\left.\begin{aligned} \mathrm{tr}\bar{\underset{\sim}{T}} &= k_1^{\varepsilon}(\mathrm{tr}\underset{\sim}{T},\ \sqrt{\mathrm{tr}\underset{\sim}{S}^2},\ \sqrt[3]{\mathrm{tr}\underset{\sim}{S}^3} \\ \sqrt{\mathrm{tr}\bar{\underset{\sim}{S}}^2} &= k_2^{\varepsilon}(\mathrm{tr}\underset{\sim}{T},\ \sqrt{\mathrm{tr}\underset{\sim}{S}^2},\ \sqrt[3]{\mathrm{tr}\underset{\sim}{S}^3}) \\ \sqrt[3]{\mathrm{tr}\bar{\underset{\sim}{S}}^3} &= k_3^{\varepsilon}(\mathrm{tr}\underset{\sim}{T},\ \sqrt{\mathrm{tr}\underset{\sim}{S}^2},\ \sqrt[3]{\mathrm{tr}\underset{\sim}{S}^3}) \end{aligned}\right\} \qquad (20)$$

where the superscript ε denotes the dependence of the functions k_i on the plastic strain.

Finally, the general rule of isotropic hardening for an initially isotropic material is given by:

$$f(\mathrm{tr}\underset{\sim}{T},\ \sqrt{\mathrm{tr}\underset{\sim}{S}^2},\ \sqrt[3]{\mathrm{tr}\underset{\sim}{S}^3}) = 0 \quad \Rightarrow \left\{\begin{aligned} & f(k_1^{\varepsilon},\ k_2^{\varepsilon},\ k_3^{\varepsilon}) = 0 \\ & k_i^{\varepsilon} = k_i^{\varepsilon}(\mathrm{tr}\underset{\sim}{T},\ \sqrt{\mathrm{tr}\underset{\sim}{S}^2},\ \sqrt[3]{\mathrm{tr}\underset{\sim}{S}^3}) \end{aligned}\right. \qquad (21)$$

The versatility of the proposed concept is manifest, when considering the following properties of rule (21):

(1) Homothetic and translational isotropic hardening rules are obviously involved as very special cases.

(2) The hardening phenomena which cannot be described by the classical formulations are taken into account (see Section 4.2 below).

(3) By rule (21), a given initial isotropic criterion can be transformed into an arbitrary subsequent isotropic criterion.

The last property shows that the proposed concept of isotropic hardening is the most general. The demonstration of this property will be given in ref.6.

3.3.2 Initially anisotropic materials

Consider now an initially anisotropic material. The invariant form of the initial criterion is given by (4). The general concept of isotropic hardening is introduced by a similar procedure: the stress tensor $\underset{\sim}{T}$ is subjected to a general isotropic transformation by the invariant form (19).

Using Rivlin's generalization of the Cayley-Hamilton theorem [20], the three isotropic invariants of $\underset{\sim}{\bar{T}}$ and the mixed (anisotropy) invariants $\bar{J}_k$ of $\underset{\sim}{\bar{T}}$ and the structural tensors $\underset{\sim}{\xi}_i$, can be expressed in terms of the corresponding invariants of the stress tensor $\underset{\sim}{T}$. Such a procedure has been developed in ref. 3.

Finally, the general rule of isotropic hardening for an initially anisotropic material is given by:

$$f(\mathrm{tr}\underset{\sim}{T}, \sqrt{\mathrm{tr}\underset{\sim}{S}^2}, \sqrt[3]{\mathrm{tr}\underset{\sim}{S}^3}, J_k) = 0 \Rightarrow f(\mathrm{tr}\underset{\sim}{\bar{T}}, \sqrt{\mathrm{tr}\underset{\sim}{\bar{S}}^2}, \sqrt[3]{\mathrm{tr}\underset{\sim}{\bar{S}}^3}, \bar{J}_k) = 0 \qquad (22)$$

Further details will be given in ref.6. We simply mention here the main properties of the general rule (22):

(1) The transformation (19) of the stress tensor being isotropic, the transformed mixed invariants $\bar{J}_k$ admit the same invariance group as the initial mixed invariants J_k; thus, the type of anisotropy of the initial criterion is preserved for the subsequent criterion.

(2) The "degree of anisotropy" varies in general. In fact, there exists a problem for a realistic definition of the "degree of anisotropy". We propose the following definition: this degree is specified by the shape, the position with respect to the deviatoric subspace and the orientation of the yield surface in the stress space. Thus, the degree is preserved if and only if the transformation of the stress tensor is that of homothetic or translational isotropic hardening rules.

(3) By rule (22), a given initial anisotropic criterion can be trans-

formed into an arbitrary subsequent anisotropic criterion of the same type of anisotropy.

3.4 Anisotropic hardening

Within the proposed concept, the anisotropic modifications of the materials' internal structure, which are induced by non-isotropic plastic strains, can be taken into account by the following transformation of the stress tensor:

$$\underset{\sim}{T} \Rightarrow \bar{\underset{\sim}{T}} = \bar{\underset{\sim}{T}}(\underset{\sim}{T}, \underset{\sim}{\xi}_i, \underset{\sim}{P}); \quad \underset{\sim}{P} - (\tfrac{1}{3}\mathrm{tr}\underset{\sim}{P})\underset{\sim}{I} = \underset{\sim}{E} \neq \underset{\sim}{0} \tag{23}$$

In (23), the tensor function $\bar{\underset{\sim}{T}}$ is isotropic with respect to its arguments $(\underset{\sim}{T}, \underset{\sim}{\xi}_i, \underset{\sim}{P})$, where $\underset{\sim}{T}$ is the stress tensor, $\underset{\sim}{P}$ the plastic strain and $\underset{\sim}{\xi}_i$ the initial structural tensors of the material. It can be shown [4] that $\bar{\underset{\sim}{T}}$ is an anisotropic function with respect to $\underset{\sim}{T}$, the type of anisotropy being specified by the invariance group of $(\underset{\sim}{\xi}_i, \underset{\sim}{P})$, which is obviously different from the initial invariance group of the material.

Following a similar procedure to that for isotropic hardening, we use the theorems of representations for tensor functions and the generalized Cayley-Hamilton theorem, in order to obtain the general rule of anisotropic hardening:

$$f(\mathrm{tr}\underset{\sim}{T}, \sqrt{\mathrm{tr}\underset{\sim}{S}^2}, \sqrt[3]{\mathrm{tr}\underset{\sim}{S}^3}, J_k) = 0 \Rightarrow f(\mathrm{tr}\bar{\underset{\sim}{T}}, \sqrt{\mathrm{tr}\bar{\underset{\sim}{S}}^2}, \sqrt[3]{\mathrm{tr}\bar{\underset{\sim}{S}}^3}, K_\ell) = 0 \tag{24}$$

where the K_ℓ are mixed invariants of $\underset{\sim}{T}$, $\underset{\sim}{\xi}_i$ and $\underset{\sim}{P}$, which take into account the initial anisotropy, the induced anisotropy and the coupling between them [8].

Further details and developements will be presented in ref. 6. The main properties of the general rule (24) are the following:

(1) A given initial anisotropic criterion can be transformed into an arbitrary subsequent anisotropic criterion, of arbitrary type and degree.

(2) The hypothesis that the hardening rule depends only on the present

plastic strain results in the fact that for an initially isotropic material, the subsequent criterion is necessarily orthotropic [8]. But, if strain history is taken into account, considering for example a sequence of small plastic strains, it can be shown that the initial isotropic condition can be transformed into an arbitrary subsequent anisotropic criterion.

4. EXAMPLES

4.1 Introduction

In developing examples of hardening rules, different specific transformations of the stress tensor may be considered. We propose a general isotropic (respectively anisotropic) hardening rule, in considering a very simple transformation of the stress tensor: the isotropic (respectively anisotropic) first order polynomial transformation.

It is worthwhile to point out that this procedure furnishes a framework which is already large enough to include the classical isotropic, kinematic and proposed anisotropic hardening rules as special cases and to account for phenomena which cannot be described by the classical formulations.

4.2 Proposed general isotropic hardening rule

When restricted to a first order polynomial transformation, the general invariant form (19) can be written in the following form:

$$\lambda\bar{\underset{\sim}{T}} = (C + \frac{A}{3}\mathrm{tr}\underset{\sim}{T})\underset{\sim}{I} + D\underset{\sim}{S} \qquad (25)$$

where the material coefficients λ, A, C, D, are scalar-valued functions of the plastic strain ε.

Consider first a material obeying in its initial state the von Mises criterion. In this case, the coefficients involved in (25) are subjected to the initial conditions: $C(0) = A(0) = 0$, $\lambda(0) = 1$, so that for $\underset{\sim}{P} = \underset{\sim}{0}$, we have $\bar{\underset{\sim}{T}}(\underset{\sim}{0}) = \underset{\sim}{S}$. Employing (25) and rule (21) of isotropic hardening, the subsequent criterion is given by:

$$\mathrm{tr}(\lambda \bar{\underset{\sim}{T}})^2 = 3C^2 + 2AC\mathrm{tr}\underset{\sim}{T} + \frac{1}{3}A^2\mathrm{tr}^2\underset{\sim}{T} + D^2\mathrm{tr}\underset{\sim}{S}^2 = 2k^2\lambda^2 \qquad (26)$$

In (26), the coefficient $D(\varepsilon)$ and $\lambda(\varepsilon)$ account for the change of size of the criterion, $C(\varepsilon)$ describes the translation parallel to the isotropic axis and $A(\varepsilon)$ stands for the change in shape. Thus, the very simple stress transformation rule (25) and the application of the von Mises criterion provide a hardening rule involving the essential features of our general concept of isotropic hardening: the initial criterion can change in size, shape and position along the isotropic axis.

The functions $A(\varepsilon)$, $C(\varepsilon)$, $D(\varepsilon)$ and $\lambda(\varepsilon)$ must be further specified from physical observations or (and) phenomenological considerations. The purpose of this paper is to show the possibilities of the proposed rule. Thus, we consider the following simple case:

$$\left.\begin{array}{lll} \varepsilon \geqslant 0 & \sqrt{3}\lambda = \sqrt{1 + \ell\varepsilon} & \sqrt{3}\,C = \sqrt{c\varepsilon} \\ & \sqrt{3}A = \sqrt{(a+2d)\varepsilon} & \sqrt{3}\,D = \sqrt{1+(a-d)\varepsilon} \end{array}\right\} \qquad (27)$$

where a, c, d, ℓ are material constants. For $\varepsilon = 0$, we obtain the von Mises criterion:

$$\varepsilon = 0 \Rightarrow \mathrm{tr}\bar{\underset{\sim}{T}}^2 = \mathrm{tr}\underset{\sim}{S}^2 = 2\,k^2 \qquad (28)$$

Employing (27) in (26) and developing with respect to the principal stresses, the subsequent criterion takes finally the form:

$$\begin{aligned} 3c\varepsilon &+ 2\sqrt{c(a+2d)}\;\varepsilon(\sigma_1+\sigma_2+\sigma_3) + (\frac{2}{3} + a\varepsilon)(\sigma_1^2+\sigma_2^2+\sigma_3^2) + \\ &+ 2(-\frac{1}{3} + d\varepsilon)(\sigma_1\sigma_2+\sigma_2\sigma_3+\sigma_3\sigma_1) = 2k^2(1+\ell\varepsilon) \end{aligned} \qquad (29)$$

In plane stress, it is easy to see that in the special case $c = 0$, the initial von Mises ellipse is transformed into a subsequent ellipse with different size and shape. Thus, the second example of isotropic hardening phenomena described in Section 2.4 is taken into account. If $c \neq 0$, the von Mises ellipse undergoes in addition a translation parallel to the isotropic axis.

Consider now a material obeying in its initial state the linear criterion:

$$\mathrm{tr}\underset{\sim}{T} - a\sqrt{\mathrm{tr}\underset{\sim}{S}^2} = c \tag{30}$$

Employing (25) and rule (21) of isotropic hardening, the subsequent criterion, when developed with respect to the principal stresses, is given by:

$$\begin{aligned}&(3A+C)(\sigma_1+\sigma_2+\sigma_3) + a\sqrt{\{3C^2 + 2AC(\sigma_1+\sigma_2+\sigma_3) +}\\ &\overline{+ \tfrac{1}{3}(A^2+2D^2)(\sigma_1^2+\sigma_2^2+\sigma_3^2) + \tfrac{2}{3}(A^2-D^2)(\sigma_1\sigma_2 + \sigma_2\sigma_3 + \sigma_3\sigma_1)\}} = \lambda c\end{aligned} \tag{31}$$

Unfortunately, there is a lack of experimental results for testing the proposed rule (31) in its general form. The increase of internal friction of granular materials, when subjected to isotropic strain ε, can be described by introducing the following forms of the coefficients involved in the stress transformation (25):

$$\begin{aligned}&\lambda \equiv 1 \quad ; \quad C \equiv 0 \quad ; \quad A \equiv 1 \quad ; \quad D = k(\varepsilon),\\ &\text{with} \qquad 0 < k(\varepsilon) \leqslant 1, \ k(0) = 1\end{aligned} \tag{32}$$

Introducing the coefficients (32) in (25) and employing rule (21) of isotropic hardening for the criterion (30), we obtain for cohesionless granular materials the following subsequent criterion:

$$\mathrm{tr}\underset{\sim}{T} - \bar{a}\sqrt{\mathrm{tr}\underset{\sim}{S}^2} = 0 \ , \quad \text{with } \bar{a} = k(\varepsilon)a \tag{33}$$

Thus, the first example of isotropic hardening phenomena described in Section 2.4 is taken into account.

4.3 Proposed general anisotropic hardening rule

4.3.1 Initially isotropic materials

Consider an initially isotropic material subjected to a plastic strain $\underset{\sim}{P}$ with a non-zero deviatoric part $\underset{\sim}{E}$. In this case, the subsequent criterion is necessarily orthotropic [8].

The proposed general hardening rule in the case of a three-dimensional stress-tensor will be developed in ref.7. We simply mention here the procedure employed and the main results obtained. Using the general invariant form of an orthotropic first-order polynomial tensor transformation, we derive the stress transformation rule, involving 15 material coefficients, which are scalar-valued functions of $\underset{\sim}{P}$. Introducing the transformed stress into the isotropic von Mises criterion, we obtain a general three-dimensional hardening rule, of which the rules proposed by Yoshimura [24] and Svensson [22] are special cases.

We develop here a similar procedure in the case of plane stress. In this case, it can be shown that the representation of the transformation $\underset{\sim}{\bar{T}} = \underset{\sim}{\bar{T}}(\underset{\sim}{T}, \underset{\sim}{P})$ takes the form:

$$\left.\begin{aligned} \underset{\sim}{\bar{T}} &= \Psi_1 \underset{\sim}{M}_1 + \Psi_2 \underset{\sim}{M}_2 + \Psi_3 \underset{\sim}{T} \\ \Psi_i &= \Psi_i(\mathrm{tr}\underset{\sim}{M}_1\underset{\sim}{T},\ \mathrm{tr}\underset{\sim}{M}_2\underset{\sim}{T},\ \mathrm{tr}\underset{\sim}{T}^2,\ \mathrm{tr}\underset{\sim}{M}_1\underset{\sim}{P},\ \mathrm{tr}\underset{\sim}{M}_2\underset{\sim}{P}) \end{aligned}\right\} \tag{34}$$

where $\underset{\sim}{M}_1 = \underset{\sim}{v}_1 \otimes \underset{\sim}{v}_1$, $\underset{\sim}{M}_2 = \underset{\sim}{v}_2 \otimes \underset{\sim}{v}_2$; $(\underset{\sim}{v}_1, \underset{\sim}{v}_2)$ being the principal directions of $\underset{\sim}{P}$ in the plane of stress.

When restricted to a first-order polynomial transformation, the invariant form (34) can be written in the form:

$$\begin{aligned} \lambda\underset{\sim}{\bar{T}} = {} & (C_1 + B_1 \mathrm{tr}\underset{\sim}{M}_1\underset{\sim}{T} + D_1 \mathrm{tr}\underset{\sim}{M}_2\underset{\sim}{T})\underset{\sim}{M}_1 + \\ & + (C_2 + B_2 \mathrm{tr}\underset{\sim}{M}_1\underset{\sim}{T} + D_2 \mathrm{tr}\underset{\sim}{M}_2\underset{\sim}{T})\underset{\sim}{M}_2 + F\underset{\sim}{T} \end{aligned} \tag{35}$$

where the material coefficients λ, C_1, C_2, B_1, B_2, D_1, D_2, F are scalar-valued functions of $\mathrm{tr}\underset{\sim}{M}_1\underset{\sim}{P} = \varepsilon_1$ and $\mathrm{tr}\underset{\sim}{M}_2\underset{\sim}{P} = \varepsilon_2$; ε_1 and ε_2 are the principal values of $\underset{\sim}{P}$.

Employing the stress transformation (35), different initial yield conditions may be considered. For a material obeying in its initial state the von Mises criterion, the subsequent criterion is given by:

$$\begin{aligned}(C_1^2 + C_2^2) &+ 2[C_1(B_1+F) + C_2B_2]\mathrm{tr}\underset{\sim}{M}_1\underset{\sim}{T} + \\ &+ 2[C_1D_1 + C_2(D_2+F)]\mathrm{tr}\underset{\sim}{M}_2\underset{\sim}{T} + \\ &+ (B_1^2+B_2^2+2FB_1)\mathrm{tr}^2\underset{\sim}{M}_1\underset{\sim}{T} + \\ &+ (D_1^2+D_2^2+2FD_2)\mathrm{tr}^2\underset{\sim}{M}_2\underset{\sim}{T} + \\ &+ 2[D_1(B_1+F) + B_2(D_2+F)]\mathrm{tr}\underset{\sim}{M}_1\underset{\sim}{T}\ \mathrm{tr}\underset{\sim}{M}_2\underset{\sim}{T} + \\ &+ F^2\ \mathrm{tr}\underset{\sim}{T}^2 = 2\ k^2\lambda^2\end{aligned} \tag{36}$$

where the initial values of the material coefficients are:

$$\underset{\sim}{P} = \underset{\sim}{0} \Rightarrow C_1 = C_2 = B_1 = D_1 = 0 \quad ; \quad B_2 = D_2 = -\frac{1}{3} \quad ; \lambda = F = 1 \tag{37}$$

so that for $\underset{\sim}{P} = \underset{\sim}{0}$, we obtain the initial isotropic von Mises criterion:

$$\underset{\sim}{P} = \underset{\sim}{0} \Rightarrow \mathrm{tr}\overline{\underset{\sim}{T}}^2 = \mathrm{tr}\underset{\sim}{S}^2 = 2\ k^2 \tag{38}$$

4.3.2 Prager's kinematic hardening rule

We consider the following very simple case for the stress transformation (35):

$$\left.\begin{aligned}&\lambda(\underset{\sim}{P}) \equiv F(\underset{\sim}{P}) \equiv 1 \quad ; \quad B_1(\underset{\sim}{P}) \equiv B_2(\underset{\sim}{P}) \equiv D_1(\underset{\sim}{P}) \equiv D_2(\underset{\sim}{P}) \equiv 0 \quad ; \\ &C_1(\underset{\sim}{P}) = c\varepsilon_1 \quad ; \quad C_2(\underset{\sim}{P}) = c\varepsilon_2 \Rightarrow \overline{\underset{\sim}{T}} = c\varepsilon_1\underset{\sim}{M}_1 + c\varepsilon_2\underset{\sim}{M}_2 + \underset{\sim}{T}\end{aligned}\right\} \tag{39}$$

where c is a material constant.

Employing (39) in (36) and developing with respect to the principal stresses, the subsequent criterion takes the particular form:

$$2c(\varepsilon_1\cos^2\theta + \varepsilon_2\sin^2\theta)\sigma_1 + 2c(\varepsilon_1\sin^2\theta + \varepsilon_2\cos^2\theta)\sigma_2+\sigma_1^2+\sigma_2^2 = \tag{40}$$

$$= 2k^2 - c^2(\varepsilon_1^2 + \varepsilon_2^2)$$

where θ is the angle between the principal directions of $\underset{\sim}{T}$ and $\underset{\sim}{P}$. Thus, we obtain Prager's kinematic hardening rule [18], which appears as a very simple particular case of the proposed general rule. We recall that by the kinematic hardening rule, the initially isotropic criterion is transformed into a particular orthotropic subsequent criterion (see Section 2.2).

Similar results are obtained for Ziegler's modification of Prager's hardening rule [25]. The Prager-Hodge hardening rule [13] is simply obtained by introducing $\lambda = \lambda(\varepsilon_1+\varepsilon_2)$ in the restrictions (39), in order to add homothetic isotropic hardening to the kinematic hardening.

4.3.3 Baltov-Sawczuk hardening rule

Several proposed anisotropic hardening rules appear as special cases of the general rule (36) in plane stress. Consider for example the Baltov-Sawczuk hardening rule [2], which can be written in the following invariant three-dimensional form:

$$\mathrm{tr}(\underset{\sim}{S} - c\underset{\sim}{E})^2 + A\ \mathrm{tr}^2[\underset{\sim}{E}(\underset{\sim}{S}-c\underset{\sim}{E})] = 2\ k^2 \tag{41}$$

where $\underset{\sim}{S}$ and $\underset{\sim}{E}$ are respectively the stress and plastic strain deviators; c and A are material constants.

The rule (41) involves a kinematic part, which is taken into account by the following stress transformation:

$$\underset{\sim}{S} \Rightarrow \underset{\sim}{\bar{S}} = \underset{\sim}{S} - c\underset{\sim}{E} \tag{42}$$

With respect to $\underset{\sim}{\bar{S}}$, the rule (41) takes the invariant form:

$$\mathrm{tr}\underset{\sim}{\bar{S}}^2 + A\ \mathrm{tr}^2\underset{\sim}{\bar{S}}\underset{\sim}{E} = 2\ k^2 \tag{43}$$

Introducing the plastic strain principal values :

$$\left.\begin{aligned} &\underset{\sim}{E} = \varepsilon_1 \underset{\sim}{M}_1 + \varepsilon_2 \underset{\sim}{M}_2 - (\varepsilon_1+\varepsilon_2)\underset{\sim}{M}_3 \\ &\underset{\sim}{M}_1 = \underset{\sim}{v}_1 \otimes \underset{\sim}{v}_1 \; ; \; \underset{\sim}{M}_2 = \underset{\sim}{v}_2 \otimes \underset{\sim}{v}_2 \; ; \; \underset{\sim}{M}_3 = \underset{\sim}{v}_3 \otimes \underset{\sim}{v}_3 \end{aligned}\right\} \quad (44)$$

where $(\underset{\sim}{v}_1, \underset{\sim}{v}_2, \underset{\sim}{v}_3)$ are the principal directions of $\underset{\sim}{E}$, we obtain finally:

$$\begin{aligned} &2A(2\varepsilon_1+\varepsilon_2)(\varepsilon_1+2\varepsilon_2)\mathrm{tr}\underset{\sim}{M}_1\overline{\underset{\sim}{S}}\ \mathrm{tr}\underset{\sim}{M}_2\overline{\underset{\sim}{S}} + \\ &\quad + A(2\varepsilon_1+\varepsilon_2)^2\ \mathrm{tr}^2\underset{\sim}{M}_1\overline{\underset{\sim}{S}} + A(\varepsilon_1+2\varepsilon_2)^2\ \mathrm{tr}\underset{\sim}{M}_2\overline{\underset{\sim}{S}} + \mathrm{tr}\overline{\underset{\sim}{S}}^2 = 2\ k^2 \end{aligned} \quad (45)$$

When applied to $\overline{S}$, the proposed anisotropic hardening rule (36), with $C_1 \equiv C_2 \equiv 0$, can be written in the form:

$$2K_1\mathrm{tr}\underset{\sim}{M}_1\overline{\underset{\sim}{S}}\ \mathrm{tr}M_2\overline{\underset{\sim}{S}} + K_2\mathrm{tr}^2\underset{\sim}{M}_1\overline{\underset{\sim}{S}} + K_3\mathrm{tr}^2\underset{\sim}{M}_2\overline{\underset{\sim}{S}} + K_4\mathrm{tr}\overline{S}^2 = 2k^2\lambda^2 \quad (46)$$

Identifying (45) with (46), we obtain:

$$\lambda \equiv K_4 \equiv 1 \; ; \; K_2 = (2\varepsilon_1+\varepsilon_2)^2 \; ; \; K_3 = (\varepsilon_1+2\varepsilon_2)^2 \; ; \; K_1^2 \equiv K_2K_3 \quad (47)$$

Thus, the Baltov-Sawczuk hardening rule appears as a special case of our proposed general rule in plane stress. The Baltov-Sawczuk rule is three-dimensional, but the involved transformation of stress has the tensorial form of a particular two-dimensional first-order polynomial transformation. A similar result is obtained for the anisotropic hardening rule proposed by Backhaus [1]. In the sense of the tensorial form of the involved stress transformation, the anisotropic hardening rules proposed by Yoshimura [24] and by Svensson [22] are more general.

4.3.4 Initially anisotropic material

Considering the procedure presented in Section 3.4, a general anisotropic hardening rule for initially anisotropic materials can be derived in a straightforward manner. In the case of an initially orthotropic material, such a procedure has already been developed and the hardening rule obtained has been tested by experimental results [10, 9]. Further details and other examples will be developed in ref. 7.

5. CONCLUSIONS

The proposed general concept results in a rational formulation of isotropic and anisotropic hardening. In the developed scheme, the generalized sense of isotropic hardening is clearly defined: a given initial isotropic criterion can be transformed into an arbitrary subsequent isotropic criterion, change of size, shape and position along the isotropic stress axis being possible for the yield surface ; likewise, a given initial anisotropic criterion can be transformed into an arbitrary subsequent anisotropic criterion, but of the same type of anisotropy. All other hardening processes are involved in our general formulation of anisotropic hardening. Thus, the proposed concepts are definitely disjointed.

In the presented examples, the proposed general isotropic and anisotropic hardening rules are based on first-order polynomial isotropic and anisotropic transformations of the stress tensor and on the application of the developed procedure to the von Mises and linear criteria. Although these stress transformations and the applied criteria are very simple, the rules obtained are already general enough to be able to include the classical hardening rules and to generalize them for the description of hardening phenomena which they cannot take into account.

More general stress transformation rules and more elaborate criteria can be introduced in the proposed concept. But even for the simple scheme developed in this work, there exists a great lack of experimental results for testing fully the proposed isotropic and anisotropic hardening rules. Well organized multiaxial experiments, guided by a sound theoretical basis are needed.

The work presented shows the usefulness and the capacity of the theory of representation for tensor functions in the developement of a rational and unified description of complex mechanical phenomena.

REFERENCES

1. BACKHAUS, G., Zur Fliessgrenze bei Allgemeiner Verfestigung, *ZAMM*, 48 (1968) : 99-108.

2. BALTOV, A. and SAWCZUK A., A Rule of Anisotropic Hardening, *Acta Mech.*, 1 (1965) : 81-92.

3. BOEHLER, J.P., Contributions Théoriques et Expérimentales à l'Etude des Milieux Plastiques Anisotropes, *Thèse de Doctorat ès-Sciences*, Grenoble, 1975.

4. BOEHLER, J.P., Lois de Comportement Anisotrope des Milieux Continus, *J. de Mécanique*, 17 (2), (1978) : 153-190.

5. BOEHLER, J.P., A Simple Derivation of Representations for Non-Polynomial Constitutive Equations in Some Cases of Anisotropy, *ZAMM*, 59 (1979) : 157-167.

6. BOEHLER, J.P., On a General Concept of Isotropic and Anisotropic Hardening. Part I: Theoretical (in preparation).

7. BOEHLER, J.P., On a General Concept of Isotropic and Anisotropic Hardening. Part II: Examples (in preparation).

8. BOEHLER, J.P. and RACLIN J., Ecrouissage Anisotrope des Matériaux Orthotropes Prédéformés, *J. Mécanique Théorique et Appliquée*, Numéro Spécial, 1982, 23-44.

9. BOEHLER, J.P. and RACLIN J., Anisotropic Hardening of Prestrained Rolled Sheet-Steel. In: *Current Advances in Mechanical Design and Production*, Proc. 2 nd Cairo Un. MDP Conf., Cairo University, 1982, 483-492.

10. BOEHLER, J.P. and RACLIN J., A Unified Theoretical and Experimental Study of Anisotropic Hardening, *Trans. 6th Int. Conf. on SMIRT*, North-Holland, Amsterdam, 1982, Invited Lecture L3/2.

11. CHABOCHE, J.L., Le Concept de Contrainte Effective Appliqué à l'Elasticité et à la Viscosité en Présence d'un Endommagement Anisotrope. In: *Mechanical Behavior of Anisotropic Solids*, *Proc. Euromech Coll. 115-CNRS Int. Coll.* 295 (Villard-de-Lans, June 1979), (Ed. J.P. Boehler), Editions du CNRS (Paris) and Martinus Nijhoff, The Hague. 1982, 737-760.

12. Hill, R., *The Mathematical Theory of Plasticity*, Oxford University Press, Oxford, 1950.

13. HODGE, P.G., Discussion on W. Prager Paper "A New Method of Analyzing Stresses and Strains in Work-Hardening Plastic Solids", *J. Appl. Mech.*, 24 (1957) : 482-483.

14. KACHANOV, L. M., Time of Rupture Process under Creep Conditions, *Izv. Akad. Nauk SSR Otd. Tekh. Nauk*, 8 (1958), 26-31 (in Russian).

15. LEMAITRE J. and CHABOCHE J.L., Aspect Phénoménologique de la Rupture par Endommagement, *J. Mécanique Appliquée*, 2 (3), 1978, 317-365.

16. MURAKAMI S. and OHNO N., A Continuum Theory of Creep and Creep Damage. In: *Creep in Structures, Proc. IUTAM Symp.* (Leicester, Sept. 1980), (Ed. A.R.S. Ponter and D.R. Hayhurst), Springer, Berlin, 1981, 422-444.

17. MURAKAMI, S. and SAWCZUK A., On Description of Rate-Independent Behavior for Prestrained Solids, *Arch. Mech.*, 32 (1979) : 251-264.

18. PRAGER, W., A New Method of Analyzing Stresses and Strains in Work-hardening Plastic Solids, *J. Appl. Mech.*, 23 (4) (1956) : 493-496.

19. REES, D.W.A., A Survey of Hardening in Metallic Materials. In: *Failure Criteria of Structured Media, Proc. CNRS Int. Coll. 351* (Villard-de-Lans, June 1983), in press.

20. RIVLIN, R.S., Further Remarks on the Stress-Deformation Relations for Isotropic Materials, *J. Rat. Mech. Anal.* 4 (1955) : 681-702.

21. SPENCER, A.J.M., The Formulation of Constitutive Equations for Anisotropic Solids. In: *Mechanical Behavior of Anisotropic Solids, Proc. Euromech. Coll. 115-CNRS Int. Coll. 295* (Villard-de-Lans, June 1979), (Ed. J.P. Boehler), Editions du CNRS (Paris) and Martinus Nijhoff, The Hague, 1982, 3-26.

22. SVENSSON, J.L., Anisotropy and the Bauschinger Effect in Cold Rolled Aluminium, *J. Mech. Engng Sci.*, 8 (1966) : 162-172.

23. TOUATI A., Comportement Mécanique des Sols Pulvérulents sous Fortes Contraintes, *Thèse de Docteur-Ingénieur*, Ecole Nationale des Ponts et Chaussées, Paris, 1982.

24. YOSHIMURA, Y., Hypothetical Theory of Anisotropy and the Bauschinger Effet Due to Plastic Strain History, Aero. Res. Inst.,Tokyo University, 349 (1959) : 224-246.

25. ZIEGLER, H., A Modification to Prager's Hardening Rule, *Q.J. Appl. Math.,* 17 (55), (1959) : 55-65.

Chapter 7

ANISOTROPIC HARDENING OF ROLLED SHEET-STEEL

J.P. Boehler
University of Grenoble, France

1. INTRODUCTION

It is well known that during manufacturing processes, the internal structure of polycrystals become strongly oriented, thus anisotropic [1]. On the macroscopic level, the oriented internal structure results in the anisotropy of the mecahnical behavior. For rolled sheet-steel, the internal structure generally presents three orthogonal planes of symmetry ; the macroscopic behavior is thus orthotropic. When sheet-steel is subjected to further irreversible processes, the anisotropy of the internal structure is modified. The modifications affect not only the degree of initial orthotropy, but also the type of anisotropy, according to the orientation of the principal directions of the irreversible deformation with respect to the rolling direction: this is the phenomenon of anisotropic hardening.

The observed experimental facts show the complexity of anisotropic phenomena and reveal the need to develop a consistent formulation of constitutive relations regarding anisotropic hardening. The aim of this Chapter is to provide a unified theoretical and experimental description of phenomenological macroscopic anisotropic behavior and of its evolution during irreversible prestrains. The proper description and the reliable assessment of anisotropic hardening of prestrained sheet-steel are of importance when developing materials with adequate mechanical properties and when designing engineering structures.

In Non-Linear Continuum Mechanics, an objective and rational formulation of the constitutive relations for anisotropic hardening materials can be developed within the framework of the theory of representations for tensor functions [2-4]. This approach has already proved fruitful in the study of the plastic behavior of anisotropic materials [4] and of prestrained initially isotropic materials [5], [6]. The same approach is developed here for initially orthotropic metals subjected to irreversible deformations. Much of this Chapter is taken from [7].

2. CONSTITUTIVE RELATION

2.1 General invariant form of the constitutive relation

We consider a plane orthotropic sheet-steel with privileged directions $\underset{\sim}{v}_1$, $\underset{\sim}{v}_2$, $\underset{\sim}{v}_3$. We define $\underset{\sim}{v}_1$ as the rolling, $\underset{\sim}{v}_2$ the transverse, and $\underset{\sim}{v}_3$ the normal direction. We suppose that the sheet-metal has been subjected to an irreversible prestraint $\underset{\sim}{P}$ with principal directions $(\underset{\sim}{E}_1, \underset{\sim}{E}_2, \underset{\sim}{v}_3)$: Ψ is the angle between the directions $\underset{\sim}{v}_1$ and $\underset{\sim}{E}_1$ (Fig.1). A plane stress $\underset{\sim}{\overline{T}}$ is then applied to the prestrained sheet. The principal directions of $\underset{\sim}{\overline{T}}$ are $(\underset{\sim}{e}_1, \underset{\sim}{e}_2, \underset{\sim}{v}_3)$; θ is the angle between the directions $\underset{\sim}{v}_1$, and $\underset{\sim}{e}_1$ (Fig.2).

We restrict our analysis to the case where the mechanical behavior of the material can be described in considering the stress tensor $\underset{\sim}{\overline{T}}$ as a tensor-valued function of a kinematic tensor $\underset{\sim}{\overline{D}}$. We suppose further that the constitutive law depends on the history of the irreversible prestraining only through the present value of the Green-Lagrange prestrain tensor $\underset{\sim}{P}$. The tensors $\underset{\sim}{\overline{T}}$ and $\underset{\sim}{\overline{D}}$ are introduced with respect to the prestrained configuration. In order to derive the constitutive law in a correct manner, we consider the two following tensors:

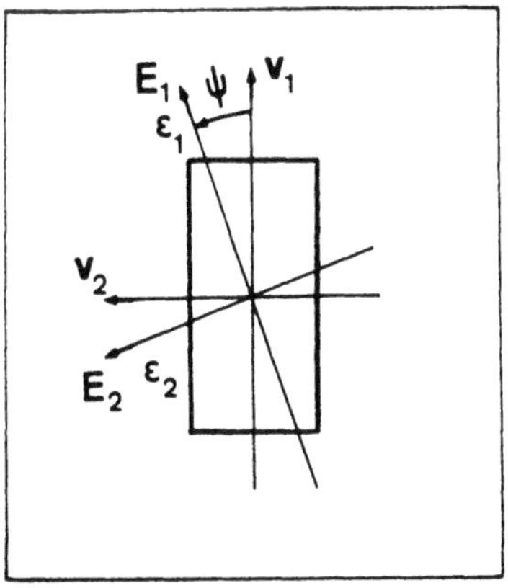

Fig.1. Irreversible prestrain

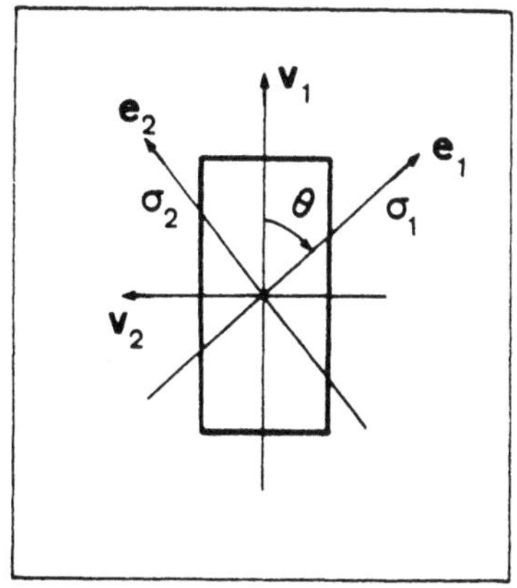

Fig.2. Plane stress

$$\underset{\sim}{T} = \underset{\sim}{R}^t \bar{\underset{\sim}{T}} \underset{\sim}{R} \quad ; \quad \underset{\sim}{D} = \underset{\sim}{R}^t \bar{\underset{\sim}{D}} \underset{\sim}{R} \qquad (1)$$

where $\underset{\sim}{R}$ is the rotation tensor of the canonical polar decomposition of the prestrain gradient tensor. It can then be shown [8] that the constitutive law of the prestrained sheet-steel takes the following form:

$$\underset{\sim}{T} = \underset{\sim}{F}(\underset{\sim}{D}, \underset{\sim}{P}, \underset{\sim}{M}) \qquad (2)$$

where $\underset{\sim}{T}$, $\underset{\sim}{D}$ and $\underset{\sim}{P}$ are the plane parts of the stress, the kinematic and the prestrain tensors respectively. The structural tensor $\underset{\sim}{M} = \underset{\sim}{v}_1 \otimes \underset{\sim}{v}_1$ accounts for the initial orthotropy of the material.

As a consequence of the Principle of Isotropy of Space, $\underset{\sim}{F}$ is an isotropic function with respect to its three arguments $\underset{\sim}{D}$, $\underset{\sim}{P}$ and $\underset{\sim}{M}$. The analysis of form (2) shows that the stress tensor $\underset{\sim}{T}$ is an orthotropic function with respect to the two mechanical arguments $\underset{\sim}{D}$ and $\underset{\sim}{P}$, [9] (see also Chapter 3).

An irreducible representation of the constitutive relation (2) is given by (see Chapter 3):

$$\left.\begin{aligned} \underset{\sim}{T} &= \alpha_o \underset{\sim}{I} + \alpha_1 \underset{\sim}{M} + \alpha_2 \underset{\sim}{D} + \alpha_3 \underset{\sim}{P} \\ \alpha_i &= \alpha_i(\mathrm{tr}\underset{\sim}{D}, \mathrm{tr}\underset{\sim}{M}\underset{\sim}{D}, \mathrm{tr}\underset{\sim}{P}, \mathrm{tr}\underset{\sim}{M}\underset{\sim}{P}, \mathrm{tr}\underset{\sim}{D}\underset{\sim}{P}, \mathrm{tr}\underset{\sim}{D}^2, \mathrm{tr}\underset{\sim}{P}^2) \end{aligned}\right\} \qquad (3)$$

The generating tensors of the representation and the scalar invariants of the functional basis are entirely determined. According to the mechanical significance of the kinematic tensor $\underset{\sim}{D}$, the invariant canonical form (6) constitutes the most general form of, for example, non-linear elasticity or non-linear plasticity of the prestrained sheet-steel.

2.2 Anisotropic hardening

For a given prestrain $\underset{\sim}{P}$, the relation between $\underset{\sim}{T}$ and $\underset{\sim}{D}$ involved in (3) is anisotropic. If $\underset{\sim}{P} = \underset{\sim}{0}$, the relation is orthotropic ; if $\underset{\sim}{P} \neq \underset{\sim}{0}$, the initial orthotropy of the material is modified and the anisotropy does not necessarily remain orthotropic: this is the phenomenon of anisotropic hardening ([7], [13]). In this Section, we analyze various modes of anisotropic hardening for initially isotropic and orthotropic materials, taking into account the orientation of the principal directions $(\underset{\sim}{E}_1, \underset{\sim}{E}_2)$ of the prestrain tensor $\underset{\sim}{P}$ with respect to the privileged directions $(\underset{\sim}{v}_1, \underset{\sim}{v}_2)$ of the material.

In order to allow a proper analysis of the different representations obtained from (3) for the various cases considered, we derive an invariant expression for the prestrain tensor $\underset{\sim}{P}$. Introducing the following two second order tensors:

$$\underset{\sim}{M}_1 = \underset{\sim}{E}_1 \otimes \underset{\sim}{E}_1 \qquad \underset{\sim}{M}_2 = \underset{\sim}{E}_2 \otimes \underset{\sim}{E}_2$$
$$M_1 = \begin{vmatrix} 1 & 0 \\ 0 & 0 \end{vmatrix} (\underset{\sim}{E}_1, \underset{\sim}{E}_2) \qquad M_2 = \begin{vmatrix} 0 & 0 \\ 0 & 1 \end{vmatrix} (\underset{\sim}{E}_1, \underset{\sim}{E}_2) \tag{4}$$

we obtain the relations :

$$\left.\begin{aligned} &\underset{\sim}{M}_1 + \underset{\sim}{M}_2 = \underset{\sim}{I} \; ; \\ &\mathrm{tr}\underset{\sim}{M}_1\underset{\sim}{P} = \varepsilon_1 \; ; \qquad \mathrm{tr}\underset{\sim}{M}_2\underset{\sim}{P} = \varepsilon_2 \; ; \\ &\mathrm{tr}\underset{\sim}{M}_1\underset{\sim}{P} + \mathrm{tr}\underset{\sim}{M}_2\underset{\sim}{P} = \mathrm{tr}\underset{\sim}{P} \end{aligned}\right\} \tag{5}$$

where ε_1 and ε_2 are the principal values of $\underset{\sim}{P}$ in the $(\underset{\sim}{v}_1, \underset{\sim}{v}_2)$ plane (cf. Fig.2).

From relations (5), the following invariant expression for $\underset{\sim}{P}$ can be established:

$$\underset{\sim}{P} = (2\mathrm{tr}\underset{\sim}{M}_1\underset{\sim}{P} - \mathrm{tr}\underset{\sim}{P})\underset{\sim}{M}_1 + (\mathrm{tr}\underset{\sim}{P} - \mathrm{tr}\underset{\sim}{M}_1\underset{\sim}{P})\underset{\sim}{I} \tag{6}$$

Employing expression (6), the invariants $\mathrm{tr}\underset{\sim}{D}\underset{\sim}{P}$ and $\mathrm{tr}\underset{\sim}{P}^2$ involved in the representation (3) can be expressed as single-valued functions of the invariants $\mathrm{tr}\underset{\sim}{P}$, $\mathrm{tr}\underset{\sim}{D}$, $\mathrm{tr}\,\underset{\sim}{M}_1\underset{\sim}{P}$ and $\mathrm{tr}\underset{\sim}{M}_1\underset{\sim}{D}$:

$$\left.\begin{aligned} \mathrm{tr}\underset{\sim}{D}\underset{\sim}{P} &= (2\mathrm{tr}\underset{\sim}{M}_1\underset{\sim}{P} - \mathrm{tr}\,\underset{\sim}{P})\mathrm{tr}\underset{\sim}{M}_1\underset{\sim}{D} + (\mathrm{tr}\underset{\sim}{P} - \mathrm{tr}\,\underset{\sim}{M}_1\underset{\sim}{P})\mathrm{tr}\underset{\sim}{D} \\ \mathrm{tr}\underset{\sim}{P}^2 &= 2(\mathrm{tr}\underset{\sim}{M}_1\underset{\sim}{P} - \mathrm{tr}\,\underset{\sim}{P})\mathrm{tr}\underset{\sim}{M}_1\underset{\sim}{P} + \mathrm{tr}^2\underset{\sim}{P} \end{aligned}\right\} \tag{7}$$

Now, we consider first an initially isotropic material, i.e. a material for which the structural tensor $\underset{\sim}{M}$ is identical to the isotropic tensor $\underset{\sim}{I}$. Introducing $\underset{\sim}{M} = \underset{\sim}{I}$ and employing relations (6) and (7), the constitutive law (3) reduces to:

$$\left.\begin{aligned} \underset{\sim}{T} &= \phi_o\underset{\sim}{I} + \phi_1\underset{\sim}{M}_1 + \phi_2\underset{\sim}{D} \\ \phi_i &= \phi_i(\mathrm{tr}\underset{\sim}{D},\ \mathrm{tr}\underset{\sim}{M}_1\underset{\sim}{D},\ \mathrm{tr}\underset{\sim}{D}^2,\ \mathrm{tr}\underset{\sim}{P},\ \mathrm{tr}\underset{\sim}{M}_1\underset{\sim}{P}) \end{aligned}\right\} \tag{8}$$

Consider now an initially orthotropic material without prestrain. An irreducible representation of the constitutive relation for such a material is obtained from (3) with the condition $\underset{\sim}{P} \equiv \underset{\sim}{O}$:

$$\left.\begin{aligned} \underset{\sim}{T} &= \alpha_o\underset{\sim}{I} + \alpha_1\underset{\sim}{M} + \alpha_2\underset{\sim}{D} \\ \alpha_i &= \alpha_i(\mathrm{tr}\underset{\sim}{D},\ \mathrm{tr}\underset{\sim}{M}\underset{\sim}{D},\ \mathrm{tr}\underset{\sim}{D}^2) \end{aligned}\right\} \tag{9}$$

Comparing the representations (8) and (9), we conclude that for the initially isotropic material the prestrain induced anisotropy is an orthotropy of privileged directions $\underset{\sim}{E}_1$, $\underset{\sim}{E}_2$. If the prestrain tensor $\underset{\sim}{P}$ evolves without rotation of its principal directions, the induced anisotropy remains an orthotropy of privileged directions $\underset{\sim}{E}_1$, $\underset{\sim}{E}_2$; the degree of the induced orthotropy changes with the values of the invariants $\mathrm{tr}\underset{\sim}{P}$, $\mathrm{tr}\underset{\sim}{M}_1\underset{\sim}{P}$.

We consider now an initially orthotropic material and we first suppose that the principal directions of $\underset{\sim}{P}$ coincide with the initially privileged directions ($\underset{\sim}{v}_1$, $\underset{\sim}{v}_2$) of the material. In this case, the second order tensor $\underset{\sim}{M}_1$ is identical to the structural tensor $\underset{\sim}{M}$. Introducing the conditions $\underset{\sim}{M}_1 = \underset{\sim}{M}$ and employing relations (6) and (7), the constitutive law (3) reduces to:

$$\left.\begin{aligned} &\underset{\sim}{T} = \beta_0 \underset{\sim}{I} + \beta_1 \underset{\sim}{M} + \beta_2 \underset{\sim}{D} \\ &\beta_i = \beta_i(\mathrm{tr}\underset{\sim}{D},\ \mathrm{tr}\underset{\sim}{M}\underset{\sim}{D},\ \mathrm{tr}\underset{\sim}{D}^2,\ \mathrm{tr}\underset{\sim}{P},\ \mathrm{tr}\underset{\sim}{M}\underset{\sim}{P}) \end{aligned}\right\} \qquad (10)$$

Comparing representation (10) with that for a non-prestrained orthotropic material (8), we conclude that the orthotropy with privileged directions ($\underset{\sim}{v}_1$, $\underset{\sim}{v}_2$) is preserved ; the degree of this orthotropy changes with the values of the invariants $\mathrm{tr}\underset{\sim}{P}$ and $\mathrm{tr}\underset{\sim}{M}\underset{\sim}{P}$.

Consider now the general case, when the principal directions ($\underset{\sim}{E}_1$, $\underset{\sim}{E}_2$) of the prestrain $\underset{\sim}{P}$ do not coincide with the initially privileged directions ($\underset{\sim}{v}_1$, $\underset{\sim}{v}_2$) of the material. Introducing the expressions (6) for $\underset{\sim}{P}$ and (7) for the invariants $\mathrm{tr}\underset{\sim}{D}\underset{\sim}{P}$ and $\mathrm{tr}\underset{\sim}{P}^2$ in the representation (3), we obtain the following equivalent representation:

$$\left.\begin{aligned} &\text{a)}\quad \underset{\sim}{T} = \gamma_0 \underset{\sim}{I} + \gamma_1 \underset{\sim}{M} + \gamma_2 \underset{\sim}{M}_1 + \gamma_3 \underset{\sim}{D} \\ &\text{b)}\quad \gamma_i = \gamma_i(\mathrm{tr}\underset{\sim}{D},\ \mathrm{tr}\underset{\sim}{D}^2,\ \mathrm{tr}\underset{\sim}{M}\underset{\sim}{D},\ \mathrm{tr}\underset{\sim}{M}_1\underset{\sim}{D},\ \mathrm{tr}\underset{\sim}{P},\ \mathrm{tr}\underset{\sim}{M}\underset{\sim}{P},\ \mathrm{tr}\underset{\sim}{M}_1\underset{\sim}{P}) \end{aligned}\right\} \qquad (11)$$

The relation between $\underset{\sim}{T}$ and $\underset{\sim}{D}$ involved in the representation (11) is anisotropic. In the general case, when the arbitrary functions γ_1 and γ_2 are independent, the type of this anisotropy is characterized by the invariance group S of the tensors $\underset{\sim}{M}$ and $\underset{\sim}{M}_1$, which is reduced to the set $\{\underset{\sim}{I}, -\underset{\sim}{I}\}$; the relation between $\underset{\sim}{T}$ and $\underset{\sim}{D}$ is then of the general anisotropic type. This signifies that in the general case, the superposition of an initial orthotropy with a prestrain induced orthotropy results in general anisotropy.

However, the arbitrary functions γ_i having to be specified from experimental investigations for each material and each class of behavior, some special cases can occur, where the resultant anisotropy reduces to an orthotropy with specific symmetry directions. Such a special case is analyzed below.

Consider the case where the ratio of the functions γ_1 and γ_2 is independent of $\underset{\sim}{D}$:

$$\gamma_2 = k(\underset{\sim}{P})\gamma_1 \tag{12}$$

where $k(\underset{\sim}{P})$ is an arbitrary scalar-valued function of $\underset{\sim}{P}$. Then, the tensorial relation (11a) reduces to:

$$\underset{\sim}{T} = \gamma_o \underset{\sim}{I} + \gamma_1[\underset{\sim}{M} + k(\underset{\sim}{P})\underset{\sim}{M}_1] + \gamma_3 \underset{\sim}{D} \tag{13}$$

In (13), we define the tensor $\underset{\sim}{m}(\underset{\sim}{P})$ by:

$$\underset{\sim}{m}(\underset{\sim}{P}) = \underset{\sim}{M} + k(\underset{\sim}{P})\underset{\sim}{M}_1 \tag{14}$$

The principal directions of the tensor $\underset{\sim}{m}$ makes an angle $\delta(\underset{\sim}{P})$ with the initially privileged directions $(\underset{\sim}{v}_1, \underset{\sim}{v}_2)$, such that:

$$\tan[2\delta(\underset{\sim}{P})] = \frac{k(\underset{\sim}{P})}{1-k(\underset{\sim}{P})} \sin 2\Psi \tag{15}$$

where Ψ is the angle between the principal directions of the prestrain tensor $\underset{\sim}{P}$ and the initially priveleged directions $(\underset{\sim}{v}_1, \underset{\sim}{v}_2)$. We further suppose that the scalar-valued functions γ_i satisfy the indentities:

$$\begin{aligned} &\gamma_i(\mathrm{tr}\underset{\sim}{D},\ \mathrm{tr}\underset{\sim}{D}^2,\ \mathrm{tr}\underset{\sim}{M}\underset{\sim}{D},\ \mathrm{tr}\underset{\sim}{M}_1\underset{\sim}{D},\ \mathrm{tr}\underset{\sim}{P},\ \mathrm{tr}\underset{\sim}{M}\underset{\sim}{P},\ \mathrm{tr}\underset{\sim}{M}_1\underset{\sim}{P}) \equiv \\ &\equiv \phi_i(\mathrm{tr}\underset{\sim}{D},\ \mathrm{tr}\underset{\sim}{D}^2,\ \mathrm{tr}\underset{\sim}{m}\underset{\sim}{D},\ \mathrm{tr}\underset{\sim}{P},\ \mathrm{tr}\underset{\sim}{m}\underset{\sim}{P}) \quad ; \quad i = 0,1,2,3. \end{aligned} \tag{16}$$

On these hypotheses, the representation (11) reduces to the following form:

$$\left.\begin{aligned} &\underset{\sim}{T} = \phi_o \underset{\sim}{I} + \phi_1 \underset{\sim}{m} + \phi_3 \underset{\sim}{D} \\ &\phi_i = \phi_i(\mathrm{tr}\underset{\sim}{D},\ \mathrm{tr}\underset{\sim}{m}\underset{\sim}{D},\ \mathrm{tr}\underset{\sim}{D}^2,\ \mathrm{tr}\underset{\sim}{P},\ \mathrm{tr}\underset{\sim}{m}\underset{\sim}{P}) \end{aligned}\right\} \tag{17}$$

Comparing (17) with the representation (8) for a non-prestrained orthotropic material, we conclude that the relation between $\underset{\sim}{T}$ and $\underset{\sim}{D}$ involved in (17) is orthotropic, the privileged directions of the orthotropy being the principal directions of the tensor $\underset{\sim}{m}$. When the prestrain tensor $\underset{\sim}{P}$ evolves, both the privileged directions and the degree of the resultant orthotropy vary.

3. PLASTIC BEHAVIOR

3.1 General invariant forms of the flow law and the yield criterion

Representation (3) can describe plastic behavior, provided that $\underset{\sim}{D}$ stands for the rate of deformation tensor and that $\underset{\sim}{T}$ is homogeneous of order zero with respect to time or, which is equivalent, with respect to the rate of deformation ([4], [9]):

$$\frac{\partial \underset{\sim}{T}}{\partial \underset{\sim}{D}} \underset{\sim}{D} = \underset{\sim}{0} \quad \text{if} \quad \frac{\partial \underset{\sim}{T}}{\partial \underset{\sim}{D}} \neq \underset{\sim}{0} \tag{18}$$

Employing condition (18) in representation (3), we obtain the most general invariant forms of both the flow law:

$$\left.\begin{aligned} &\frac{\underset{\sim}{D}}{\mathrm{tr}^{1/2} \underset{\sim}{D}^2} = \Psi_0 \underset{\sim}{I} + \Psi_1 \underset{\sim}{M} + \Psi_2 \underset{\sim}{T} + \Psi_3 \underset{\sim}{P} \\ &\Psi_i = \Psi_i(\mathrm{tr}\underset{\sim}{T},\ \mathrm{tr}\underset{\sim}{M}\underset{\sim}{T},\ \mathrm{tr}\underset{\sim}{P},\ \mathrm{tr}\underset{\sim}{M}\underset{\sim}{P},\ \mathrm{tr}\underset{\sim}{T}\underset{\sim}{P},\ \mathrm{tr}\underset{\sim}{T}^2,\ \mathrm{tr}\underset{\sim}{P}^2) \end{aligned}\right\} \tag{19}$$

and the yield condition:

$$f(\mathrm{tr}\underset{\sim}{T},\ \mathrm{tr}\underset{\sim}{M}\underset{\sim}{T},\ \mathrm{tr}\underset{\sim}{P},\ \mathrm{tr}\underset{\sim}{M}\underset{\sim}{P},\ \mathrm{tr}\underset{\sim}{T}\underset{\sim}{P},\ \mathrm{tr}\underset{\sim}{T}^2,\ \mathrm{tr}\underset{\sim}{P}^2) = 0 \tag{20}$$

The general invariant forms (19) and (20) in the plane stress case are derived by a similar procedure than the corresponding forms in the three-dimensional case (see Chapter 5).

The flow law (19) specifies the rate of deformation tensor only to within a scalar multiplier. The relation between $\underset{\sim}{D}$ and $\underset{\sim}{T}$ is anisotropic.

In general, the principal directions of the tensors $\underset{\sim}{D}$ and $\underset{\sim}{T}$ do not coincide, their relative orientation depending on the relative orientations of the mechanical tensors $\underset{\sim}{D}$ and $\underset{\sim}{P}$ and the structural tensor $\underset{\sim}{M}$. It is worthwile to point out that the flow law (19) and the yield condition (20) are not necessarily related by the rule of plastic potential. Relation (20) indicates the type and the minimal number of the basic invariants involved in the most general form of the yield condition. The essential variables to be observed in experiments are thus specified.

3.2 Proposed criterion

The general invariant form (20) of the yield criterion has to be specified for each class of orthotropic materials through further hypotheses. For prestrained rolled sheet-metal, we have proposed [8] a new generalization of the Von Mises condition in developing a homogeneous form of order two in stress and employing the simplest combinations of the basic invariants involved in (20):

$$A_1 \mathrm{tr}^2\underset{\sim}{T} + A_2 \mathrm{tr}\underset{\sim}{T}^2 + (A_3 \mathrm{tr}\underset{\sim}{M}\underset{\sim}{T} + A_4 \mathrm{tr}\underset{\sim}{T})\mathrm{tr}\underset{\sim}{M}\underset{\sim}{T} + (A_5 \mathrm{tr}\underset{\sim}{T} + A_6 \mathrm{tr}\underset{\sim}{M}\underset{\sim}{T})\mathrm{tr}\underset{\sim}{T}\underset{\sim}{P} = 2A_7 k^2 \quad (21)$$

where the A_i are polynomials of order two with respect to the invariants $\mathrm{tr}\underset{\sim}{P}$, $\mathrm{tr}^{1/2}\underset{\sim}{P}^2$ and $\mathrm{tr}\underset{\sim}{M}\underset{\sim}{P}$. In criterion (21), the terms $A_1 \mathrm{tr}^2\underset{\sim}{T}$ and $A_2 \mathrm{tr}\underset{\sim}{T}^2$ account for the isotropic part of the initial criterion and the isotropic hardening induced by $\underset{\sim}{P}$, the terms $A_3 \mathrm{tr}^2\underset{\sim}{M}\underset{\sim}{T}$ and $A_4 \mathrm{tr}\underset{\sim}{T}\mathrm{tr}\underset{\sim}{M}\underset{\sim}{T}$ account for the orthotropic part of the initial criterion and the evolution of its degree with $\underset{\sim}{P}$, the term $A_5 \mathrm{tr}\underset{\sim}{T}\mathrm{tr}\underset{\sim}{T}\underset{\sim}{P}$ describes the new anisotropy induced by $\underset{\sim}{P}$ and the term $A_6 \mathrm{tr}\underset{\sim}{M}\underset{\sim}{T}\mathrm{tr}\underset{\sim}{T}\underset{\sim}{P}$ stands for the coupling between the initial anisotropy and the induced anisotropy.

For the non-prestrained material ($\underset{\sim}{P} = \underset{\sim}{0}$), criterion (21) reduces to the yield condition for orthotropic materials submitted to plane stress [4]. Thus, the proposed criterion constitutes a double generalization of the isotropic Von Mises criterion : generalization for initially orthotropic materials ; generalization for the description of deformation induced anisotropy.

The proposed criterion (21) describes both isotropic and anisotropic hardening induced by $\underset{\sim}{P}$. In order to give a simple example of anisotropic hardening, we consider an initially isotropic material submitted to a prestraining of principal directions $\underset{\sim}{v}_1$, $\underset{\sim}{v}_2$, $\underset{\sim}{v}_3$, ($\Psi = 0$, cf. Figure 1). Before prestraining, the material obeys the isotropic Von Mises condition. After prestraining, the criterion is orthotropic and is represented in the principal stress plane by a family of ellipses parametrized by the orientation angle θ (cf. Fig.2). For a prestraining $\underset{\sim}{P}$ such that $\varepsilon_1 = -\varepsilon_2 = 0.1$, the obtained family of ellipses is represented in Figure 3.

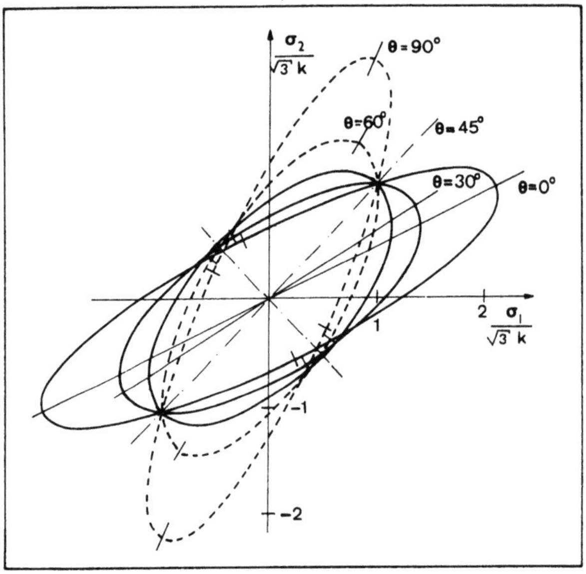

Fig. 3. - Anisotropic hardening of an initially isotropic metal

4. EXPERIMENTAL BEHAVIOR OF ROLLED SHEET-STEEL

4.1 Experimental procedure

A number of tests were performed on large sheets of mild steel obtained by cold rolling with three different reductions in thickness. For each rolling reduction, the large sheets were submitted to four different uniaxial prestrainings in the rolling direction, namely the zero prestrain, a prestrain the closest possible to the limiting curve of appearance of localized deformation and two intermediate prestrains (Fig.4).

For each rolling reduction and each prestrain, small specimens were cut out from the large sheets in seven different orientations θ with respect to the principal axis of prestrain: $\theta = 0°, 15°, 30°, 45°, 60°, 75°$ and $90°$. The small specimens obtained in this way were then tested under simple tensile stress. The size and the shape of the specimens are presented in Fig.5.

A preliminary experimental study was developed in order to improve the interpretation of the off-axis tests. For such tests, the principal directions of the mechanical sollicitation do not coincide with the material symmetry axes of the medium. In the classical testing procedure, with rigidly clamped heads, non homogeneous states of strain and stress develop during the axial straining ([14], [15]) and the specimens tend to assume a S-shape (Fig. 6a). This effect is characteristic for anisotropic solids (see also Chapter 5). In order to avoid such effects, the ideal procedure should assure a uniform stress distribution on the ends of the specimens and non-constrained deformation (Fig. 6b). Such an ideal procedure cannot be achieved practically. In order to approach the ideal case, we propose a new testing device consisting of hinged fixtures with knife-edges permetting the ends of the specimens to rotate. The optimate position for the knife-edges is at the ends of the specimens, wich results in a quasi-homogeneous state of strain in the working length (Fig.7).

At different stages of the axial straining, the developed displacement fields in the specimens were measured by the method of stereo-photogrammetry with deformation parallaxes [11]. Examples of measurements ([15], [16]) are given in Fig. 8, which shows the axial and transverse isodisplacement lines of specimens oriented at $\theta = 60°$ to the direction of prestraining and subjected to tensile axial deformation up to 1 %. Fig.8a shows the S-shaped deformation obtained with the classical testing device. Fig. 8b shows that the new testing device proposed allows a homogeneous state of strain in nearly all of the working length to be obtained.

4.2 Experimental results and comparison with theoretical predictions

In order to apply the proposed criterion to the performed experimen-

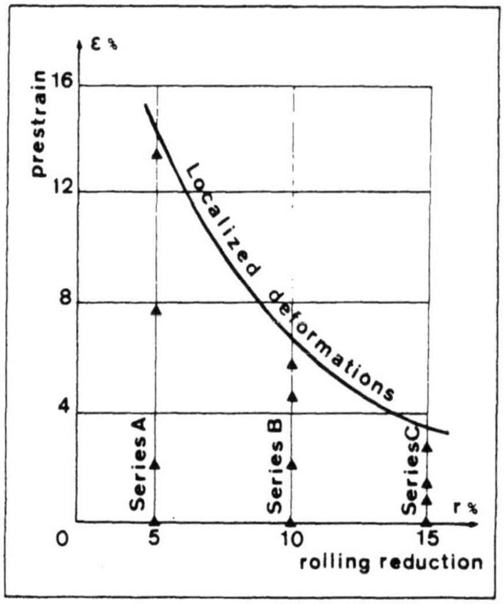

Fig. 4 - Applied irreversible prestrains

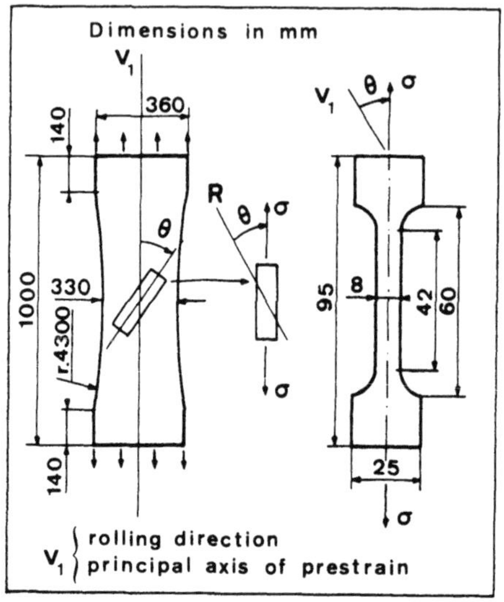

Fig. 5 - Shape and size of specimens

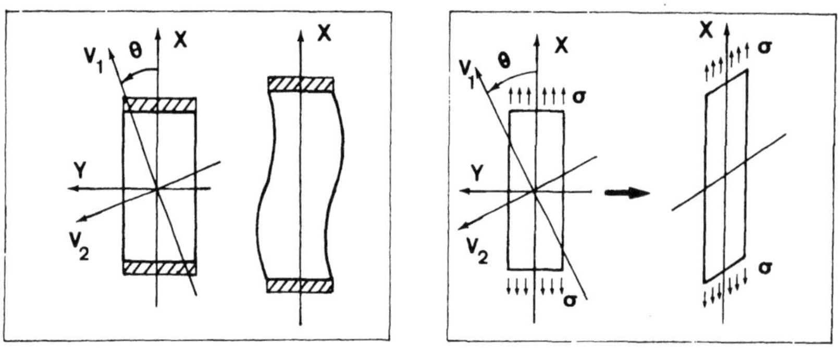

Fig. 6 - a) Classical testing device b) Ideal testing device

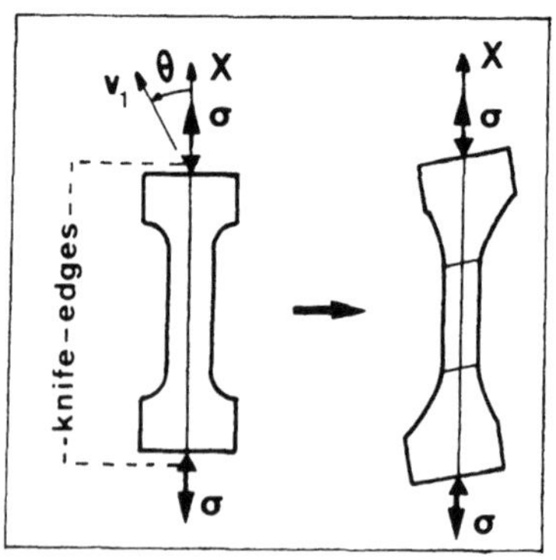

Fig. 7 - New testing device for "off-axis" anisotropic specimens

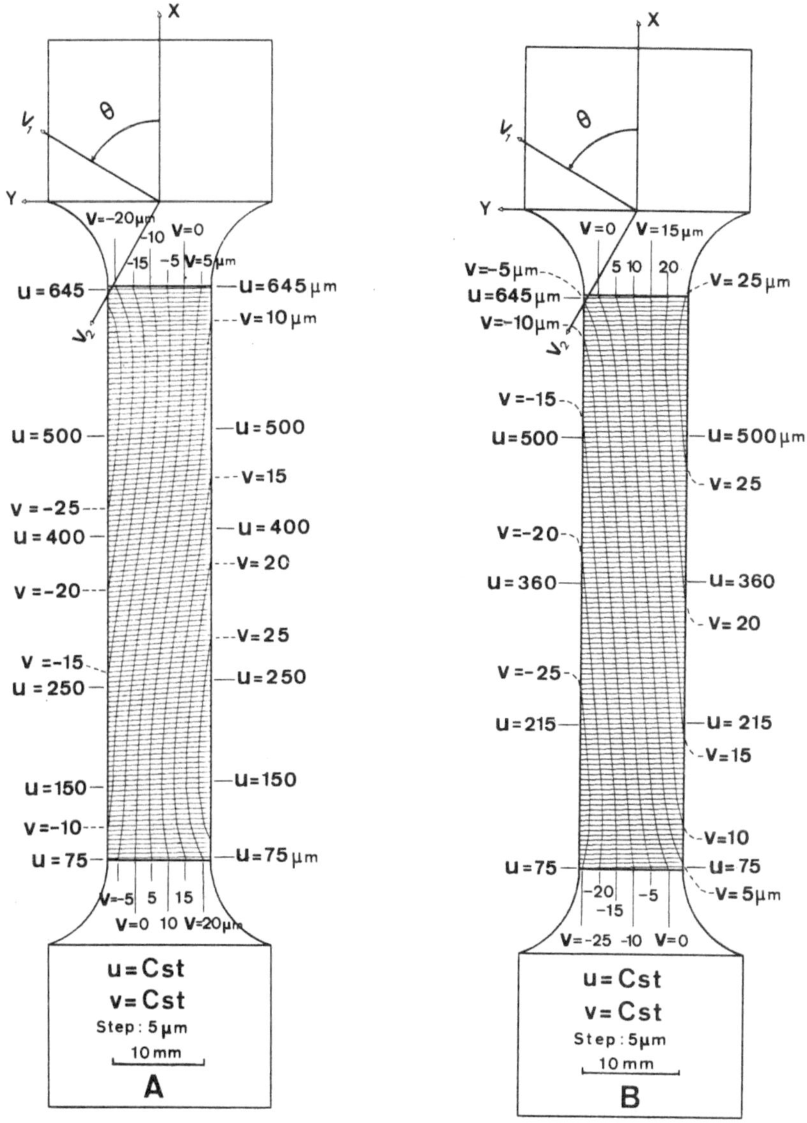

Fig. 8 - Anisotropic steel subjected to "off-axis" simple tension ($\mathbf{v}_1$ = rolling direction ; θ = 60°) : axial and transverse isodisplacement lines.

A : classical testing device ; B : proposed new testing device tal programm, we first analyze the flow law for the non-prestrained material ($\underset{\sim}{P} = \underset{\sim}{0}$). In the case of the associated flow law (normality rule), criterion (21) results in the following expression:

$$\frac{\underset{\sim}{D}}{\mathrm{tr}^{1/2}\underset{\sim}{D}^2} = (-\frac{2}{3}\,\mathrm{tr}\underset{\sim}{T} + d_o\mathrm{tr}\underset{\sim}{M}\underset{\sim}{T})\underset{\sim}{I} + (d_o\mathrm{tr}\underset{\sim}{T} + c_o\mathrm{tr}\underset{\sim}{M}\underset{\sim}{T})\underset{\sim}{M} + 2\underset{\sim}{T} \tag{22}$$

where c_o and d_o are constants. In our tests, the irreversible prestrainings were applied in the $\underset{\sim}{v}_1$ direction. The stress and rate of deformation tensors are thus expressed in the ($\underset{\sim}{v}_1$, $\underset{\sim}{v}_2$) plane by:

$$\underset{\sim}{T} \Rightarrow \begin{vmatrix} \sigma & 0 \\ 0 & 0 \end{vmatrix} \quad ; \quad \underset{\sim}{D} \Rightarrow \begin{vmatrix} \dot{\varepsilon}_1 & 0 \\ 0 & \dot{\varepsilon}_2 \end{vmatrix} \tag{23}$$

Introducing the expression (23) in the flow law (22), one obtains:

$$\dot{\varepsilon}_2 = \lambda\dot{\varepsilon}_1 \qquad \text{with} \qquad \lambda = \frac{3d_o - 2}{2(3d_o + 3c_o + 2)} \tag{24}$$

Relation (24) is only valid for incipient flow. The evolution of this relation during the prestraining can be obtained from the flow law associated with criterion (21) through the normality rule. However, the prestrains applied in our tests are small ; thus, in a first approximation, we can admit that relation (24) is valid during the entire prestraining process. In the case of an incompressible material, the principal values of the prestraining tensor $\underset{\sim}{P}$ are then given by:

$$\varepsilon_1 = \varepsilon \quad ; \quad \varepsilon_2 = \lambda\varepsilon \quad ; \quad \varepsilon_3 = -(1 + \lambda)\varepsilon \tag{25}$$

Making use of the expressions (25), the application of the proposed criterion (21) to oriented simple tensile tests results in:

$$R^2_\theta(\varepsilon) = \frac{2\,k^2\,(1 + H\varepsilon + L\varepsilon^2)}{(\frac{2}{3} + A_1\varepsilon + A_2\varepsilon^2) + (B_o + B_1\varepsilon + B_2\varepsilon^2)\cos^2\theta + (C_o + C_1\varepsilon + C_2\varepsilon^2)\cos^4\theta} \tag{26}$$

where R_θ is the simple tensile strength in the direction θ and the coeffificients H, L, A_i, B_i, C_i are constants related to isotropic and anisotropic hardening.

The eleven constants appearing in (26) were calculated from our experimental data for the three rolling conditions. The numerical values are given in [12]. The obtained results are presented in Fig.9, where the limit stresses in tensile axial straining are given with respect to the orientation angle θ, for each rolling reduction r. The points indicate the tests' data ; the curves parametrized by the prestrain ε represent the theoretical predictions obtained from the proposed criterion.

Considering that the further the yield curve moves away from a horizontal straigth line (corresponding to the isotropic case), the greater the increase in the degree of anisotropy, we observe two principal phenomena: for a given rolling reduction, the initial anisotropy of the sheet-steel increases with the irreversible prestrain; the initial anisotropy, as well as the anisotropic hardening during irreversible prestrains, are markedly affected by the rolling reduction.

For each rolling reduction, we have four different prestrains and seven different orientations, thus 28 experimental data. Eleven data are needed to calculate the constants of criterion (26); thus, seventeen data test the reliability of the proposed theory for each rolling reduction. In spite of the complexity of the observed phenomena, the agreement between theoretical predictions and experimental data is very good.

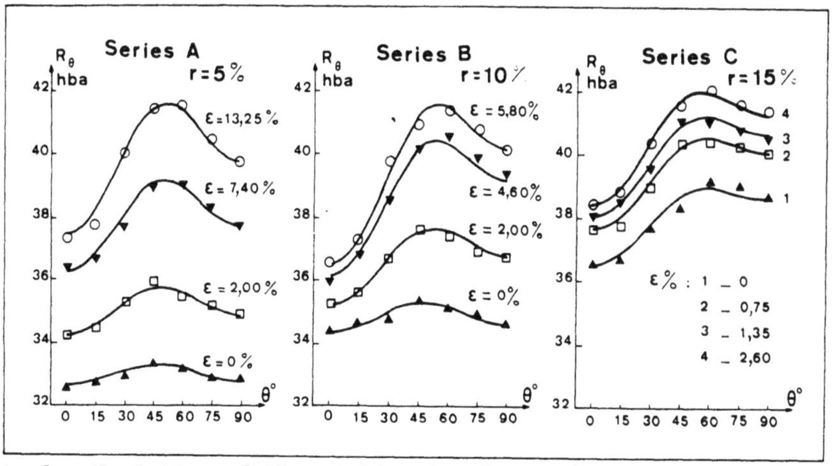

Fig.9 - Evolution of the yield criterion with respect to rolling reduction, irreversible prestrain and orientation of the specimens (after [7]).

REFERENCES

1. PARNIERE, P., Relations entre Textures et Comportement Mécanique Anisotrope des Métaux, in: *Mechanical Behavior of Anisotropic Solids*, Proc. Euromech Colloquium - Colloque International 295 du CNRS, J.P. BOEHLER ed., Editions du CNRS (Paris) and Martinus Nijhoff Publishers (The Hague), 1982 : 303-332.

2. ADKINS, J.E., Symmetry Relations for Orthotropic and Transversely Isotropic Materials, *Arch. Rat. Mech. An.*, 4 (1960) : 192-213.

3. SPENCER, A.J.M., Theory of Invariants, in: *Continuum Physics*, C. ERINGEN ed., Academic Press, New-York, (1971) : 239-353.

4. BOEHLER, J.P., Lois de Comportement Anisotrope des Milieux Continus, *Journal de Mécanique*, 17,2 (1978) : 183-190.

5. RACLIN, J., Remarques sur les Conditions de Plasticité des Milieux Ecrouissables, *Troisième Congrès Français de Mécanique*, Grenoble, (1977).

6. MURAKAMI, S. and SAWCZUK, A., On Description of Rate-Independent Behaviour for Prestrained Solids, *Archives of Mechanics*, 32 (1979) : 251-264.

7. BOEHLER, J.P. and RACLIN, J., Anisotropic Hardening of Prestrained Sheet-Steel, in: *Current Advances in Mechanical Design and Production*, Proc. Second Cairo University MDP Conference, Dpt. of MDP, Faculty of Engineering, Cairo University, (1982) : 483-492.

8. BOEHLER, J.P. and RACLIN, J., A Unified Theoretical and Experimental Study of Anisotropic Hardening, *6th Int. Conference on SMIRT*, North-Holland, Amsterdam, (1981), Invited Lecture L 3/2.

9. BOEHLER, J.P. and RACLIN, J., Ecrouissage Anisotrope des Matériaux Orthotropes Prédéformés, *Journal de Mécanique Théorique et Appliquée*, Numéro Spécial (1982) : 23-44.

10. BOEHLER, J.P. and SAWCZUK, A., On Yielding of Oriented Solids, *Acta Mechanica*, 27 (1977) : 185-204.

11. BENEFICE, P. and BOEHLER, J.P., The Applications of Stereophotogrammetry to the Analysis of Displacement Fields in Solid Mechanics, *Proc. IUTAM Symposium Optical Methods in Mechanics of Solids*, A. LAGARDE ed., Sijthoff and Noordhoff, Alphen aan den Rijn, (1981) : 657-668.

12. BOEHLER, J.P., RACLIN, J. and BENEFICE, P., Etude des Champs Cinématiques Non Homogènes, de l'Ecrouissage et de la Fissuration des Milieux Anisotropes, *Rapport Contrat ATP 3669 CNRS*, Grenoble (1981).

13. RACLIN, J., Contributions Théoriques et Expérimentales à l'Etude de la Plasticité, de l'Ecrouissage et de la Rupture des Solides Anisotropes, *Thèse de Doctorat es-Sciences*, Grenoble, 1984.

14. BOEHLER, J.P., Anisotropie et Comportement Rhéologique des Matériaux, Conférence Générale, *19ème Colloque du Groupe Français de Rhéologie*, Novembre 1984 (in print).

15. BOEHLER, J.P., EL AOUFI, L. and RACLIN, J., On Experimental Testing Methods for Anisotropic Materials, *Res Mechanica* (in print).

16. EL AOUFI, L., Contributions à l'Etude Expérimentale et Numérique du Comportement Mécanique des Matériaux Anisotropes, *Thèse de Doctorat de Spécialité*, Grenoble, 1984.

Chapter 8

ISOTROPIC POLYNOMIAL INVARIANTS AND TENSOR FUNCTIONS

A.J.M. Spencer
The University of Nottingham, England

1. INTRODUCTION. NOTATION AND DEFINITIONS

Some of the circumstances in which invariance problems arise in continuum mechanics are described elsewhere in this book. In Chapters 8 and 9 we consider the purely algebraic problem of determining systems of polynomial invariants and tensor polynomial functions for a given set of vectors and tensors, for some of the transformation groups which are of importance in continuum mechanics.

For the most part we consider vectors and tensors in three-dimensional Euclidian space; occasionally it will be useful to refer to results in two-dimensional space. Unless otherwise stated, 'tensor' means a second-order tensor. Rectangular cartesian coordinates are used throughout. Results may, if desired, be expressed in terms of curvilinear coordinate systems by standard techniques of tensor analysis. Index notation and summation convention are employed; lower case Latin indices take the values 1,2,3 unless otherwise stated.

General vectors are denoted by $\underset{\sim}{u}$, $\underset{\sim}{v}$, $\underset{\sim}{w}$ or $\underset{\sim}{u}_R$ and their components in a fixed rectangular cartesian coordinate system $Ox_1x_2x_3$ by $u_i, v_i, w_i, \ldots$ or $u_i^{(R)}$. The symbols $\underset{\sim}{v}_1$, $\underset{\sim}{v}_2$, $\underset{\sim}{v}_3$ will be reserved for special vectors which represent directions associated with material anisotropy. For some purposes it is convenient to associate with a vector $\underset{\sim}{u}$ (say) an anti-symmetric tensor $\underset{\sim}{U}$, with components U_{ij}, where

$$U_{ij} = e_{ijk}u_k \, , \tag{1}$$

and e_{ijk} are components of the third-order alternating tensor. Then

$$u_i = \tfrac{1}{2}e_{ijk}U_{jk}. \tag{2}$$

Components in the $Ox_1x_2x_3$ coordinate system of general (not necessarily symmetric or anti-symmetric) second-order tensors $\underset{\sim}{P}$, $\underset{\sim}{Q}$, ..., or $\underset{\sim}{P}_R$, are denoted by P_{ij}, Q_{ij}, ..., or $P_{ij}^{(R)}$; of symmetric second-order tensors $\underset{\sim}{A}$, $\underset{\sim}{B}$, $\underset{\sim}{C}$, ..., or $\underset{\sim}{A}_R$, by A_{ij}, B_{ij}, C_{ij}, ..., or $A_{ij}^{(R)}$; of anti-symmetric tensors $\underset{\sim}{X}$, $\underset{\sim}{Y}$, $\underset{\sim}{Z}$, or $\underset{\sim}{X}_R$, by X_{ij}, Y_{ij}, Z_{ij}, or $X_{ij}^{(R)}$. With these second-order tensors we associate 3×3 matrices, also denoted by $\underset{\sim}{P}$, $\underset{\sim}{Q}$, $\underset{\sim}{P}_R$; $\underset{\sim}{A}$, $\underset{\sim}{B}$, $\underset{\sim}{C}$, $\underset{\sim}{A}_R$; $\underset{\sim}{X}$, $\underset{\sim}{Y}$, $\underset{\sim}{Z}$, $\underset{\sim}{X}_R$, where, for example, $\underset{\sim}{P} = (P_{ij})$. The use of the same notation to denote tensors and the associated matrices of tensor components should not generally cause confusion. When the meaning is not clear from the context we shall state explicitly whether matrix or tensor notation is being employed.

We shall use both tensor component notation and direct tensor notation as convenient, with a bias towards direct notation. Thus, for example, $\underset{\sim}{u}.\underset{\sim}{v}$ denotes the scalar product u_iv_i; $\underset{\sim}{u}\underset{\sim}{P}$ and $\underset{\sim}{P}\underset{\sim}{u}$ are vectors with components u_iP_{ij} and $P_{ij}u_j$ respectively; $\underset{\sim}{P}\underset{\sim}{Q}$ is the second-order tensor with components $P_{ij}Q_{jk}$ and $\underset{\sim}{P}\underset{\sim}{Q}^T$ is the second-order tensor with components $P_{ij}Q_{kj}$; and the dyadic product $\underset{\sim}{u}\otimes\underset{\sim}{v}$ is the second-order tensor with components u_iv_j. The calculation of the components of such products is most easily performed by matrix multiplications. Thus if $\underset{\sim}{u}$, $\underset{\sim}{v}$ are interpreted as 3×1 column matrices and $\underset{\sim}{P}$, $\underset{\sim}{Q}$ as 3×3 matrices of tensor components, then the scalar $\underset{\sim}{u}.\underset{\sim}{v}$ is the single element of the matrix product $\underset{\sim}{u}^T\underset{\sim}{v}$; the components of the vectors $\underset{\sim}{u}\underset{\sim}{P}$ and $\underset{\sim}{P}\underset{\sim}{u}$ are the elements of the matrix products $\underset{\sim}{u}^T\underset{\sim}{P}$ and $\underset{\sim}{P}\underset{\sim}{u}$ respectively; the components of the tensor $\underset{\sim}{P}\underset{\sim}{Q}$ are the elements of the matrix product $\underset{\sim}{P}\underset{\sim}{Q}$; and the components of $\underset{\sim}{u} \otimes \underset{\sim}{v}$ are elements of the matrix product $\underset{\sim}{u}\underset{\sim}{v}^T$.

The transformation

$$\bar{x}_i = M_{ij} x_j \qquad \text{or (in matrix notation)} \qquad \bar{\underset{\sim}{x}} = \underset{\sim}{M}\underset{\sim}{x} \tag{3}$$

determines a new set of rectangular cartesian coordinates $O\bar{x}_1\bar{x}_2\bar{x}_3$ if

$$M_{ik} M_{jk} = \delta_{ij}, \qquad \text{or} \quad \underset{\sim}{M}\underset{\sim}{M}^T = \underset{\sim}{I}. \tag{4}$$

Here δ_{ij} is the Kronecker delta symbol, $\underset{\sim}{I}$ is the unit tensor or matrix (so that $(\delta_{ij}) = \underset{\sim}{I}$) and $\underset{\sim}{M}^T$ is the transpose of $\underset{\sim}{M}$. Then also

$$M_{ki} M_{kj} = \delta_{ij}\,, \qquad \underset{\sim}{M}^T\underset{\sim}{M} = \underset{\sim}{I} \qquad \text{and} \qquad \det \underset{\sim}{M} = \pm 1\,. \tag{5}$$

The orthogonal matrix $\underset{\sim}{M}$ defines an orthogonal transformation of coordinates. The set of orthogonal transformations form a group (the full orthogonal group in three dimensions). The set of matrices $\underset{\sim}{M}$ is a matrix representation of this group.

Under an orthogonal transformation, an *absolute vector* $\underset{\sim}{u}$ with components u_i in the $Ox_1x_2x_3$ system has components $\bar{u}_i$ in the $O\bar{x}_1\bar{x}_2\bar{x}_3$ system where

$$\bar{u}_i = M_{ij} u_j\,, \qquad \text{or} \qquad \bar{\underset{\sim}{u}} = \underset{\sim}{M}\underset{\sim}{u}\,. \tag{6}$$

The components of an *axial vector* $\underset{\sim}{u}$ transform according to the rule

$$\bar{u}_i = (\det \underset{\sim}{M}) M_{ij} u_j\,, \qquad \text{or (matrix notation)} \quad \bar{\underset{\sim}{u}} = (\det \underset{\sim}{M})\underset{\sim}{M}\underset{\sim}{u}. \tag{7}$$

The components of an (absolute) second-order tensor $\underset{\sim}{P}$ transform according to the rule

$$\bar{P}_{ij} = M_{ir} M_{js} P_{rs}\,, \qquad \text{or (matrix notation)} \quad \bar{\underset{\sim}{P}} = \underset{\sim}{M}\underset{\sim}{P}\underset{\sim}{M}^T\,. \tag{8}$$

If

$$U_{ij} = e_{ijk} u_k\,, \qquad \bar{U}_{ij} = e_{ijk} \bar{u}_k \tag{9}$$

and U_{ij} transform as a second-order tensor, so that

$$\bar{U}_{ij} = M_{ir}M_{js}U_{rs}$$

then we find that

$$\bar{u}_i = (\det \underset{\sim}{M})M_{ik}u_k \tag{10}$$

so that u_i are components of an axial vector. It is therefore possible, and often convenient, to identify an axial vector with an anti-symmetric tensor, by (9), and it is not then necessary to consider axial vectors as separate entities. Henceforth we shall do this and, in the case of the full orthogonal group, deal only with absolute vectors and second-order tensors.

If attention is confined to *proper orthogonal transformations* (i.e. rotations of the coordinate system) for which $\det \underset{\sim}{M} = +1$, then it is not necessary to distinguish between absolute and axial vectors.

A function $f(u_i, v_i, \ldots, P_{ij}, Q_{ij}, \ldots)$ of the components of a set of vectors and tensors in an *invariant* under a given group of transformations if

$$f(u_i, v_i, \ldots, P_{ij}, Q_{ij}, \ldots) = (\det \underset{\sim}{M})^\rho f(\bar{u}_i, \bar{v}_i, \ldots, \bar{P}_{ij}, \bar{Q}_{ij}, \ldots)$$

for every transformation M_{ij} of the group. Since, in our case, $\det \underset{\sim}{M} = \pm 1$, we may take $\rho = 0$ or $\rho = 1$. If $\rho = 0$, then f is an *absolute* or *even* invariant; if $\rho = 1$, then f is a *relative* or *odd* invariant. We suppose invariants to be absolute unless otherwise stated.

The following are three simple and well-known examples of invariants under the full orthogonal group.

(a) The scalar product $\underset{\sim}{u}.\underset{\sim}{v}$ of two absolute (or two axial) vectors:

$$\bar{u}_i\bar{v}_i = M_{ip}u_p M_{iq}v_q = \delta_{pq}u_p v_q = u_p v_p = u_i v_i \ .$$

(b) The trace $\operatorname{tr}\underset{\sim}{P}$ of the matrix $\underset{\sim}{P}$:

$$\bar{P}_{ii} = M_{ip}M_{iq}P_{pq} = \delta_{pq}P_{pq} = P_{pp} = P_{ii} \ .$$

(c) The determinant of $\underset{\sim}{P}$:

$$\det \underset{\sim}{\bar{P}} = \det(\underset{\sim}{M}\underset{\sim}{P}\underset{\sim}{M}^T) = (\det \underset{\sim}{M})^2 \det \underset{\sim}{P} = \det \underset{\sim}{P} .$$

The central problem in invariant theory is as follows. For a given set of vectors and tensors and a given transformation group, determine a set of invariants (a *basis*) from which all other invariants can be generated. For polynomial invariants the basis is called an *integrity basis*. A polynomial invariant is *reducible* if it can be expressed as a polynomial in other polynomial invariants, otherwise it is *irreducible*. A basis which contains only irreducible invariants is *minimal*. Polynomial relations between invariants which do not permit any one invariant to be expressed as a polynomial in the others are *syzygies*.

2. RESULTS FROM CLASSICAL THEORY

Classical invariant theory, as described in, for example, Grace and Young [1], Elliott [2], Turnbull [3], Weyl [4] and Gurevich [5], was mainly motivated by projective geometry and is usually concerned with invariance under the full linear group of transformations. Consequently its numerous results are often not directly applicable to invariance under orthogonal transformations. However, there are important general theorems.

(a) *Hilbert's Theorem.* For any finite set or vectors and tensors of any order, there exists a finite integrity basis. (Proof in Gurevich [5]).

(b) *Peano's Theorem.* Consider a set of R tensors $\underset{\sim}{A}_p$ (of any order) all of which are similar in the sense that they have the same order and have the same index symmetries. Suppose each has ν distinct components which, arranged in some definite order, are denoted $A_i^{(P)}$ $(i = 1,2,\dots,\nu;\ P = 1,2,\dots,R)$. The *polarization operator* D_{PQ} is defined by

$$D_{PQ} = A_i^{(P)} \partial/\partial A_i^{(Q)} .$$

Then Peano's theorem states that, with the possible exception of the νth order determinants in which a typical column is comprised of the elements $A_i^{(P)}$, every polynomial invariant of the tensors can be expressed as a polynomial in invariants of $\nu-1$ of the tensors and invariants formed from them by polarization. It is therefore sufficient to consider $\nu-1$ of the R tensors. A proof of the theorem is given (under the name Pascal's Theorem) in Weyl [4].

(c) Every *algebraic invariant* is a solution of an algebraic equation whose coefficients are *rational invariants.*

(d) Every *rational invariant* is a quotient of two *polynomial invariants* (which may be relative invariants).

(e) Every *polynomial invariant* is a sum of *homogeneous polynomial invariants.*

The results *(c)*, *(d)* and *(e)* (proofs in Gurevich [5] and Spencer [6]) show that the study of algebraic invariants reduces to that of homogeneous polynomial invariants.

3. ORTHOGONAL TRANSFORMATION GROUPS

Orthogonal transformations in three dimensions comprise rotations and reflections. A rotation θ about an axis defined by a unit vector $\underset{\sim}{v}_1$ with components $v_i^{(1)}$ is characterized by the matrix $\underset{\sim}{M}_\theta = (M_{ij}^{(\theta)})$, where

$$M_{ij}^{(\theta)} = \delta_{ij} \cos\theta + e_{ijk} v_k^{(1)} \sin\theta + (1-\cos\theta) v_i^{(1)} v_j^{(1)} .$$

In particular, if $\underset{\sim}{v}_1$ coincides with the base vector $\underset{\sim}{e}_1$ of a system of rectangular cartesian coordinates $Ox_1x_2x_3$, then

$$\underset{\sim}{M}_\theta = \begin{pmatrix} 1 & 0 & 0 \\ 0 & \cos\theta & \sin\theta \\ 0 & -\sin\theta & \cos\theta \end{pmatrix} .$$

A reflection in the planes normal to $\underset{\sim}{v}_1$ is characterized by the matrix $\underset{\sim}{R}_1 = (R_{ij}^{(1)})$, where

$$R_{ij}^{(1)} = \delta_{ij} - 2v_i^{(1)}v_j^{(1)} .$$

If $\underset{\sim}{v}_1$ coincides with $\underset{\sim}{e}_1$, then

$$\underset{\sim}{R}_1 = \begin{pmatrix} -1 & 0 & 0 \\ 0 & 1 & 0 \\ 0 & 0 & 1 \end{pmatrix}.$$

For the remainder of this section we choose $\underset{\sim}{v}_1$, $\underset{\sim}{v}_2$, $\underset{\sim}{v}_3$ to be a right-handed set of mutually orthogonal unit vectors, and define $\underset{\sim}{R}_2$ and $\underset{\sim}{R}_3$ in the analogous manner to $\underset{\sim}{R}_1$.

Particular orthogonal transformations of special interest are those represented by the following matrices:

$\underset{\sim}{I}$, (the identity transformation);

$\underset{\sim}{C} = -\underset{\sim}{I} = \underset{\sim}{R}_1\underset{\sim}{R}_2\underset{\sim}{R}_3$, which represents a central inversion;

$\underset{\sim}{D}_1 = \underset{\sim}{R}_2\underset{\sim}{R}_3 = \underset{\sim}{M}_\pi$, which represents a rotation π about the $\underset{\sim}{v}_1$ axis. If $\underset{\sim}{v}_1$, $\underset{\sim}{v}_2$, $\underset{\sim}{v}_3$ coincide with $\underset{\sim}{e}_1$, $\underset{\sim}{e}_2$, $\underset{\sim}{e}_3$, then

$$\underset{\sim}{D}_1 = \begin{pmatrix} 1 & 0 & 0 \\ 0 & -1 & 0 \\ 0 & 0 & -1 \end{pmatrix}.$$

The matrices $\underset{\sim}{D}_2$ and $\underset{\sim}{D}_3$ are defined similarly.

Material symmetries require invariance under certain groups of orthogonal transformations. The material symmetries of concern to us are the following.

(a) Isotropy with a centre of symmetry. In this case the isotropy group is the full orthogonal group, which consists of all orthogonal transformations, represented by the orthogonal matrices $\underset{\sim}{M}$ with $\det \underset{\sim}{M} = \pm 1$.

(b) Isotropy without a centre of symmetry. In this case the isotropy group is the proper orthogonal group, or rotation group, which group

is represented by the proper orthogonal matrices $\underset{\sim}{M}$ with $\det \underset{\sim}{M} = +1$. Material isotropy with no centre of symmetry requires invariance under this group.

(c) Transverse isotropy. Material with a single preferred direction is transversely isotropic. The constitutive equations are invariant under rotations about the preferred direction. If the $\underset{\sim}{v}_1$ direction is chosen as the preferred direction, we have invariance under the rotations $\underset{\sim}{M}_\theta$. There are five cases, whose transformation groups consist of the following matrices:

(i) $\underset{\sim}{M}_\theta$; *(ii)* $\underset{\sim}{M}_\theta, \underset{\sim}{R}_2, \underset{\sim}{R}_3$; *(iii)* $\pm\underset{\sim}{M}_\theta, \underset{\sim}{R}_1, \underset{\sim}{C}$;

(iv) $\underset{\sim}{M}_\theta, \underset{\sim}{D}_2, \underset{\sim}{D}_3$; *(v)* $\pm\underset{\sim}{M}_\theta, \underset{\sim}{R}_1, \underset{\sim}{R}_2, \underset{\sim}{R}_3, \underset{\sim}{D}_2, \underset{\sim}{D}_3, \underset{\sim}{C}$.

Clearly, all the groups include $\underset{\sim}{M}_0 = \underset{\sim}{I}$ and $\underset{\sim}{M}_\pi = \underset{\sim}{D}_1$.

The groups *(a)*, *(b)*, *(c)* are continuous groups, with an infinite number of elements. Finite sub-groups of the full orthogonal group are also of interest. In particular crystal symmetries are represented by thirty-two groups arranged in seven systems. These groups are known as the crystal classes. The transformations which form these groups are not listed here. They can be found in Green and Adkins [7], Spencer [6] or in texts on crystal physics. We consider only two cases, which may also be regarded as symmetries appropriate to fibre-reinforced materials reinforced by two families of continuously distributed fibres.

(d) Prismatic symmetry. The transformation group consists of the matrices

$$\underset{\sim}{I}, \underset{\sim}{C}, \underset{\sim}{R}_1, \underset{\sim}{D}_1.$$

This is the appropriate symmetry group for a material reinforced by two families of non-orthogonal fibres lying in planes x_1 = constant, the two families having distinct mechanical properties (e.g. an unbalanced angle-ply laminate).

(e) Orthogonal (rhombic) symmetry. The transformation group consists of

$$\underset{\sim}{I}, \underset{\sim}{R}_1, \underset{\sim}{R}_2, \underset{\sim}{R}_3, \underset{\sim}{C}, \underset{\sim}{D}_1, \underset{\sim}{D}_2, \underset{\sim}{D}_3.$$

This is the symmetry group for an orthotropic material, which has symmetry under reflections in each of the coordinate planes. It is appropriate for a material reinforced by two families of orthogonal fibres, with distinct mechanical properties, directed parallel to the x_2 and x_3 axes (e.g. a cross-ply laminate), or for a material reinforced by two mechanically equivalent families of fibres whose bisectors are parallel to the x_2 and x_3 axes (e.g. a balanced angle-ply laminate).

4. INTEGRITY BASES FOR VECTORS

Consider a set of M absolute vectors $\underset{\sim}{u}^{(P)}$ (P = 1,2,...,M) in three dimensions. Then it is easily verified that the following scalar products are invariant under the full orthogonal group

$$u_i^{(P)} u_i^{(Q)} = \underset{\sim}{u}_P \cdot \underset{\sim}{u}_Q \qquad (P,Q = 1,2,\dots,M). \tag{11}$$

It can be shown that this set is complete, and that (11) form an integrity basis under the full orthogonal group for the set of vectors. The result is due to Cauchy; a proof based on Peano's theorem is given in Weyl [4] and Spencer [6].

For the proper orthogonal group, there are additional invariants of the form

$$I_{PQR} = e_{ijk} u_i^{(P)} u_j^{(Q)} u_k^{(R)} \qquad (P \neq Q \neq R \neq P) \tag{12}$$

and the integrity basis comprises the invariants (11) and (12). The expressions (12) are relative invariants for the full orthogonal group. We note the syzygies

$$I_{PQR} I_{STU} = \begin{vmatrix} \underset{\sim}{u}_P \cdot \underset{\sim}{u}_S & \underset{\sim}{u}_P \cdot \underset{\sim}{u}_T & \underset{\sim}{u}_P \cdot \underset{\sim}{u}_U \\ \underset{\sim}{u}_Q \cdot \underset{\sim}{u}_S & \underset{\sim}{u}_Q \cdot \underset{\sim}{u}_T & \underset{\sim}{u}_Q \cdot \underset{\sim}{u}_U \\ \underset{\sim}{u}_R \cdot \underset{\sim}{u}_S & \underset{\sim}{u}_R \cdot \underset{\sim}{u}_T & \underset{\sim}{u}_R \cdot \underset{\sim}{u}_U \end{vmatrix}. \tag{13}$$

If the set of vectors contains axial vectors then some modification is required. However, as remarked in Section 1, it is usually preferable to identify an axial vector with an anti-symmetric tensor.

5. ISOTROPIC TENSORS

An isotropic tensor is defined to be a tensor whose components are unchanged by any proper orthogonal transformation of coordinates. Isotropic tensors which are unchanged under both proper and improper orthogonal transformations can be obtained by deleting from the list of isotropic tensors those which are not invariant under the central inversion $\underset{\sim}{C}$. The following method of deriving isotropic tensors is due to Smith and Rivlin [8]. We consider isotropic tensors in three dimensions.

Let $\alpha_{i_1 i_2 \ldots i_\mu}$ be an isotropic tensor of order μ. Then for any orthogonal matrix $\underset{\sim}{M}$

$$\bar{\alpha}_{i_1 i_2 \ldots i_\mu} = M_{i_1 j_1} M_{i_2 j_2} \ldots M_{i_\mu j_\mu} \alpha_{j_1 j_2 \ldots j_\mu} = \alpha_{i_1 i_2 \ldots i_\mu} . \tag{14}$$

Consider the scalar expression

$$J = \alpha_{i_1 i_2 \ldots i_\mu} u_{i_1}^{(1)} u_{i_2}^{(2)} \ldots u_{i_\mu}^{(\mu)} , \tag{15}$$

where $u_i^{(P)}$ are components of absolute vectors. Then

$$\alpha_{i_1 i_2 \ldots i_\mu} = \frac{\partial^\mu J}{\partial u_{i_1}^{(1)} \partial u_{i_2}^{(2)} \ldots \partial u_{i_\mu}^{(\mu)}} . \tag{16}$$

It is easily verified that J is invariant under orthogonal transformations. It can therefore be expressed as a polynomial in expressions of the types (11) and (12). It is also multilinear in the components of each of the vectors $\underset{\sim}{u}_P$ $(P = 1,2,\ldots,\mu)$. Thus a typical term of J is

$$\beta (u_i^{(A)} u_i^{(B)}) (u_j^{(C)} u_j^{(D)}) \ldots (e_{pqr} u_p^{(G)} u_q^{(H)} u_r^{(K)}) \ldots , \tag{17}$$

where β is a numerical coefficient and A, B, C, D, ..., G, H, K, ... is a permutation of $1,2,\ldots,\mu$. Since

$$\frac{\partial^2 (u_i^{(A)} u_i^{(B)})}{\partial u_j^{(A)} \partial u_k^{(B)}} = \delta_{jk} \,, \qquad \frac{\partial^3 (e_{pqr} u_p^{(G)} u_q^{(H)} u_r^{(K)})}{\partial u_s^{(G)} \partial u_t^{(H)} \partial u_u^{(K)}} = e_{stu} \,,$$

it follows from (16) and (17) that $\alpha_{i_1 i_2 \ldots i_\mu}$ can be expressed as a linear combination of terms each of which is an outer product of Kronecker deltas and third order alternating tensors. Hence:

> Every isotropic tensor is a linear combination of outer products of Kronecker deltas δ_{ij} and alternating tensors e_{stu}.

Since

$$e_{ijk} e_{rst} = \begin{vmatrix} \delta_{ir} & \delta_{is} & \delta_{it} \\ \delta_{jr} & \delta_{js} & \delta_{jt} \\ \delta_{kr} & \delta_{ks} & \delta_{kt} \end{vmatrix} \,, \tag{18}$$

$\alpha_{i_1 i_2 \ldots i_\mu}$ can be expressed so that each of its terms contains at most one alternating tensor.

To obtain the isotropic tensors for the full orthogonal group, we delete the tensors of the form e_{stu}. Hence for the full orthogonal group every isotropic tensor is a linear combination of products of Kronecker deltas.

6. ISOTROPIC INVARIANTS OF VECTORS AND SECOND ORDER TENSORS - GENERAL FORM

Let I be a homogeneous polynomial invariant of ν vectors $\underset{\sim}{u}_R$ and λ tensors $\underset{\sim}{P}_S$ in three dimensions. Thus

$$I = \beta_{i_1 i_2 \ldots i_N j_1 k_1 j_2 k_2 \ldots j_M k_M} u_{i_1}^{(R_1)} u_{i_2}^{(R_2)} \ldots u_{i_N}^{(R_N)} P_{j_1 k_1}^{(S_1)} P_{j_2 k_2}^{(S_2)} \ldots P_{j_M k_M}^{(S_M)} \,, \tag{19}$$

where $R_1, R_2, \ldots, R_N$ are integers (not necessarily different) chosen from $1,2,\ldots,\nu$; $S_1, S_2, \ldots, S_M$ are integers (not necessarily different) from $1,2 \ldots,\lambda$, and the βs are numerical coefficients. Under an orthogonal transformation $\underset{\sim}{M}$, the components of $\underset{\sim R}{u}$ and $\underset{\sim S}{P}$ transform to

$$\bar{u}_p^{(R)} = M_{pi} u_i^{(R)} , \qquad \bar{P}_{qr}^{(S)} = M_{qj} M_{rk} P_{jk}^{(S)} \tag{20}$$

and since I is invariant, it is also equal to (19) with each u_i and P_{jk} replaced by the corresponding $\bar{u}_i$ and $\bar{P}_{jk}$. Hence, using (20) and equating coefficients

$$\begin{aligned}\beta_{i_1 i_2 \ldots i_N j_1 k_1 j_2 k_2 \ldots j_M k_M} = {} & M_{p_1 i_1} M_{p_2 i_2} \ldots M_{p_N i_N} M_{q_1 j_1} M_{r_1 k_1} \times \\ & \times M_{q_2 j_2} M_{r_2 k_2} \ldots M_{q_M j_M} M_{r_M k_M} \times \\ & \times \beta_{p_1 p_2 \ldots p_N q_1 r_1 q_2 r_2 \ldots q_M r_M} .\end{aligned}$$

Thus the βs are components of an isotropic tensor and so can be expressed as polynomials in the isotropic tensors δ_{ij} and e_{ijk} (with indices chosen from $i_1, i_2, \ldots, i_N, j_1, j_2, \ldots, j_M$ and $k_1, k_2, \ldots, k_M$). On substituting such a polynomial in (19), it follows that I can be expressed as a polynomial in expressions of the following types.

$$(a) \quad \pi_{ii}^{(1)} = \operatorname{tr} \underset{\sim}{\pi}_1 , \tag{21a}$$

$$(b) \quad u_i^{(R_1)} \pi_{ij}^{(2)} u_j^{(R_2)} = \underset{\sim}{u}_{R_1} \underset{\sim}{\pi}_2 \underset{\sim}{u}_{R_2} , \tag{21b}$$

$$(c) \quad e_{ijk} \pi_{iq}^{(3)} u_q^{(R_1)} \pi_{jk}^{(4)} , \tag{22a}$$

$$(d) \quad e_{ijk} \pi_{iq}^{(5)} u_q^{(R_1)} \pi_{jr}^{(6)} u_r^{(R_2)} \pi_{ks}^{(7)} u_s^{(R_3)} , \tag{22b}$$

where $\pi_{ij}^{(1)}, \ldots, \pi_{ij}^{(7)}$ are expressions of the form

$$\pi_{ij} = P_{ik}^{(S_1)} P_{k\ell}^{(S_2)} \ldots P_{mj}^{(S_R)} , \qquad \text{or} \qquad \pi_{ij} = \delta_{ij} , \tag{23}$$

or expressions obtained from (23) by replacing some or all of $P_{ij}^{(S)}$ by

$P_{ji}^{(S)}$. In each term of I at most one expression of type *(c)* or *(d)* need appear. R_1, R_2, R_3 are integers (not necessarily different) chosen from $1,2,\ldots,\nu$, and S_1, S_2, ... are integers (not necessarily different) chosen from $1,2,\ldots,\lambda$. Expressions of the form (21) are absolute invariants for the full and proper orthogonal groups; expressions of the form (22) are absolute invariants for the proper orthogonal group and relative invariants for the full orthogonal group.

The expressions (21) and (22) give an integrity basis for $\underset{\sim}{u}_R$ and $\underset{\sim}{P}_S$ but the basis is not minimal or even finite. It is therefore necessary to eliminate the redundant members.

Since any tensor can be expressed as the sum of a symmetric tensor and an anti-symmetric tensor, it is sufficient to consider invariants of symmetric tensors, anti-symmetric tensors, absolute vectors and axial vectors. Since an anti-symmetric tensor can be associated with an axial vector, it is sufficient to consider invariants of symmetric tensors, anti-symmetric tensors and absolute vectors. For invariance under the proper orthogonal group and its sub-groups, it is not necessary to distinguish between absolute and axial vectors.

7. TRACES OF MATRIX PRODUCTS AND MATRIX POLYNOMIALS

In this section, all matrices are 3×3 matrices. A *matrix product* of degree N formed from a set of matrices $\underset{\sim}{P}_1$, $\underset{\sim}{P}_2$, ..., $\underset{\sim}{P}_\lambda$ is a product

$$\underset{\sim}{\pi} = \underset{\sim}{P}_{S_1}\underset{\sim}{P}_{S_2}\ldots\underset{\sim}{P}_{S_N} \tag{24}$$

or, in components,

$$\pi_{ij} = P_{ik}^{(S_1)}P_{k\ell}^{(S_2)}\ldots P_{rj}^{(S_N)} ,$$

where S_1, S_2, ..., S_N are integers (not necessarily different) chosen from the integers $1,2,\ldots,\lambda$. The *trace* of $\underset{\sim}{\pi}$ is denoted $\mathrm{tr}\,\underset{\sim}{\pi}$ and is a scalar:

$$\operatorname{tr} \underset{\sim}{\Pi} = \operatorname{tr} \underset{\sim}{P}_{S_1} \underset{\sim}{P}_{S_2} \ldots \underset{\sim}{P}_{S_N} = P_{ik}^{(S_1)} P_{k\ell}^{(S_2)} \ldots P_{ri}^{(S_N)} . \tag{25}$$

Note that the trace of a matrix product is the trace of its transpose

$$\operatorname{tr} \underset{\sim}{\Pi} = \operatorname{tr} \underset{\sim}{\Pi}^T , \qquad \operatorname{tr} \underset{\sim}{P}\underset{\sim}{Q} \ldots \underset{\sim}{S} = \operatorname{tr} \underset{\sim}{S}^T \ldots \underset{\sim}{Q}^T \underset{\sim}{P}^T , \tag{26}$$

and the trace of a matrix product is unchanged by cyclic interchange of its factors

$$\operatorname{tr} \underset{\sim}{P}\underset{\sim}{Q} \ldots \underset{\sim}{R}\underset{\sim}{S} = \operatorname{tr} \underset{\sim}{Q} \ldots \underset{\sim}{R}\underset{\sim}{S}\underset{\sim}{P} = \ldots = \operatorname{tr} \underset{\sim}{S}\underset{\sim}{P}\underset{\sim}{Q} \ldots \underset{\sim}{R} . \tag{27}$$

A *matrix polynomial* is a sum of matrix products with coefficients which are polynomials in traces of matrix products.

We use the notation

$$\operatorname{tr} \underset{\sim}{\Pi} \equiv 0 , \tag{28}$$

to indicate that $\operatorname{tr} \underset{\sim}{\Pi}$ can be expressed as a polynomial in traces of matrix products of degree lower than the degree of $\underset{\sim}{\Pi}$. If $\underset{\sim}{\Pi}_1$ and $\underset{\sim}{\Pi}_2$ are of the same degree, and $\operatorname{tr}(\underset{\sim}{\Pi}_1 - \underset{\sim}{\Pi}_2) \equiv 0$, we say that $\underset{\sim}{\Pi}_1$ and $\underset{\sim}{\Pi}_2$ are *equivalent*.

The Cayley-Hamilton theorem for a 3×3 matrix is the identity

$$\underset{\sim}{P}^3 - \underset{\sim}{P}^2 \operatorname{tr} \underset{\sim}{P} + \tfrac{1}{2}\underset{\sim}{P}\{(\operatorname{tr} \underset{\sim}{P})^2 - \operatorname{tr} \underset{\sim}{P}^2\} - \underset{\sim}{I} \det \underset{\sim}{P} = \underset{\sim}{0} . \tag{29}$$

Hence

$$\operatorname{tr} \underset{\sim}{P}^3 - \tfrac{3}{2} \operatorname{tr} \underset{\sim}{P}^2 \operatorname{tr} \underset{\sim}{P} + \tfrac{1}{2}(\operatorname{tr} \underset{\sim}{P})^3 - 3 \det \underset{\sim}{P} = 0 \tag{30}$$

so that $\det \underset{\sim}{P}$ can be expressed as a polynomial in $\operatorname{tr} \underset{\sim}{P}$, $\operatorname{tr} \underset{\sim}{P}^2$ and $\operatorname{tr} \underset{\sim}{P}^3$.

If in (29) we replace $\underset{\sim}{P}$ by $\lambda\underset{\sim}{P}+\mu\underset{\sim}{Q}+\nu\underset{\sim}{R}$, where λ, μ and ν are arbitrary scalars, and equate to zero the coefficient of $\lambda\mu\nu$, we obtain an important identity due to Rivlin [9], namely

$$\underset{\sim}{P}\underset{\sim}{Q}\underset{\sim}{R} + \underset{\sim}{P}\underset{\sim}{R}\underset{\sim}{Q} + \underset{\sim}{Q}\underset{\sim}{P}\underset{\sim}{R} + \underset{\sim}{Q}\underset{\sim}{R}\underset{\sim}{P} + \underset{\sim}{R}\underset{\sim}{P}\underset{\sim}{Q} + \underset{\sim}{R}\underset{\sim}{Q}\underset{\sim}{P}$$

$$- (\underset{\sim}{Q}\underset{\sim}{R}+\underset{\sim}{R}\underset{\sim}{Q})\,\mathrm{tr}\,\underset{\sim}{P} - (\underset{\sim}{R}\underset{\sim}{P}+\underset{\sim}{P}\underset{\sim}{R})\,\mathrm{tr}\,\underset{\sim}{Q} - (\underset{\sim}{P}\underset{\sim}{Q}+\underset{\sim}{Q}\underset{\sim}{P})\,\mathrm{tr}\,\underset{\sim}{R}$$

$$- \underset{\sim}{P}(\mathrm{tr}\,\underset{\sim}{Q}\underset{\sim}{R} - \mathrm{tr}\,\underset{\sim}{Q}\,\mathrm{tr}\,\underset{\sim}{R}) - \underset{\sim}{Q}(\mathrm{tr}\,\underset{\sim}{R}\underset{\sim}{P} - \mathrm{tr}\,\underset{\sim}{R}\,\mathrm{tr}\,\underset{\sim}{P}) - \underset{\sim}{R}(\mathrm{tr}\,\underset{\sim}{P}\underset{\sim}{Q} - \mathrm{tr}\,\underset{\sim}{P}\,\mathrm{tr}\,\underset{\sim}{Q})$$

$$- \underset{\sim}{I}(\mathrm{tr}\,\underset{\sim}{P}\underset{\sim}{Q}\underset{\sim}{R} + \mathrm{tr}\,\underset{\sim}{R}\underset{\sim}{Q}\underset{\sim}{P} - \mathrm{tr}\,\underset{\sim}{P}\,\mathrm{tr}\,\underset{\sim}{Q}\underset{\sim}{R} - \mathrm{tr}\,\underset{\sim}{Q}\,\mathrm{tr}\,\underset{\sim}{R}\underset{\sim}{P} - \mathrm{tr}\,\underset{\sim}{R}\,\mathrm{tr}\,\underset{\sim}{P}\underset{\sim}{Q} + \mathrm{tr}\,\underset{\sim}{P}\,\mathrm{tr}\,\underset{\sim}{Q}\,\mathrm{tr}\,\underset{\sim}{R})$$

$$= \underset{\sim}{0}\,. \qquad (31)$$

The identity (29) is recovered from (31) by setting $\underset{\sim}{P} = \underset{\sim}{Q} = \underset{\sim}{R}$.

By multiplying (29) on the right by $\underset{\sim}{S}$ and taking the trace, we obtain

$$\mathrm{tr}\,\underset{\sim}{P}^3\underset{\sim}{S} - \mathrm{tr}\,\underset{\sim}{P}^2\underset{\sim}{S}\,\mathrm{tr}\,\underset{\sim}{P} + \tfrac{1}{2}\,\mathrm{tr}\,\underset{\sim}{P}\underset{\sim}{S}\{(\mathrm{tr}\,\underset{\sim}{P})^2 - \mathrm{tr}\,\underset{\sim}{P}^2\} - \mathrm{tr}\,\underset{\sim}{S}\,\det\underset{\sim}{P} = 0\,,$$

or,

$$\mathrm{tr}\,\underset{\sim}{P}^3\underset{\sim}{S} \equiv 0\,. \qquad (32)$$

Similarly, from (31),

$$\mathrm{tr}(\underset{\sim}{P}\underset{\sim}{Q}\underset{\sim}{R}+\underset{\sim}{P}\underset{\sim}{R}\underset{\sim}{Q}+\underset{\sim}{Q}\underset{\sim}{P}\underset{\sim}{R}+\underset{\sim}{Q}\underset{\sim}{R}\underset{\sim}{P}+\underset{\sim}{R}\underset{\sim}{P}\underset{\sim}{Q}+\underset{\sim}{R}\underset{\sim}{Q}\underset{\sim}{P})\underset{\sim}{S} \equiv 0\,. \qquad (33)$$

In (33) set $\underset{\sim}{P} = \underset{\sim}{Q}$. This gives

$$\mathrm{tr}(\underset{\sim}{P}^2\underset{\sim}{R}+\underset{\sim}{P}\underset{\sim}{R}\underset{\sim}{P}+\underset{\sim}{R}\underset{\sim}{P}^2)\underset{\sim}{S} \equiv 0\,. \qquad (34)$$

Replace $\underset{\sim}{S}$ in (34) by $\underset{\sim}{P}\underset{\sim}{S}$, and we obtain

$$\mathrm{tr}(\underset{\sim}{P}^2\underset{\sim}{R}\underset{\sim}{P}+\underset{\sim}{P}\underset{\sim}{R}\underset{\sim}{P}^2+\underset{\sim}{R}\underset{\sim}{P}^3)\underset{\sim}{S} \equiv 0\,.$$

But $\mathrm{tr}\,\underset{\sim}{R}\underset{\sim}{P}^3\underset{\sim}{S} = \mathrm{tr}\,\underset{\sim}{P}^3\underset{\sim}{S}\underset{\sim}{R} \equiv 0$, so

$$\mathrm{tr}(\underset{\sim}{P}^2\underset{\sim}{R}\underset{\sim}{P}+\underset{\sim}{P}\underset{\sim}{R}\underset{\sim}{P}^2)\underset{\sim}{S} \equiv 0\,. \qquad (35)$$

Next consider $\mathrm{tr}\,\underset{\sim}{P}\underset{\sim}{Q}\underset{\sim}{P}\underset{\sim}{R}\underset{\sim}{Q}\underset{\sim}{S}$. Then from relations of the type (34)

$$\begin{aligned}\mathrm{tr}(\underset{\sim}{P}\underset{\sim}{Q}\underset{\sim}{P})\underset{\sim}{R}\underset{\sim}{Q}\underset{\sim}{S} &\equiv -\,\mathrm{tr}(\underset{\sim}{P}^2\underset{\sim}{Q}+\underset{\sim}{Q}\underset{\sim}{P}^2)\underset{\sim}{R}\underset{\sim}{Q}\underset{\sim}{S} \\ &= -\,\mathrm{tr}\,\underset{\sim}{P}^2(\underset{\sim}{Q}\underset{\sim}{R}\underset{\sim}{Q})\underset{\sim}{S} - \mathrm{tr}(\underset{\sim}{Q}\underset{\sim}{P}^2\underset{\sim}{R}\underset{\sim}{Q})\underset{\sim}{S} \\ &\equiv \mathrm{tr}\,\underset{\sim}{P}^2(\underset{\sim}{Q}^2\underset{\sim}{R}+\underset{\sim}{R}\underset{\sim}{Q}^2)\underset{\sim}{S} + \mathrm{tr}(\underset{\sim}{Q}^2\underset{\sim}{P}^2\underset{\sim}{R}+\underset{\sim}{P}^2\underset{\sim}{R}\underset{\sim}{Q}^2)\underset{\sim}{S}\,.\end{aligned} \qquad (36)$$

Also, from relations of the form (34),

$$\begin{aligned}\operatorname{tr}\mathbf{P}(\mathbf{QPRQ})\mathbf{S} &\equiv -\operatorname{tr}\mathbf{P}(\mathbf{Q}^2\mathbf{PR}+\mathbf{PRQ}^2)\mathbf{S}\\ &= -\operatorname{tr}(\mathbf{PR}^2\mathbf{P})\mathbf{RS}-\operatorname{tr}\mathbf{P}^2\mathbf{RQ}^2\mathbf{S}\\ &\equiv \operatorname{tr}(\mathbf{P}^2\mathbf{Q}^2+\mathbf{Q}^2\mathbf{P}^2)\mathbf{RS}-\operatorname{tr}\mathbf{P}^2\mathbf{RQ}^2\mathbf{S}\,. \qquad (37)\end{aligned}$$

Comparing (36) and (37) gives

$$\operatorname{tr}\mathbf{P}^2\mathbf{RQ}^2\mathbf{S}\equiv 0\,. \qquad (38)$$

By replacing $\mathbf{P}$ in turn by $\mathbf{P}+\mathbf{T}$ and $\mathbf{P}-\mathbf{T}$, there follows

$$\operatorname{tr}(\mathbf{PT}+\mathbf{TP})\mathbf{RQ}^2\mathbf{S}\equiv 0\,, \qquad (39)$$

and, in a similar manner,

$$\operatorname{tr}(\mathbf{PT}+\mathbf{TP})\mathbf{R}(\mathbf{QK}+\mathbf{KQ})\mathbf{S}\equiv 0\,. \qquad (40)$$

Now, in turn, in (40), replace $\mathbf{S}$ by $\mathbf{LS}$ and $\mathbf{Q}$ by $\mathbf{QL}$

$$\operatorname{tr}(\mathbf{PR}+\mathbf{TP})\mathbf{R}(\mathbf{QK}+\mathbf{KQ})\mathbf{LS}\equiv 0\,, \qquad \operatorname{tr}(\mathbf{PT}+\mathbf{TP})\mathbf{R}(\mathbf{QLK}+\mathbf{KQL})\mathbf{S}\equiv 0\,.$$

Hence, by subtracting these two relations

$$\operatorname{tr}(\mathbf{PT}+\mathbf{TP})\mathbf{RQ}(\mathbf{KL}-\mathbf{LK})\mathbf{S}\equiv 0\,. \qquad (41)$$

Also, in (40), replace $\mathbf{R}$ by $\mathbf{RQ}$ and $\mathbf{Q}$ by $\mathbf{L}$

$$\operatorname{tr}(\mathbf{PT}+\mathbf{TP})\mathbf{RQ}(\mathbf{LK}+\mathbf{KL})\mathbf{S}\equiv 0\,. \qquad (42)$$

From (41) and (42)

$$\operatorname{tr}(\mathbf{PT}+\mathbf{TP})\mathbf{RQKLS}\equiv 0\,.$$

It follows that *the trace of a matrix product of degree seven is equivalent to the negative of the trace of the matrix product obtained from it by interchanging two adjacent factors.*

Now, in (33) replace $\mathbf{P}$, $\mathbf{Q}$, $\mathbf{R}$, $\mathbf{S}$ by $\mathbf{PQ}$, $\mathbf{RS}$, $\mathbf{TK}$, $\mathbf{L}$ respectively. Then

$$\operatorname{tr}(\mathbf{PQRSTK}+\mathbf{PQTKRS}+\mathbf{RSPQTK}+\mathbf{RSTKPQ}+\mathbf{TKPQRS}+\mathbf{TKRSPQ})\mathbf{L}\equiv 0\,. \qquad (43)$$

Each term in (43) can be derived from $\operatorname{tr}\mathbf{PQRSTKL}$ by an *even* number of interchanges of adjacent factors. Hence each term is equivalent to $\operatorname{tr}\mathbf{PQRSTKL}$, and hence

$$\operatorname{tr}\mathbf{PQRSTKL}\equiv 0\,. \qquad (44)$$

Thus the trace of any matrix product of degree seven (or more) can be expressed as a polynomial in traces of matrix products of degree less than seven.

To summarise the main results of this section we have the following *theorem:*

The trace of any matrix polynomial in μ 3×3 matrices $\underset{\sim}{P}_R$ can be expressed as a polynomial in traces of matrix products, each of which products satisfies:

(a) it is a product of factors of the form $\underset{\sim}{P}_R$, $\underset{\sim}{P}_R^2$, $\underset{\sim}{P}_R^3$;

(b) the first and last factors are not powers of the same matrix;

(c) if $\underset{\sim}{P}_R^3$ is a factor, there are no other factors;

(d) no two factors are the same;

(e) $\underset{\sim}{P}_R$ precedes $\underset{\sim}{P}_R^2$ in any product with both $\underset{\sim}{P}_R$ and $\underset{\sim}{P}_R^2$ as factors;

(f) if $\underset{\sim}{P}_R^2$ and $\underset{\sim}{P}_S^2$, $\underset{\sim}{R} \neq \underset{\sim}{S}$, are both factors, they are consecutive factors;

(g) no product has total degree greater than six.

8. INVARIANTS OF SYMMETRIC SECOND-ORDER TENSORS

For invariants of second-order tensors only, it is not necessary to distinguish between the full and the proper orthogonal groups. From Section 6, equation (21a), any polynomial invariant of a set of symmetric second-order tensors is a polynomial in expressions of the form $\mathrm{tr}\,\underset{\sim}{\pi}$, where $\underset{\sim}{\pi}$ is a matrix product in the matrices of the tensor components. For a finite set of matrices there is only a finite number of matrix products which satisfy the theorem proved in Section 7. Hence the matrix products which satisfy the theorem yield a *finite* integrity basis. However, in general this basis is not minimal.

An invariant of the basis cannot be of degree greater than six, and so cannot involve more than six distinct tensors. It is therefore sufficient to consider invariants of up to six distinct tensors; an integrity basis for more than six tensors is obtained by combining the bases for the tensors taken six at a time, in all possible combinations.

The procedure is to list the matrix products which satisfy the theorem, for one to six matrices in turn, and express as many as possible of these invariants in terms of the remainder. We eventually arrive thus at a minimal basis; proof of minimality is discussed briefly in Section 7 of Chapter 9. An alternative approach to the construction of minimal integrity bases, which is based on group-theoretical methods, is described by Smith [10, 11].

In addition to the results of Section 7, we make extensive use of the fact that the trace of a matrix product formed from *symmetric* matrices is equal to the trace of the matrix product with its factors written in reverse order; that is, if $\underset{\sim}{A}$, $\underset{\sim}{B}$, ..., $\underset{\sim}{F}$ are symmetric matrices

$$\operatorname{tr} \underset{\sim}{A}\underset{\sim}{B}\ldots\underset{\sim}{F} = \operatorname{tr} \underset{\sim}{F}\ldots\underset{\sim}{B}\underset{\sim}{A} \; . \tag{45}$$

The results for up to three tensors are as follows.

One matrix $\underset{\sim}{A}$

A minimal integrity basis is:

$$\operatorname{tr} \underset{\sim}{A}, \quad \operatorname{tr} \underset{\sim}{A}^2, \quad \operatorname{tr} \underset{\sim}{A}^3.$$

Two matrices $\underset{\sim}{A}$, $\underset{\sim}{B}$

The following invariants satisfy the theorem:

$$\operatorname{tr} \underset{\sim}{A}\underset{\sim}{B}, \quad \operatorname{tr} \underset{\sim}{A}\underset{\sim}{B}^2, \quad \operatorname{tr} \underset{\sim}{B}\underset{\sim}{A}^2, \quad \operatorname{tr} \underset{\sim}{A}^2\underset{\sim}{B}^2, \quad \operatorname{tr} \underset{\sim}{A}\underset{\sim}{B}\underset{\sim}{A}^2\underset{\sim}{B}^2.$$

However, $\operatorname{tr} \underset{\sim}{A}\underset{\sim}{B}\underset{\sim}{A}^2\underset{\sim}{B}^2 \equiv -\operatorname{tr} \underset{\sim}{A}^2\underset{\sim}{B}\underset{\sim}{A}\underset{\sim}{B}^2 = -\operatorname{tr} \underset{\sim}{A}\underset{\sim}{B}\underset{\sim}{A}^2\underset{\sim}{B}$. Hence $\operatorname{tr} \underset{\sim}{A}\underset{\sim}{B}\underset{\sim}{A}^2\underset{\sim}{B}^2 \equiv 0$, and this invariant can be omitted from the basis.

Three matrices $\underset{\sim}{A}$, $\underset{\sim}{B}$, $\underset{\sim}{C}$

We have to consider

$$\operatorname{tr} \underset{\sim}{A}\underset{\sim}{B}\underset{\sim}{C}, \quad \operatorname{tr} \underset{\sim}{A}^2\underset{\sim}{B}\underset{\sim}{C}, \quad \operatorname{tr} \underset{\sim}{A}^2\underset{\sim}{B}^2\underset{\sim}{C}, \quad \operatorname{tr} \underset{\sim}{A}\underset{\sim}{B}\underset{\sim}{A}^2\underset{\sim}{C}, \quad \operatorname{tr} \underset{\sim}{A}^2\underset{\sim}{B}^2\underset{\sim}{C}^2, \quad \operatorname{tr} \underset{\sim}{A}^2\underset{\sim}{B}^2\underset{\sim}{A}\underset{\sim}{C} \tag{46}$$

and invariants derived from these by permuting $\underset{\sim}{A}$, $\underset{\sim}{B}$, $\underset{\sim}{C}$.

By cyclic interchanges and transposition

$$\operatorname{tr} \underset{\sim}{A}\underset{\sim}{B}\underset{\sim}{C} = \operatorname{tr} \underset{\sim}{B}\underset{\sim}{C}\underset{\sim}{A} = \operatorname{tr} \underset{\sim}{C}\underset{\sim}{A}\underset{\sim}{B} = \operatorname{tr} \underset{\sim}{C}\underset{\sim}{B}\underset{\sim}{A} = \operatorname{tr} \underset{\sim}{A}\underset{\sim}{C}\underset{\sim}{B} = \operatorname{tr} \underset{\sim}{B}\underset{\sim}{A}\underset{\sim}{C} \, ,$$

so of these we retain only $\operatorname{tr} ABC$.

Similarly $\operatorname{tr} A^2BC = \operatorname{tr} A^2CB$, etc., so we retain

$$\operatorname{tr} A^2BC, \quad \operatorname{tr} B^2CA, \quad \operatorname{tr} C^2AB .$$

By similar arguments, we retain

$$\operatorname{tr} A^2B^2C, \quad \operatorname{tr} B^2C^2A, \quad \operatorname{tr} C^2A^2B .$$

The remaining invariants in (46) are reducible, as follows.

$$\operatorname{tr} ABA^2C \equiv -\operatorname{tr} A^2BAC = -\operatorname{tr} CABA^2 = -\operatorname{tr} ABA^2C ,$$

so $\operatorname{tr} ABA^2C \equiv 0$. Replacing B by B^2 gives $\operatorname{tr} AB^2A^2C \equiv 0$. Finally

$$2 \operatorname{tr} A^2B^2C^2 = \operatorname{tr} A^2(B^2C^2+C^2B^2) \equiv -\operatorname{tr} A^2BC^2B = -\operatorname{tr}(A^2BC^2)B \equiv 0.$$

As the number of tensors increases, the number of invariants needing consideration and the number of relations between them grow rapidly, and we omit the rather lengthy details. The final results are shown in Table 1. The most difficult case is that of six tensors, where we have initially to consider $6! = 720$ invariants of the type $\operatorname{tr} ABCDEF$. These reduce to 60 by using the results that the invariants are unchanged by factor reversal and cyclic interchange. Between these 60 invariants there are various relations which can be reduced to *(a)* 60 relations of the type

$$\operatorname{tr} ABC(DEF+DFE+EDF+EFD+FDE+FED) \equiv 0 ,$$

and *(b)* 90 relations of the type

$$\operatorname{tr} AB\{(CD)EF+(CD)FE+E(CD)F+EF(CD)+F(CD)E+FE(CD)\} \equiv 0 .$$

Of these 150 relations, it can be shown that 50 are independent; they may be used to express 50 of the invariants in terms of the remaining ten, for example the ten listed in Table 1.

9. INVARIANTS OF SECOND-ORDER TENSORS AND VECTORS; PROPER ORTHOGONAL GROUP

For the proper orthogonal group it is not necessary to distinguish between absolute and axial vectors. With each vector $\underset{\sim}{u}$ we can associate an anti-symmetric tensor $\underset{\sim}{U}$ by the relations

$$U_{ij} = e_{ijk}u_k , \qquad u_k = \tfrac{1}{2}e_{ijk}U_{ij} \tag{47}$$

It is therefore sufficient to consider invariants of symmetric and anti-symmetric second-order tensors. Hence the forms of invariants we have to consider are, from (21a),

$$\operatorname{tr} \underset{\sim}{\pi} ,$$

where $\underset{\sim}{\pi}$ is a matrix product formed from symmetric matrices $\underset{\sim}{A}$, $\underset{\sim}{B}$, ..., and anti-symmetric matrices $\underset{\sim}{U}$, $\underset{\sim}{V}$, ..., $\underset{\sim}{X}$, $\underset{\sim}{Y}$, ...

Equivalently, we recall that the required invariants are of the forms (21a,b) and (22a,b). By making substitutions of the forms (47) and employing the identity (18) it can be shown that each of the invariants (21b), (22a) and (22b) can be expressed as a polynomial in traces of matrix products formed from symmetric and anti-symmetric matrix products (details are given in Spencer and Rivlin [12] and Spencer [6, 13]). Hence it is sufficient to consider the invariants $\operatorname{tr} \underset{\sim}{\pi}$ which are traces of these matrix products.

The same methods as were used to reduce the integrity basis for symmetric tensors can now be used for symmetric and anti-symmetric tensors. In addition to the results used earlier based on the Cayley-Hamilton theorem for general matrices we have available the following results for traces of matrix products involving anti-symmetric matrices $\underset{\sim}{X}$, $\underset{\sim}{Y}$, $\underset{\sim}{Z}$ and symmetric matrices $\underset{\sim}{A}$, $\underset{\sim}{B}$, $\underset{\sim}{C}$, ... (which may be the unit matrix $\underset{\sim}{I}$):

$$\operatorname{tr} \underset{\sim}{X} = 0 , \qquad \operatorname{tr} \underset{\sim}{X}\underset{\sim}{A} = 0 , \tag{48}$$

$$\operatorname{tr} \underset{\sim}{X}\underset{\sim}{A}\underset{\sim}{B}\ldots\underset{\sim}{C} = -\operatorname{tr} \underset{\sim}{C}\ldots\underset{\sim}{B}\underset{\sim}{A}\underset{\sim}{X} , \tag{49}$$

$$\operatorname{tr} \underset{\sim}{X}\underset{\sim}{A}\ldots\underset{\sim}{Y}\underset{\sim}{B}\ldots\underset{\sim}{C} = \operatorname{tr} \underset{\sim}{C}\ldots\underset{\sim}{B}\underset{\sim}{Y}\ldots\underset{\sim}{A}\underset{\sim}{X} , \tag{50}$$

$$\operatorname{tr} \underset{\sim}{X}\underset{\sim}{A}\ldots\underset{\sim}{Y}\underset{\sim}{B}\ldots\underset{\sim}{Z}\underset{\sim}{C}\ldots\underset{\sim}{D} = -\operatorname{tr} \underset{\sim}{D}\ldots\underset{\sim}{C}\underset{\sim}{Z}\ldots\underset{\sim}{B}\underset{\sim}{Y}\ldots\underset{\sim}{A}\underset{\sim}{X} , \tag{51}$$

where the dots represent symmetric matrices. In addition we have the

TABLE 1
INVARIANTS OF VECTORS AND TENSORS - PROPER ORTHOGONAL GROUP

MATRICES	MATRIX PRODUCTS
A	$A;\ A^2;\ A^3$
A,B	$AB;\ AB^2{*};\ A^2B^2$
A,B,C	$ABC;\ A^2BC{*};\ A^2B^2C{*}$
A,B,C,D	$ABCD,\ ABDC;\ A^2BCD{*},\ A^2BDC{*};\ A^2B^2CD,\ A^2C^2BD,\ A^2D^2BC,\ B^2C^2AD,\ B^2D^2AC,\ C^2D^2AB;\ A^2BACD{*}$
A,B,C,D,E	$ABCDE,\ ABDEC,\ ABECD,\ ACBED,\ ACDBE,\ ADBCE;\ A^2BCDE{*},\ A^2BCED{*},\ A^2CBDE{*},\ A^2BEDC{*}$
A,B,C,D,E,F	$ACFEBD,\ ADCBFE,\ ADCFBE,\ ADFBCE,\ ADFCBE,\ AEBDCF,\ AECBDF,\ AECDBF,\ AEDBCF,\ AEDCBF$
X	X^2
X,A	$X^2A;\ X^2A^2;\ X^2AXA^2$
X,A,B	$XAB;\ XA^2B{*};\ XA^2B^2;\ XA^2BA{*};\ XA^2B^2A{*},\ X^2AB;\ X^2A^2B{*};\ X^2AXB;\ X^2AXB^2{*}$
X,A,B,C	$XABC,\ XACB,\ XBAC;\ XA^2BC{*},\ XA^2CB{*},\ XBA^2C{*};\ XA^2BCA{*};\ XA^2B^2C{*},\ XB^2A^2C;\ X^2ABC,\ X^2BCA;\ X^2A^2BC{*};\ X^2AXBC$
X,A,B,C,D	$XABCD,\ XABDC,\ XACDB,\ XBACD,\ XBADC,\ XCABD;\ XBA^2CD{*},\ XCA^2DB{*},\ XDA^2BC{*};\ X^2ACDB,\ X^2ABDC,\ X^2ABCD$
X,A,B,C,D,E	$XBDACE,\ XACBDE,\ XABCED,\ XABDCE,\ XACEBD$
X,Y	XY
X,Y,A	$XYA;\ XYA^2;\ X^2YA^{\dagger},\ X^2YA^{2\dagger};\ X^2AYA^{2\dagger}$
X,Y,A,B	$XYAB,\ XYBA;\ XYA^2B{*},\ XYBA^2{*};\ XYA^2B^2;\ XYA^2BA{*};\ X^2YAB^{\dagger};\ X^2AYB^{\dagger};\ X^2BYA^2{*}^{\dagger}$
X,Y,A,B,C	$XYABC,\ XYACB,\ XYBAC,\ XYBCA,\ XYCAB;\ XYA^2BC{*},\ XYA^2CB{*},\ XYBCA^2{*};\ X^2AYBC^{\dagger}$
X,Y,A,B,C,D	$XYACDB,\ XYABDC,\ XYABCD,\ XYBCDA,\ XYBADC,\ XYCBDA$
X,Y,Z	XYZ
X,Y,Z,A	$XYZA,\ XZYA;\ XYZA^2,\ XZYA^2;\ XYAZA^2$
X,Y,Z,A,B	$XYZAB,\ YZXAB,\ XYAZB;\ XYBZA^2{*}$
X,Y,Z,A,B,C	$XYAZBC$

Spencer-Rivlin identity [12]

$$\underset{\sim}{X}\underset{\sim}{P}\underset{\sim}{Y} + \underset{\sim}{P}^T\underset{\sim}{X}\underset{\sim}{Y} + \underset{\sim}{X}\underset{\sim}{Y}\underset{\sim}{P}^T = \underset{\sim}{X}\underset{\sim}{Y}\,\mathrm{tr}\,\underset{\sim}{P} + \tfrac{1}{2}\underset{\sim}{P}^T\,\mathrm{tr}\,\underset{\sim}{X}\underset{\sim}{Y} + \underset{\sim}{I}(\mathrm{tr}\,\underset{\sim}{X}\underset{\sim}{Y}\underset{\sim}{P}^T - \tfrac{1}{2}\,\mathrm{tr}\,\underset{\sim}{X}\underset{\sim}{Y}\,\mathrm{tr}\,\underset{\sim}{P}) \,, \tag{52}$$

where $\underset{\sim}{P}$ is a general 3×3 matrix. This identity is independent of the Cayley-Hamilton theorem, although it reduces to the Cayley-Hamilton theorem for anti-symmetric 3×3 matrices when $\underset{\sim}{X} = \underset{\sim}{Y} = \underset{\sim}{P}$. From (52) it follows that

$$\mathrm{tr}(\underset{\sim}{X}\underset{\sim}{Y}\underset{\sim}{P}+\underset{\sim}{X}\underset{\sim}{P}^T\underset{\sim}{Y}+\underset{\sim}{P}\underset{\sim}{X}\underset{\sim}{Y})\underset{\sim}{Q} \equiv 0\,, \qquad \text{for any matrix } \underset{\sim}{Q}. \tag{53}$$

Rivlin and Smith [14] have proved that there are no other distinct matrix identities except (31) and (52) for symmetric and anti-symmetric 3×3 matrices.

We omit details of the reductions, which are similar to those employed in the case of symmetric tensors only. The results are given in Table 1. In Table 1, the invariants are the traces of the tabulated matrix products, where $\underset{\sim}{A}$, $\underset{\sim}{B}$, $\underset{\sim}{C}$, $\underset{\sim}{D}$, $\underset{\sim}{E}$, $\underset{\sim}{F}$ are symmetric matrices associated with symmetric tensors, and $\underset{\sim}{X}$, $\underset{\sim}{Y}$, $\underset{\sim}{Z}$ are anti-symmetric matrices associated with anti-symmetric tensors or with vectors (by (47)). The integrity basis for a given set of not more than six vectors and tensors is obtained by adding to the invariants tabulated for the given set the invariants for all the subsets of the given set. The integrity bases for more than six vectors and tensors is the sum of the integrity bases for the vectors and tensors taken six at a time in all combinations. An asterisk (*) indicates that in addition to the listed invariant, the basis also includes invariants formed by cyclic permutation of symmetric matrices, and a dagger (†) that it also includes invariants formed by cyclic permutation of skew-symmetric matrices.

10. INVARIANTS OF SECOND-ORDER TENSORS AND VECTORS; FULL ORTHOGONAL GROUP

For the full orthogonal group it is necessary to distinguish axial from absolute vectors. We associate axial vectors with second-order anti-symmetric tensors (which transform in the correct manner; see (9), (10)) but do not make this association for absolute vectors. Invariants of the forms (21a,b) are absolute invariants under the full group, and those of the forms (22a,b) are only relative invariants under the full group, where now $\pi^{(1)}_{ij}$, $\pi^{(2)}_{ij}$, ..., $\pi^{(7)}_{ij}$ are components of matrix products in symmetric matrices and in anti-symmetric matrices associated with anti-symmetric tensors and axial vectors. General tensors can be dealt with by decomposing them into sums of symmetric and anti-symmetric tensors.

To obtain an integrity basis for the full group it is not sufficient to eliminate odd invariants from a basis for the proper group, because some invariants which are reducible when regarded as invariants of the proper group are irreducible when regarded as invariants of the full group.

We have to consider invariants of the forms (21a) and (21b). Invariants of the form (21a) are as for the proper orthogonal group. For invariants of the form (21b), that is $u_i \pi^{(2)}_{ij} v_j$ or, as a special case, $u_i \pi^{(2)}_{ij} u_j$, we set

$$\underset{\sim}{A} = \underset{\sim}{A}^T = (u_i u_j), \qquad \underset{\sim}{C} = \underset{\sim}{C}^T = (u_i v_j + u_j v_i), \qquad \underset{\sim}{X} = -\underset{\sim}{X}^T = \tfrac{1}{2}(u_i v_j - u_j v_i). \tag{54}$$

Then

$$u_i \pi^{(2)}_{ij} v_j = \mathrm{tr}\,(\underset{\sim}{C}+\underset{\sim}{X})\,\underset{\sim}{\pi}_2\,, \qquad u_i \pi^{(2)}_{ij} u_j = \mathrm{tr}\,\underset{\sim}{A}\underset{\sim}{\pi}_2\,, \tag{55}$$

so that it is sufficient to consider invariants of the forms

$$\mathrm{tr}\,\underset{\sim}{A}\underset{\sim}{\pi}\,, \qquad \mathrm{tr}\,\underset{\sim}{C}\underset{\sim}{\pi}\,, \qquad \mathrm{tr}\,\underset{\sim}{X}\underset{\sim}{\pi}\,, \tag{56}$$

where $\underset{\sim}{A}$ and $\underset{\sim}{C}$ are symmetric, and $\underset{\sim}{X}$ is anti-symmetric. However, invariants of the forms (56) were considered in Section 9, and listed in Table 1. Hence the required results can be read off from Table 1. The complete list of invariants is long; it is given in Smith [15], but is not reproduced here.

The integrity bases derived in Sections 8, 9 and 10 by reduction of traces of matrix products, using matrix identities based on the Cayley-

Hamilton theorem and the Spencer-Rivlin identity, are in fact minimal bases in the sense that none of their members is reducible. However, we have so far given no proof of this minimality property. We refer to this problem in Section 7 of Chapter 9.

11. ISOTROPIC TENSOR POLYNOMIAL FUNCTIONS OF VECTORS AND TENSORS

A problem which arises frequently in continuum physics is the following. Let $F_{i_1 i_2 \ldots i_r}$ be components of a tensor of order r, which are functions of the components $u_{ij}^{(1)}, u_{ij}^{(2)}, \ldots, u_{ij}^{(\nu)}$ of a set of ν vectors and the components $P_{mn}^{(1)}, P_{mn}^{(2)}, \ldots, P_{mn}^{(\lambda)}$ of a set of λ second-order tensors. For example, in the cases r = 1 and r = 2 we have

$$f_i = \phi_i(u_j^{(p)}, P_{mn}^{(q)}), \qquad F_{ij} = \Phi_{ij}(u_j^{(p)}, P_{mn}^{(q)}),$$

$$(p = 1,2,\ldots\nu;\ q = 1,2,\ldots\lambda). \tag{57}$$

Under a coordinate transformation $\bar{\underset{\sim}{x}} = \underset{\sim}{M}\underset{\sim}{x}$, the components of the vectors and tensors become

$$\bar{f}_i = M_{ik} f_k, \qquad \bar{F}_{ij} = M_{ir} M_{js} F_{rs},$$

$$\bar{u}_i^{(p)} = M_{ij} u_j^{(p)}, \qquad \bar{P}_{ij}^{(q)} = M_{im} M_{jm} P_{mn}^{(q)}.$$

We say that f_i, F_{ij} are form-invariant under a group of transformations $\underset{\sim}{M}$ if, for each $\underset{\sim}{M}$ of the group,

$$\begin{aligned} M_{ij}\phi_j(u_j^{(p)}, P_{mn}^{(q)}) &= \phi_i(\bar{u}_j^{(p)}, \bar{P}_{mn}^{(q)}), \\ M_{ir} M_{js} \Phi_{rs}(u_j^{(p)}, P_{mn}^{(q)}) &= \Phi_{ij}(\bar{u}_j^{(p)}, \bar{P}_{mn}^{(q)}), \end{aligned} \tag{58}$$

The extension to tensors of higher order is evident. If the transformation group is the full or proper orthogonal group, we say that f_i and F_{ij} are isotropic vector and tensor functions respectively of their arguments. We restrict ourselves to polynomial functions. If f is a scalar, then the equivalent statement to (58) is that f is an invariant of $u_j^{(p)}$ and $P_{mn}^{(q)}$; the concept of a form-invariant tensor is therefore a generalization of that of a scalar invariant.

To determine canonical forms for form-invariant tensor polynomials we reduce the problem to the problem of determining integrity bases, which has already been considered. The method was first described by Pipkin and Rivlin [16]. Let v_i, w_i be components of arbitrary vectors, and form

$$J_1 = v_i f_i , \qquad J_2 = v_i w_j F_{ij} . \tag{59}$$

Evidently

$$f_i = \partial J_1/\partial v_i , \qquad F_{ij} = \partial^2 J_2/\partial v_i \partial w_j . \tag{60}$$

It is easily verified that J_1 is an invariant of v_i, $u_k^{(p)}$ and $P_{mn}^{(q)}$, and linear in v_i; also that J_2 is an invariant of v_i, w_i, $u_k^{(p)}$ and $P_{mn}^{(q)}$, and bilinear in v_i and w_i. Hence J_1 and J_2 can be expressed in the forms

$$J_1 = \sum_\alpha I_\alpha J_\alpha^{(1)} , \qquad J_2 = \sum_\alpha I_\alpha J_\alpha^{(2)} ,$$

where I_α are polynomials in invariants of the vectors $u_k^{(p)}$ and $P_{mn}^{(q)}$; $J_\alpha^{(1)}$ are invariants of v_i, $u_k^{(p)}$ and $P_{mn}^{(q)}$ which are linear in v_i; and $J_\alpha^{(2)}$ are invariants of v_i, w_i, $u_k^{(p)}$ and $P_{mn}^{(q)}$ which are bilinear in v_i and w_i. Hence, from (59) and (60),

$$f_i = \sum_\alpha I_\alpha \partial J_\alpha^{(1)}/\partial v_i , \qquad F_{ij} = \sum_\alpha I_\alpha \partial^2 J_\alpha^{(2)}/\partial v_i \partial w_j . \tag{61}$$

This gives canonical forms for f_i and F_{ij} although, because of the possible existence of syzygies, the expressions may contain redundant terms even if $J_\alpha^{(1)}$ and $J_\alpha^{(2)}$ are elements of an irreducible integrity basis. The extension to tensors of higher order is obvious. The use of the differentiation operation is only for convenience; the process may be described easily in purely algebraic terms.

The method needs modification if F_{ij} (or a tensor of higher order) has index symmetries. If, for example, F_{ij} is symmetric, there are several possible ways to proceed.

(a) Construct F_{ij} as in (61); then interchange i and j to form F_{ji}, and take

$$F_{ij} = \tfrac{1}{2}(F_{ij}+F_{ji}) .$$

This expression will often contain redundant terms.

(b) Instead of forming J_2 as in (59), form

$$J_2' = v_i v_j F_{ij} .$$

Then J_2' is an invariant of v_i, $u_k^{(p)}$, $P_{mn}^{(q)}$, and quadratic in v_i, and

$$F_{ij} = \tfrac{1}{2}\partial^2 J'/\partial v_i \partial v_j \tag{62}$$

and F_{ij} is symmetric with respect to interchanges of i and j.

(c) Insteady of forming J_2 as in (59), form

$$J_2'' = K_{ij} F_{ij} ,$$

where K_{ij} are components of an arbitrary symmetric tensor, and J_2'' is expressed as a symmetric function of K_{ij} and K_{ji}. Then J_2'' is an invariant of K_{ij}, $u_k^{(p)}$, $P_{mn}^{(q)}$, and linear in K_{ij}. Hence

$$F_{ij} = \partial J_2''/K_{ij} = \partial J_2''/\partial K_{ji} \tag{63}$$

so F_{ij} has the required index symmetry.

Methods analogous to *(a)* and *(c)*, but not to *(b)*, can be exployed when F_{ij} is anti-symmetric.

As a simple example let us use method *(c)* above to find the form-invariant symmetric tensor function of a single symmetric tensor $\underset{\sim}{A}$. From the tables, the invariants of $\underset{\sim}{A}$ and $\underset{\sim}{K}$ are

$$\mathrm{tr}\,\underset{\sim}{A},\ \mathrm{tr}\,\underset{\sim}{A}^2,\ \mathrm{tr}\,\underset{\sim}{A}^3,\ \mathrm{tr}\,\underset{\sim}{A}\underset{\sim}{K},\ \mathrm{tr}\,\underset{\sim}{A}^2\underset{\sim}{K},\ \mathrm{tr}\,\underset{\sim}{K},\ \mathrm{tr}\,\underset{\sim}{K}^2,\ \mathrm{tr}\,\underset{\sim}{K}^3,\ \mathrm{tr}\,\underset{\sim}{A}\underset{\sim}{K}^2,\ \mathrm{tr}\,\underset{\sim}{A}^2\underset{\sim}{K}^2 .$$

Since we require J_2'' to be linear in $\underset{\sim}{K}$, we retain only the first six of these. Hence J_2'' has the form

$$J_2'' = I_1\,\mathrm{tr}\,\underset{\sim}{K} + I_2\,\mathrm{tr}\,\underset{\sim}{A}\underset{\sim}{K} + I_3\,\mathrm{tr}\,\underset{\sim}{A}^2\underset{\sim}{K} ,$$

where I_1, I_2 and I_3 are polynomials in $\mathrm{tr}\,\underset{\sim}{A}$, $\mathrm{tr}\,\underset{\sim}{A}^2$ and $\mathrm{tr}\,\underset{\sim}{A}^3$. Using (63) then gives

$$F_{ij} = I_1\delta_{ij} + I_2A_{ij} + I_3A_{ik}A_{kj} ,$$

or

$$\underset{\sim}{F} = I_1\underset{\sim}{I} + I_2\underset{\sim}{A} + I_3\underset{\sim}{A}^2 , \tag{64}$$

which is a well-known result.

In a similar manner we can form symmetric tensor functions of any number of symmetric tensors. Full results are given in [6]. For two symmetric tensors $\underset{\sim}{A}$ and $\underset{\sim}{B}$ the result is

$$\underset{\sim}{F} = I_1\underset{\sim}{I} + I_2\underset{\sim}{A} + I_3\underset{\sim}{A}^2 + I_4\underset{\sim}{B} + I_5\underset{\sim}{B}^2 + I_6(\underset{\sim}{A}\underset{\sim}{B}+\underset{\sim}{B}\underset{\sim}{A})$$
$$+ I_7(\underset{\sim}{A}^2\underset{\sim}{B}+\underset{\sim}{B}\underset{\sim}{A}^2) + I_8(\underset{\sim}{A}\underset{\sim}{B}^2+\underset{\sim}{B}^2\underset{\sim}{A}) + I_9(\underset{\sim}{A}^2\underset{\sim}{B}^2+\underset{\sim}{B}^2\underset{\sim}{A}^2) \tag{65}$$

where I_α are invariants of $\underset{\sim}{A}$ and $\underset{\sim}{B}$. This result was obtained directly by using results in matrix polynomial theory by Rivlin [9], and its extension for an arbitrary number of tensors by Spencer and Rivlin [17, 18] and Spencer [19].

As a further example, consider an anti-symmetric tensor function $\underset{\sim}{X}$ of two symmetric tensors $\underset{\sim}{A}$ and $\underset{\sim}{B}$, for the proper orthogonal group. We form

$$K = 2X_{ij}W_{ij} = X_{ij}(W_{ij}-W_{ji}) ,$$

where W_{ij} are components of an arbitrary anti-symmetric tensor and K is expressed as a symmetric function of W_{ij} and $-W_{ji}$. Then, from tables of invariants of $\underset{\sim}{W}$, $\underset{\sim}{A}$ and $\underset{\sim}{B}$, K is a linear combination of the invariants

$$\mathrm{tr}\,\underset{\sim}{W}\underset{\sim}{A}\underset{\sim}{B},\quad \mathrm{tr}\,\underset{\sim}{W}\underset{\sim}{A}^2\underset{\sim}{B},\quad \mathrm{tr}\,\underset{\sim}{W}\underset{\sim}{B}^2\underset{\sim}{A},\quad \mathrm{tr}\,\underset{\sim}{W}\underset{\sim}{A}^2\underset{\sim}{B}^2,\quad \mathrm{tr}\,\underset{\sim}{W}\underset{\sim}{A}^2\underset{\sim}{B}\underset{\sim}{A},$$
$$\mathrm{tr}\,\underset{\sim}{W}\underset{\sim}{B}^2\underset{\sim}{A}\underset{\sim}{B},\quad \mathrm{tr}\,\underset{\sim}{W}\underset{\sim}{A}^2\underset{\sim}{B}^2\underset{\sim}{A},\quad \mathrm{tr}\,\underset{\sim}{W}\underset{\sim}{B}^2\underset{\sim}{A}^2\underset{\sim}{B}$$

with coefficients I_α which are invariants of $\underset{\sim}{A}$ and $\underset{\sim}{B}$. Hence $\underset{\sim}{X}$ has the form

$$\underset{\sim}{X} = I_1(\underset{\sim}{A}\underset{\sim}{B}-\underset{\sim}{B}\underset{\sim}{A}) + I_2(\underset{\sim}{A}^2\underset{\sim}{B}-\underset{\sim}{B}\underset{\sim}{A}^2) + I_3(\underset{\sim}{B}^2\underset{\sim}{A}-\underset{\sim}{A}\underset{\sim}{B}^2) + I_4(\underset{\sim}{A}^2\underset{\sim}{B}^2-\underset{\sim}{B}^2\underset{\sim}{A}^2)$$
$$+ I_5(\underset{\sim}{A}^2\underset{\sim}{B}\underset{\sim}{A}-\underset{\sim}{A}\underset{\sim}{B}\underset{\sim}{A}^2) + I_6(\underset{\sim}{B}^2\underset{\sim}{A}\underset{\sim}{B}-\underset{\sim}{B}\underset{\sim}{A}\underset{\sim}{B}^2) + I_7(\underset{\sim}{A}^2\underset{\sim}{B}^2\underset{\sim}{A}-\underset{\sim}{A}\underset{\sim}{B}^2\underset{\sim}{A}^2)$$
$$+ I_8(\underset{\sim}{B}^2\underset{\sim}{A}^2\underset{\sim}{B}-\underset{\sim}{B}\underset{\sim}{A}^2\underset{\sim}{B}^2) \; . \qquad (66)$$

To obtain a vector function $\underset{\sim}{u}$ of $\underset{\sim}{A}$ and $\underset{\sim}{B}$ we simply associate $\underset{\sim}{u}$ with the anti-symmetric tensor $\underset{\sim}{X}$, and set

$$u_i = \tfrac{1}{2}e_{ijk}X_{jk} \; .$$

Several other examples of symmetric tensor, anti-symmetric tensor and vector functions of symmetric tensors, anti-symmetric tensors and vectors, under both the full and the proper orthogonal groups, are given in [6].

REFERENCES

1. GRACE, J.H. and YOUNG, A., *The Algebra of Invariants*, Cambridge University Press, 1903
2. ELLIOTT, E.B., *Algebra of Quantics*, Oxford University Press, 1913
3. TURNBULL, H.W., *The Theory of Determinants, Matrices and Invariants*, 3rd edition, Dover Publications, 1960
4. WEYL, H., *The Classical Groups*, Princeton University Press, 1939
5. GUREVICH, G.B., *Foundations of the Theory of Algebraic Invariants* (English translation), Noordhoff, 1964
6. SPENCER, A.J.M., Theory of Invariants, in: *Continuum Physics, Vol.I* (ed. A.C. Eringen), Academic Press, 1971: 239-353
7. GREEN, A.E. and ADKINS, J.E., *Large Elastic Deformations*, Oxford University Press, 1960
8. SMITH, G.F. and RIVLIN, R.S., The anisotropic tensors, *Q.Appl.Math.* 15 (1957): 309-314
9. RIVLIN, R.S., Further remarks on the stress-deformation relations for isotropic materials, *J.Rat.Mech.Anal.* 4 (1955): 681-701
10. SMITH, G.F., On the generation of integrity bases, *Atti Accad.Naz. Lincei*, Ser.8,9 (1968): 51-101
11. SMITH, G.F., The generation of integrity bases, in: *Non-linear Continuum Theories in Mechanics and Physics and their Applications* (Coordinator R. S. Rivlin), Edizioni Cremonese, Roma, 1970: 311-351
12. SPENCER, A.J.M. and RIVLIN, R.S., Isotropic integrity bases for vectors and second-order tensors, Part I, *Arch.Rat.Mech.Anal.* 9 (1962): 45-63

13. SPENCER, A.J.M., Isotropic integrity bases for vectors and second-order tensors, Part II, *Arch.Rat.Mech.Anal.* 18 (1965): 51-82

14. RIVLIN, R.S. and SMITH, G.F., On identities for 3 × 3 matrices, *Rend. di Matematica Serie VI*, 8 (1975): 345-353

15. SMITH, G.F., On isotropic integrity bases, *Arch.Rat.Mech.Anal.* 18 (1965): 282-292

16. PIPKIN, A.C. and RIVLIN, R.S., The formulation of constitutive equations in continuum physics I, *Arch.Rat.Mech.Anal.* 4 (1969): 129-144

17. SPENCER, A.J.M. and RIVLIN, R.S., The theory of matrix polynomials and its application to the mechanics of isotropic continua, *Arch.Rat. Mech.Anal.* 2 (1959): 309-336

18. SPENCER, A.J.M. and RIVLIN, R.S., Further results in the theory of matrix polynomials, *Arch.Rat.Mech.Anal.* 4 (1960): 214-230

19. SPENCER, A.J.M., The invariants of six symmetric 3 × 3 matrices, *Arch. Rat.Mech.Anal.* 7 (1961): 64-77

Chapter 9

ANISOTROPIC INVARIANTS AND ADDITIONAL RESULTS FOR INVARIANT AND TENSOR REPRESENTATIONS

A.J.M. Spencer
The University of Nottingham, England

1. TRANSVERSE ISOTROPY

The transformation groups which characterize transverse isotropy were listed in *(c)* of Section 3 of Chapter 8. For simplicity, we consider Case *(ii)*, which is invariance under the group generated by the rotations $\underset{\sim}{M}_\theta$ and the reflection $\underset{\sim}{R}_3$. If only second-order tensors are considered, it is not necessary to distinguish between the five cases listed under *(c)* of Section 3 of Chapter 8.

The isotropic tensors δ_{ij} and e_{ijk} played an important part in establishing isotropic integrity bases. Smith and Rivlin [1] constructed the corresponding *anisotropic tensors* for other groups of transformations. For transverse isotropy $\underset{\sim}{\alpha}$ satisfies (14) of Chapter 8, but now $\underset{\sim}{M}$ is not a general orthogonal matrix, but is $\underset{\sim}{R}_3$ or any matrix $\underset{\sim}{M}_\theta$ which represents a rotation through θ about the axis $\underset{\sim}{v}_1$. By the same argument as was used in Section 3 of Chapter 8, the tensor $\underset{\sim}{\alpha}$ is given by (16) of Chapter 8, where now J is a transversely isotropic invariant of the vectors $\underset{\sim}{u}_p$, and is multilinear in their components. The transversely isotropic integrity basis for a set of vectors is known; it is essentially the integrity basis for two-dimensional vectors under rotations and reflections in two dimensions, and consists of the scalar products $\underset{\sim}{u}_P . \underset{\sim}{u}_Q$ together with the resolved components $\underset{\sim}{v}_1 . \underset{\sim}{u}_P$ of the vectors $\underset{\sim}{u}_P$ in the direction $\underset{\sim}{v}_1$. Hence a typical term of J is

$$(u_i^{(A)} u_i^{(B)})(u_j^{(C)} u_j^{(D)}) \ldots (v_r^{(1)} u_r^{(G)})(v_s^{(1)} u_s^{(H)}) \ldots$$

and it follows that the components of $\underset{\sim}{\alpha}$ are of the form

$$\alpha_{i_1 i_2 \ldots i_p} = \sum_n \beta_n \delta_{k_1 \ell_1} \delta_{k_2 \ell_2} \ldots \delta_{k_q \ell_q} v_{m_1}^{(1)} v_{m_2}^{(1)} \ldots v_{m_r}^{(1)}$$

where β_n are numerical coefficients and $k_1, \ell_1, k_2, \ell_2, \ldots, k_q, \ell_q, m_1, m_2, \ldots, m_r$ are permutations of $i_1, i_2, \ldots, i_p$.

Now let I be a homogeneous polynomial transversely isotropic invariant of ν vectors $\underset{\sim}{u}_R$ and λ tensors $\underset{\sim}{P}_S$, so that I has the form (19) of Chapter 8. Then by arguments similar to those of Section 6 of Chapter 8 it follows that the βs are now components of a transversely isotropic tensor. Consequently I can be expressed as a polynomial in expressions of the following types:

$$\begin{array}{llll} (a) & \pi_{ii}^{(1)}\ ; & (b) & u_i^{(R_1)} \pi_{ij}^{(2)} u_j^{(R_2)}\ ; \\ (c) & v_i^{(1)} \pi_{ij}^{(3)} u_j^{(R_3)}\ ; & (d) & v_i^{(1)} \pi_{ij}^{(4)} v_j^{(1)}\ ; \end{array} \qquad (1)$$

where $\pi_{ij}^{(\alpha)}$ are expressions of the form (23) of Chapter 8. However, (1) simply states that I is a polynomial in the members of an isotropic integrity basis for $\underset{\sim}{u}_R$, $\underset{\sim}{P}_S$ and $\underset{\sim}{v}_1$; thus an isotropic integrity basis for $\underset{\sim}{u}_R$, $\underset{\sim}{P}_S$ and $\underset{\sim}{v}_1$ is a transversely isotropic intergrity basis for $\underset{\sim}{u}_R$ and $\underset{\sim}{P}_S$. It must be noted, however, that because of the possible existence of syzygies, and because $\underset{\sim}{v}_1$ is required to be a unit vector, an irreducible isotropic integrity basis for $\underset{\sim}{u}_R$, $\underset{\sim}{P}_S$ and $\underset{\sim}{v}_1$ is not necessarily irreducible when it is regarded as a transversely isotropic integrity basis.

As an example we determine the transversely isotropic integrity basis for a single symmetric tensor $\underset{\sim}{A}$. This is the isotropic integrity basis for the tensor $\underset{\sim}{A}$ and the vector $\underset{\sim}{v}_1$, namely

$$\mathrm{tr}\,\underset{\sim}{A}, \quad \mathrm{tr}\,\underset{\sim}{A}^2, \quad \mathrm{tr}\,\underset{\sim}{A}^3, \quad \underset{\sim}{v}_1\underset{\sim}{A}\underset{\sim}{v}_1, \quad \underset{\sim}{v}_1\underset{\sim}{A}^2\underset{\sim}{v}_1\,, \tag{2}$$

and this basis is irreducible.

An alternative derivation of (1) can be based on physical arguments. For definiteness, suppose I is a transversely isotropic polynomial invariant of a tensor $\underset{\sim}{A}$. Then I must be a polynomial in the components of $\underset{\sim}{A}$ and $\underset{\sim}{v}_1$. But I must be unchanged if it is referred to a new coordinate system $\bar{\underset{\sim}{x}} = \underset{\sim}{M}\underset{\sim}{x}$. Thus

$$I(\underset{\sim}{A},\underset{\sim}{v}_1) = I(\underset{\sim}{M}\underset{\sim}{A}\underset{\sim}{M}^T,\underset{\sim}{M}\underset{\sim}{v}_1)\,.$$

That is, I is an isotropic invariant of $\underset{\sim}{A}$ and $\underset{\sim}{v}_1$. This approach is developed further in Chapter 10 .

A complete list of invariants of an arbitrary number of vectors, symmetric tensors and anti-symmetric tensors, for each of the five transformation groups defined in *(c)* of Section 3 of Chapter 8, has been given by Smith [2]. His method of derivation differs from the one described here.

Tensor polynomial functions of vectors and tensors, which are form-invariant under the transformations which describe transverse isotropy, may be derived by methods analogous to those used in Section 11 of Chapter 8 . For example, to determine such a form-invariant symmetric tensor function $\underset{\sim}{F}$ of a symmetric tensor $\underset{\sim}{A}$, we write down the most general isotropic invariant of $\underset{\sim}{A}$, $\underset{\sim}{v}_1$ and a symmetric tensor $\underset{\sim}{K}$, which is linear in $\underset{\sim}{K}$. This invariant (written as an invariant which is symmetric in K_{ij} and K_{ji}) is

$$J = I_1\,\mathrm{tr}\,\underset{\sim}{K} + I_2\,\mathrm{tr}\,\underset{\sim}{A}\underset{\sim}{K} + I_3\,\mathrm{tr}\,\underset{\sim}{A}^2\underset{\sim}{K} + I_4\underset{\sim}{v}_1\underset{\sim}{K}\underset{\sim}{v}_1$$

$$+ I_5\underset{\sim}{v}_1(\underset{\sim}{A}\underset{\sim}{K}+\underset{\sim}{K}\underset{\sim}{A})\underset{\sim}{v}_1 + I_6\underset{\sim}{v}_1(\underset{\sim}{A}^2\underset{\sim}{K}+\underset{\sim}{K}\underset{\sim}{A}^2)\underset{\sim}{v}_1\,,$$

where I_α are invariants of $\underset{\sim}{A}$ and $\underset{\sim}{v}_1$. Then $F_{ij} = \partial J/\partial K_{ij}$, which gives

$$\underset{\sim}{F} = I_1\underset{\sim}{I} + I_2\underset{\sim}{A} + I_3\underset{\sim}{A}^2 + I_4\underset{\sim}{v}_1 \otimes \underset{\sim}{v}_1$$
$$+ I_5(\underset{\sim}{v}_1 \otimes \underset{\sim}{A}\underset{\sim}{v}_1 + \underset{\sim}{v}_1\underset{\sim}{A} \otimes \underset{\sim}{v}_1) + I_6(\underset{\sim}{v}_1 \otimes \underset{\sim}{A}^2\underset{\sim}{v}_1 + \underset{\sim}{v}_1\underset{\sim}{A}^2 \otimes \underset{\sim}{v}_1) . \qquad (3)$$

2. ORTHOTROPIC SYMMETRY

The symmetry group for orthotropic symmetry with respect to three orthogonal planes normal to the three unit vectors $\underset{\sim}{v}_1$, $\underset{\sim}{v}_2$, $\underset{\sim}{v}_3$ is given in *(e)* of Section 3 of Chapter 8. By arguments similar to those used in Section 1, the anisotropic tensors for this symmetry group are generated by the tensors

$$\underset{\sim}{v}_1 \otimes \underset{\sim}{v}_1 , \qquad \underset{\sim}{v}_2 \otimes \underset{\sim}{v}_2 , \qquad \underset{\sim}{v}_3 \otimes \underset{\sim}{v}_3 . \qquad (4)$$

Since $\underset{\sim}{v}_1$, $\underset{\sim}{v}_2$ and $\underset{\sim}{v}_3$ are mutually orthogonal, we have

$$\underset{\sim}{v}_1 \otimes \underset{\sim}{v}_1 + \underset{\sim}{v}_2 \otimes \underset{\sim}{v}_2 + \underset{\sim}{v}_3 \otimes \underset{\sim}{v}_3 = \underset{\sim}{I} ,$$

and so we may discard one of the set (4), say $\underset{\sim}{v}_3 \otimes \underset{\sim}{v}_3$, in favour of the unit tensor $\underset{\sim}{I}$ and take the basic set of anisotropic tensors to be

$$\underset{\sim}{v}_1 \otimes \underset{\sim}{v}_1 , \qquad \underset{\sim}{v}_2 \otimes \underset{\sim}{v}_2 , \qquad \underset{\sim}{I} .$$

It follows by arguments analogous to those used in Section 1 that any polynomial invariant, under this transformation group, of a set of vectors $\underset{\sim}{u}_R$ and tensors $\underset{\sim}{P}_S$, is an isotropic invariant of $\underset{\sim}{u}_R$, $\underset{\sim}{P}_S$, $\underset{\sim}{v}_1$ and $\underset{\sim}{v}_2$, and of even degree in the components of $\underset{\sim}{v}_1$ and in the components of $\underset{\sim}{v}_2$. Thus in this case also the tables of isotropic invariants can be used to generate integrity bases for the symmetry group under consideration. For example, for a single symmetric tensor $\underset{\sim}{A}$, an integrity basis consists of

$$\operatorname{tr}\underset{\sim}{A}, \quad \operatorname{tr}\underset{\sim}{A}^2, \quad \operatorname{tr}\underset{\sim}{A}^3, \quad \underset{\sim}{v}_1\underset{\sim}{A}\underset{\sim}{v}_1, \quad \underset{\sim}{v}_1\underset{\sim}{A}^2\underset{\sim}{v}_1, \quad \underset{\sim}{v}_2\underset{\sim}{A}\underset{\sim}{v}_2, \quad \underset{\sim}{v}_2\underset{\sim}{A}^2\underset{\sim}{v}_2, \qquad (5)$$

together with squares and products in pairs of

$$\underset{\sim}{v}_1 \cdot \underset{\sim}{v}_2, \qquad \underset{\sim}{v}_1\underset{\sim}{A}\underset{\sim}{v}_2, \qquad \underset{\sim}{v}_1\underset{\sim}{A}^2\underset{\sim}{v}_2 .$$

However, since $\underset{\sim}{v}_1$ and $\underset{\sim}{v}_2$ are orthogonal, we have $\underset{\sim}{v}_1 \cdot \underset{\sim}{v}_2 = 0$, and it can be shown that $(\underset{\sim}{v}_1\underset{\sim}{A}\underset{\sim}{v}_2)^2$, $(\underset{\sim}{v}_1\underset{\sim}{A}^2\underset{\sim}{v}_2)^2$ and $(\underset{\sim}{v}_1\underset{\sim}{A}\underset{\sim}{v}_2)(\underset{\sim}{v}_1\underset{\sim}{A}^2\underset{\sim}{v}_2)$ are reducible, and so (5) constitutes the required integrity basis.

To obtain the most general tensor function $\underset{\sim}{F}$ of a symmetric tensor $\underset{\sim}{A}$ which is form-invariant under the transformations which characterize orthotropic symmetry, we first write down the most general isotropic invariant of $\underset{\sim}{A}$, $\underset{\sim}{v}_1 \otimes \underset{\sim}{v}_1$, $\underset{\sim}{v}_2 \otimes \underset{\sim}{v}_2$ and and arbitrary symmetric tensor $\underset{\sim}{K}$, which is linear in $\underset{\sim}{K}$. This (written as an invariant which is symmetric in K_{ij} and K_{ji}) is

$$
\begin{aligned}
J = {} & I_1 \operatorname{tr} \underset{\sim}{K} + I_2 \operatorname{tr} \underset{\sim}{A}\underset{\sim}{K} + I_3 \operatorname{tr} \underset{\sim}{A}^2\underset{\sim}{K} + I_4 \underset{\sim}{v}_1\underset{\sim}{K}\underset{\sim}{v}_1 \\
& + I_5 \underset{\sim}{v}_1(\underset{\sim}{A}\underset{\sim}{K}+\underset{\sim}{K}\underset{\sim}{A})\underset{\sim}{v}_1 + I_6 \underset{\sim}{v}_1(\underset{\sim}{A}^2\underset{\sim}{K}+\underset{\sim}{K}\underset{\sim}{A}^2)\underset{\sim}{v}_1 \\
& + I_7 \underset{\sim}{v}_2\underset{\sim}{K}\underset{\sim}{v}_2 + I_8 \underset{\sim}{v}_2(\underset{\sim}{A}\underset{\sim}{K}+\underset{\sim}{K}\underset{\sim}{A})\underset{\sim}{v}_2 + I_9 \underset{\sim}{v}_2(\underset{\sim}{A}^2\underset{\sim}{K}+\underset{\sim}{K}\underset{\sim}{A}^2)\underset{\sim}{v}_2 \,, \qquad (6)
\end{aligned}
$$

where I_1, I_2, ..., I_9 are polynomials in the invariants (5). In constructing (6) it has been taken into account that $\underset{\sim}{v}_1$ and $\underset{\sim}{v}_2$ are orthogonal unit vectors. Then, as before, $F_{ij} = \partial J/\partial K_{ij}$, and hence

$$
\begin{aligned}
\underset{\sim}{F} = {} & I_1\underset{\sim}{I} + I_2\underset{\sim}{A} + I_3\underset{\sim}{A}^2 + I_4\underset{\sim}{v}_1 \otimes \underset{\sim}{v}_1 + I_5(\underset{\sim}{v}_1 \otimes \underset{\sim}{A}\underset{\sim}{v}_1 + \underset{\sim}{v}_1\underset{\sim}{A} \otimes \underset{\sim}{v}_1) \\
& + I_6(\underset{\sim}{v}_1 \otimes \underset{\sim}{A}^2\underset{\sim}{v}_1 + \underset{\sim}{v}_1\underset{\sim}{A}^2 \otimes \underset{\sim}{v}_1) + I_7\underset{\sim}{v}_2 \otimes \underset{\sim}{v}_2 \\
& + I_8(\underset{\sim}{v}_2 \otimes \underset{\sim}{A}\underset{\sim}{v}_2 + \underset{\sim}{v}_2\underset{\sim}{A} \otimes \underset{\sim}{v}_2) + I_9(\underset{\sim}{v}_2 \otimes \underset{\sim}{A}^2\underset{\sim}{v}_2 + \underset{\sim}{v}_2\underset{\sim}{A}^2 \otimes \underset{\sim}{v}_2) \,. \qquad (7)
\end{aligned}
$$

Some other examples of the construction of tensor polynomial functions appropriate for transversely isotropic or orthotropic symmetry will be described in Chapter 10.

3. CRYSTAL SYMMETRIES

The same general procedures can be used to determine integrity bases under the symmetries which correspond to the thirty-two crystal classes. Extensive results have been derived by G. F. Smith, Rivlin and M. M. Smith. The number of cases which require attention for a full treatment is very large, and in many of these cases the results are very complicated, so no attempt is made to summarize them here. References to results obtained prior to 1968 are given in Spencer [3].

For a given crystal class, the first task is to establish an integrity basis for an arbitrary number of vectors. This was achieved for all the crystal classes by Smith and Rivlin [4, 5]. To do this it is necessary to consider polynomials, in the components of the vectors, which remain invariant under the transformations of the group under consideration. The process is very lengthy, and involves the use of several theorems in the theory of symmetric polynomials.

Having established integrity bases for vectors, it is then possible to construct the anisotropic tensors for each group [1, 4, 5]. Knowledge of the anisotropic tensors then enables general forms of invariants of vectors and tensors to be written down. In special cases, it is then possible to reduce these to minimal integrity bases. Full results are known for vectors [4, 5], a single symmetric tensor (Smith and Rivlin [6]), a single symmetric tensor and a single vector (Smith, Smith & Rivlin [7]), and for N symmetric tensors (Smith and Kiral [8]). Further results have been derived by Kiral and Smith [9].

In principle, a knowledge of the appropriate integrity bases leads to the determination of form invariant tensor functions by the procedure described in Section 11 of Chapter 8 and also used in Sections 1 and 2 of this chapter.

4. TENSORS OF THIRD AND HIGHER ORDER

Relatively little is known about integrity bases for tensors of order higher than two, for any of the transformation groups of interest in continuum mechanics. An exception is the set of results presented by Betten in Chapter 11. Apart from these, there appears to be no thorough systematic treatment, although some isolated results appear in the literature.

For definiteness let us consider the full orthogonal group, for which all isotropic tensors can be expressed in terms of the Kronecker delta δ_{ij}. By procedures analogous to those described in Section 6 of Chapter 8, all invariants can be obtained by forming inner products of Kronecker deltas with the tensors under consideration. For example, for a fourth-order tensor $\underset{\sim}{A}$, with components $A_{ijk\ell}$, we may form the following invariants of degree one in the components of $\underset{\sim}{A}$:

$$\delta_{ij}\delta_{k\ell}A_{ijk\ell} = A_{iikk}\,, \qquad \delta_{ik}\delta_{j\ell}A_{ijk\ell} = A_{ijij}\,, \qquad \delta_{i\ell}\delta_{jk}A_{ijk\ell} = A_{ijji}$$

Thus invariants are (in this case) obtained by contracting indices in pairs until a scalar is obtained. Similarly, the invariants which are bilinear in the components of two third-order tensors $\underset{\sim}{A}$ and $\underset{\sim}{B}$, with components A_{ijk} and B_{ijk}, are

$$\begin{array}{lll}
A_{iij}B_{jkk}\,, & A_{iij}B_{kjk}\,, & A_{iij}B_{kkj}\,, \\
A_{iji}B_{jkk}\,, & A_{iji}B_{kjk}\,, & A_{iji}B_{kkj}\,, \\
A_{ijj}B_{ikk}\,, & A_{ijk}B_{ijk}\,, & A_{ijk}B_{ikj}\,, \\
A_{ijj}B_{kik}\,, & A_{ijk}B_{jik}\,, & A_{ijk}B_{kij}\,, \\
A_{ijj}B_{kki}\,, & A_{ijk}B_{jki}\,, & A_{ijk}B_{kji}\,.
\end{array}$$

Clearly, the existence of any index symmetries will reduce the number of independent invariants.

As in the case of vectors and second-order tensors, the operation of inner multiplication by isotropic tensors yields an infinite number of invariants. By Hilbert's theorem, only a finite number of invariants can be independent. Presumably the reduction to a finite and even a minimal set of polynomial invariants may in principle be effected by application of identities of the form

$$\begin{vmatrix} \delta_{ip} & \delta_{iq} & \delta_{ir} & \delta_{is} \\ \delta_{jp} & \delta_{jq} & \delta_{jr} & \delta_{js} \\ \delta_{kp} & \delta_{kq} & \delta_{kr} & \delta_{ks} \\ \delta_{\ell p} & \delta_{\ell q} & \delta_{\ell r} & \delta_{\ell s} \end{vmatrix} = 0 ,$$

but the details are certain to be algebraically complicated.

In a similar manner, infinite integrity bases can be generated by using the appropriate isotropic or anisotropic tensors, for invariance under any other group of orthogonal transformations. The extension to the construction of form-invariant tensor functions is also straightforward in principle, but sure to be complicated in practice.

5. REDUCTION OF A GENERAL TENSOR TO A SUM OF TRACELESS SYMMETRIC TENSORS

In constructing integrity bases and in related problems, it is often desirable to express tensor quantities in their simplest form, that is with the smallest possible number of independent components. Thus it is best to decompose a general second-order tensor, with 9 components, into a symmetric tensor (6 components) and an anti-symmetric tensor (3 components). Since the trace of a symmetric tensor is a scalar, a symmetric tensor can be reduced to a scalar and a traceless (deviatoric) tensor with 5 independent components. Also, an anti-symmetric tensor can be associated with a vector. Thus, for the components P_{ij} of a general tensor $\underset{\sim}{P}$, we may write

$$P_{ij} = A_{ij} + e_{ijk}u_k + w\delta_{ij} ,$$

where

$$A_{ij} = \tfrac{1}{2}(P_{ij}+P_{ji}) - \tfrac{1}{3}P_{kk}\delta_{ij} , \qquad \text{(5 independent components)},$$

$$u_k = \tfrac{1}{2}e_{rsk}P_{rs} , \qquad \text{(3 components)},$$

$$w = \tfrac{1}{3}P_{kk} , \qquad \text{(1 scalar)},$$

which express $\underset{\sim}{P}$ in terms of the traceless symmetric tensor $\underset{\sim}{A}$, the vector $\underset{\sim}{u}$, and the scalar w.

This process of reducing tensors to their simplest forms is even more desirable when dealing with tensors of higher order. An elementary method of reducing tensors of any order to sums of traceless symmetric tensors is described in Spencer [10]. A related problem is treated in a more general way by Hannabuss [11].

We first express the tensor as a sum of fully symmetric tensors, i.e. tensors which are symmetric with respect to interchange of any pair of indices. Vectors and scalars are regarded as fully symmetric tensors of orders one and zero. We use bracketed indices to denote the index symmetrization operation, so that, for example,

$$A_{(ij)k} = \tfrac{1}{2}(A_{ijk}+A_{jik}) ,$$

$$A_{(ijk)} = \tfrac{1}{6}(A_{ijk}+A_{ikj}+A_{jik}+A_{jki}+A_{kij}+A_{kji}) .$$

Then $A_{(ij)k}$ is symmetric for interchanges of (ij) and $A_{(ijk)}$ is symmetric for any permutation of (ijk).

Suppose $A_{ij\ldots k\ell pq\ldots s}$ are components of a tensor of order n, and that ℓ is the (r-1)*th* index and p the r*th* index. Then

$$A_{(ij\ldots k\ell)pq\ldots s} = rA_{(ij\ldots k\ell p)q\ldots s} - A_{(j\ldots k\ell p)iq\ldots s}$$
$$- A_{(i\ldots k\ell p)jq\ldots s} - \ldots - A_{(ij\ldots kp)\ell q\ldots s} .$$

Now add $(r-1)A_{(ij\ldots k\ell)pq\ldots s}$ to each side, and note that for any pair of indices t and u

$$A_{i\ldots t\ldots u\ldots s} - A_{i\ldots u\ldots t\ldots s} = e_{tuv}e_{xyv}A_{i\ldots x\ldots y\ldots s} .$$

We also denote

$$B_{ij\ldots kq\ldots sw} = e_{uvw}A_{(uij\ldots k)vq\ldots s} .$$

Then $B_{ij\ldots w}$ are components of a tensor $\underset{\sim}{B}$ of order n-1 which is symmetric in its first (r-2) indices. Hence

$$rA_{(ij\ldots k\ell)pq\ldots s} = rA_{(ij\ldots k\ell p)q\ldots s} + e_{ipw}B_{j\ldots k\ell q\ldots sw}$$

$$+ e_{jpw}B_{i\ldots k\ell q\ldots sw} + \ldots + e_{\ell pw}B_{ij\ldots q\ldots sw}$$

Thus the tensor $\underset{\sim}{A}$ symmetrized in its first r-1 indices is the sum of $\underset{\sim}{A}$ symmetrized in its first r indices, and tensors formed by contracting $\underset{\sim}{B}$ with the third order alternating tensor. By repeated application of this process, $\underset{\sim}{A}$ can be expressed entirely in terms of symmetric tensors of order n or less, and isotropic tensors.

For example, the components A_{ijk} of a third-order tensor can be expressed in this way as

$$12A_{ijk} = 12A_{(ijk)} + 4e_{ikp}C_{(jp)} + 4e_{jkp}C_{(ip)} + 6e_{ijp}B_{(kp)}$$

$$+ 3(\delta_{jk}u_i - \delta_{ik}u_j) + \delta_{jk}v_i + \delta_{ik}v_j - 2\delta_{ij}v_k ,$$

where

$$B_{kp} = e_{prs}A_{rsk} , \qquad C_{ip} = e_{rsp}A_{(ri)s} \qquad (C_{ii} = 0) ,$$

$$u_p = e_{rsp}B_{rs} = A_{prr} - A_{rpr} , \qquad v_p = e_{rsp}C_{rs} = A_{(pr)r} - A_{rrp} .$$

In the most general case, A_{ijk} has 27 independent components, $A_{(ijk)}$ has 10, $B_{(ij)}$ has 6, $C_{(ij)}$ has 5, u_i has 3, v_i has 3. The decomposition is not unique.

A symmetric tensor of any order is traceless if the value zero is obtained as a result of contracting on any pair of indices. Any symmetric tensor may be decomposed into traceless symmetric tensors. For example, for a third-order symmetric tensor $\underset{\sim}{A}$

$$A_{ijk} = A'_{ijk} + \tfrac{1}{5}(A_{ppi}\delta_{jk} + A_{ppj}\delta_{ki} + A_{ppk}\delta_{ij}) .$$

This defines the tensor $\underset{\sim}{A}'$ which has the property $A'_{iik} = 0$. Then $\underset{\sim}{A}'$ has seven independent components. In general a traceless symmetric tensor of order n has 2n+1 independent components.

These decompositions can be extended to tensors of any order. Details are given in [10].

6. LINEARLY INDEPENDENT INVARIANTS - GENERATING FUNCTIONS

The procedure followed in Sections 4-10 of Chapter 8 was to construct integrity bases which usually contain redundant elements, and then by examination of particular cases to reduce as many as possible of the elements. In this way we hope to arrive at a minimal basis but (except in the simplest cases) the procedure gives no assurance that the basis is minimal. For the integrity bases derived in these lectures, minimality can be established by methods of the theory of group representations. Space does not permit a detailed description of these methods, but we state some of the principal results. For details we refer to Smith [12, 13].

Consider first a single traceless symmetric tensor $\tilde{A}$ of order n under the proper orthogonal transformation group in three dimensions. The tensor has $2n+1 = p$ independent components. These components in a coordinate system $\tilde{x}$, arranged in some definite order, are regarded as elements of a p-dimensional vector $\tilde{\xi}_n$.

Suppose a proper orthogonal transformation of coordinates

$$\bar{\tilde{x}} = \tilde{M}\tilde{x}$$

induces in the vector $\tilde{\xi}_n$ the transformation

$$\bar{\tilde{\xi}}_n = \tilde{L}_n \tilde{\xi}_n ,$$

where $\tilde{L}_1 = \tilde{M}$ and $\tilde{L}_n$ is a $p \times p$ matrix.

Now form all the distinct products of degree m in the components of $\tilde{A}$, and arrange these in some definite order as components of a vector $\tilde{\xi}_n^{(m)}$. The coordinate transformation induces a transformation of $\tilde{\xi}_n^{(m)}$, which is of the form

$$\bar{\xi}_n^{(m)} = L_n^{(m)} \xi_n^{(m)} .$$

Then it is known that the number P(m) of linearly independent invariants of degree m in the components of A is given by

$$P(m) = \frac{1}{2\pi} \int_0^{2\pi} \operatorname{tr} L_n^{(m)} (1 - \cos\theta)\, d\theta ,$$

where θ is the rotation angle represented by the matrix M.

It is also known that $\operatorname{tr} L_n^{(m)}$ is the coefficient of a^m in the formal expansion in positive powers of a of $\phi(a, e^{i\theta}, n)$, where

$$\phi(a,t,n) = \{(1-at^{-n})(1-at^{-n+1})\ldots(1-at^{-1})(1-a)(1-at)\ldots$$
$$\ldots(1-at^{n-1})(1-at^{n})\}^{-1}$$
$$= \prod_{s=-n}^{n} \{(1-at^{s})^{-1}\} .$$

Hence

$$\sum_{m=0}^{\infty} \operatorname{tr} L_n^{(m)} a^m = \phi(a,e^{i\theta},n) .$$

It follows that

$$\sum_{m=0}^{\infty} P(m) a^m = \frac{1}{2\pi} \int_0^{2\pi} \phi(a,e^{i\theta},n)(1 - \cos\theta)\, d\theta$$
$$= \frac{1}{2\pi} \int_0^{2\pi} \prod_{s=-n}^{n} \{(1-ae^{is\theta})^{-1}\}(1 - \cos\theta)\, d\theta ,$$

and hence the integral on the right-hand side above is a generating function for the number P(m) of linearly independent invariants of A which have degree m.

These results extend to sets of tensors. If $A_1, A_2, \ldots, A_\mu$ are a set of traceless symmetric tensors of orders $n_1, n_2, \ldots, n_\mu$, then the number of their linearly independent invariants of degree m_1 in the components of

$\underset{\sim}{A}_1$, m_2 in the components of $\underset{\sim}{A}_2$, and so on, is

$$P(m_1,m_2,\ldots,m_\mu) = \frac{1}{2\pi}\int_0^{2\pi} \operatorname{tr} \underset{\sim}{L}_{n_1}^{(m_1)} \operatorname{tr} \underset{\sim}{L}_{n_2}^{(m_2)} \ldots \operatorname{tr} \underset{\sim}{L}_{n_\mu}^{(m_\mu)} (1-\cos\theta)\, d\theta ,$$

and it follows that

$$\sum_{m_r=0}^{\infty} P(m_1,m_2,\ldots,m_\mu)\, a_1^{m_1} a_2^{m_2} \ldots a_\mu^{m_\mu}$$

$$= \frac{1}{2\pi}\int_0^{2\pi} \prod_{r=1}^{\mu} \{\phi(a_r, e^{i\theta}, n_r)\}(1-\cos\theta)\, d\theta$$

$$= \frac{1}{2\pi}\int_0^{2\pi} \prod_{r=1}^{\mu} \left[\prod_{s=-n}^{n} \{(1-a_r e^{is\theta})^{-1}\}\right](1-\cos\theta)\, d\theta .$$

Some examples of these generating functions are as follows:

(a) *One vector* $\underset{\sim}{u}$. Let $\underset{\sim}{A} = \underset{\sim}{u}$. Then

$$\sum_0^\infty P(m)\, u^m = \frac{1}{2\pi}\int_0^{2\pi} \frac{(1-\cos\theta)\, d\theta}{(1-ue^{i\theta})(1-u)(1-ue^{-i\theta})} .$$

(b) *One traceless symmetric second-order tensor* $\underset{\sim}{A}$. Then

$$\sum_0^\infty P(m)\, a^m = \frac{1}{2\pi}\int_0^{2\pi} \frac{(1-\cos\theta)\, d\theta}{(1-ae^{2i\theta})(1-ae^{i\theta})(1-a)(1-ae^{-i\theta})(1-ae^{-2i\theta})} .$$

(c) *Two vectors* $\underset{\sim}{u}$, $\underset{\sim}{v}$. Let $\underset{\sim}{A}_1 = \underset{\sim}{u}$, $\underset{\sim}{A}_2 = \underset{\sim}{v}$. Then

$$\sum_0^\infty P(m_1,m_2)\, u^{m_1} v^{m_2} = \frac{1}{2\pi}\int_0^{2\pi} \frac{(1-\cos\theta)\, d\theta}{(1-ue^{i\theta})(1-u)(1-ue^{-i\theta})(1-ve^{i\theta})(1-v)(1-ve^{-i\theta})} .$$

(d) A vector $\underset{\sim}{u} = \underset{\sim}{A}_1$ *and a traceless symmetric tensor* $\underset{\sim}{A} = \underset{\sim}{A}_2$. Then

$$\sum_{0}^{\infty} P(m_1,m_2)u^{m_1}a^{m_2}$$

$$= \frac{1}{2\pi}\int_0^{2\pi} \frac{(1-\cos\theta)\,d\theta}{(1-ue^{i\theta})(1-u)(1-ue^{-i\theta})(1-ae^{2i\theta})(1-ae^{i\theta})(1-a)(1-ae^{-i\theta})(1-ae^{-2i\theta})}$$

Generating functions are particularly useful because they indicate the existing of syzygies. For example, for three vectors in two dimensions, the generating function can be evaluated as

$$\frac{1-x^2y^2z^2}{(1-x^2)(1-y^2)(1-z^2)(1-zy)(1-xz)(1-yz)} .$$

The denominator indicates the existence of invariants of degrees (2,0,0), (0,2,0), (0,0,2), (1,1,0), (1,0,1) and (0,1,1) (these are the six scalar products). The numerator reveals a syzygy of degree (2,2,2), which is in fact (for vectors in two dimensions)

$$\begin{vmatrix} \underset{\sim}{u}.\underset{\sim}{u} & \underset{\sim}{u}.\underset{\sim}{v} & \underset{\sim}{u}.\underset{\sim}{w} \\ \underset{\sim}{v}.\underset{\sim}{u} & \underset{\sim}{v}.\underset{\sim}{v} & \underset{\sim}{v}.\underset{\sim}{w} \\ \underset{\sim}{w}.\underset{\sim}{u} & \underset{\sim}{w}.\underset{\sim}{v} & \underset{\sim}{w}.\underset{\sim}{w} \end{vmatrix} = 0 .$$

The natural method of evaluating the integral generating functions is by contour integration, but except in the simplest cases the evaluation and summation of the residues is very tedious. Many further results are available because of an analogy between the linear group in two dimensions and the orthogonal group in three dimensions. Classical invariant theory yields generating functions for numbers of invariants under the two dimensional linear group and many cases were worked out, at great length, by Sylvester and Franklin (for references, see Spencer [14]). These generating functions are algebraic, and involve no integration, but the effort needed to apply them is similar to that required to use the integral generating function. However, the known results can be interpreted in terms of the three-dimensional orthogonal groups. The connection between the types of generating functions is explored in [14].

7. MINIMALITY OF AN INTEGRITY BASIS

For a given set of vectors and tensors, and for invariants of specified degrees in these vectors and tensors, let

k = number of linearly independent invariants;

β = number of invariants in a minimal integrity basis;

β' = number of invariants in an integrity basis under consideration;

θ = number of invariants which can be contructed from invariants of lower degree from a minimal basis;

η = number of syzygies between the θ invariants;

η' = number of known syzygies between the θ invariants.

Then $\beta \leqslant \beta'$, $\eta' \leqslant \eta$, $k = \beta + \theta - \eta$. Hence

$$\beta' \geqslant \beta = k - \theta + \eta \geqslant k - \theta + \eta' .$$

Now β' and η' are known, k can be calculated by the methods outlined in Section 6 of Chapter 9 and, working by ascending degrees, θ can be calculated. Thus if

$$\beta' = k - \theta + \eta'$$

then $\beta' = \beta$ and $\eta' = \eta$, and the minimality of the basis is proved for invariants of the specified degrees. The process is applied successively to invariants of increasing degrees. For a finite basis the process eventually terminates.

Minimality of the integrity basis listed in Table 1 has been proved by Smith [12, 13] in this manner.

REFERENCES

1. SMITH, G.F. and RIVLIN, R.S., The anisotropic tensors, *Q.Appl.Math.*, 15 (1957): 309-314

2. SMITH, G.F., On transversely isotropic functions of vectors, symmetric

second-order tensors and skew-symmetric second-order tensors, *Q.Appl. Math.* 39 (1982): 509-516

3. SPENCER, A.J.M., Theory of Invariants, in *Continuum Physics, Vol.I* (ed. A.C. Eringen), Academic Press, 1971

4. SMITH, G.F. and RIVLIN, R.S., Integrity bases for vectors - the crystal classes, *Arch.Rat.Mech.Anal.* 15 (1964): 169-221

5. SMITH, G.F., Tensor and integrity bases for the gyroidal crystal classes, *Q.Appl.Math.* 25 (1967): 218-221

6. SMITH, G.F. and RIVLIN, R.S., The strain-energy function for anisotropic elastic materials, *Trans.Amer.Math.Soc.* 88 (1958): 175-193

7. SMITH, G.F., SMITH, M.M. and RIVLIN, R.S., Integrity bases for a symmetric tensor and a vector - the crystal classes, *Arch.Rat.Mech. Anal.* 12 (1963): 93-133

8. SMITH, G.F. and KIRAL, E., Integrity bases for N symmetric second-order tensors - the crystal classes, *Rend.Circ.Mat.Palermo* II, Ser.18 (1969): 5-22

9. KIRAL, E. and SMITH, G.F., On the constitutive relations for anisotropic materials - triclinic, monoclinic, rhombic, tetragonal and hexagonal crystal systems, *Int.J.Engng.Sci.* 12 (1974):471-490

10. SPENCER, A.J.M., A note on the decomposition of tensors into traceless symmetric tensors, *Int.J.Engng. Sci.* 8 (1970):489-505

11. HANNABUSS, K.C., The irreducible components of homogeneous functions and symmetric tensors, *J.Inst.Math.Applics.* 14 (1974): 83-88

12. SMITH, G.F., On isotropic integrity bases, *Arch.Rat.Mech.Anal.* 18 (1965): 282-292

13. SMITH, G.F., On the minimality of integrity bases for symmetric 3×3 matrices, *Arch.Rat.Mech.Anal.* 5 (1960): 382-389

14. SPENCER, A.J.M., On generating functions for the number of invariants of orthogonal tensors, *Mathematika* 17 (1970): 275-286

Chapter 10

KINEMATIC CONSTRAINTS, CONSTITUTIVE EQUATIONS AND FAILURE RULES FOR ANISOTROPIC MATERIALS

A.J.M. Spencer
The University of Nottingham, England

1. KINEMATIC CONSTRAINTS

It is common in many branches of continuum mechanics to treat material as though it is incompressible. Although no material is truly incompressible, there are many materials in which the ability to resist volume changes greatly exceeds the ability to resist shearing deformations; examples are liquids with low viscosity, like water, and some natural and artificial rubbers. For such materials, the assumption of incompressibility is a good approximation in many circumstances, and often greatly simplifies the solution of specific problems. It should be noted, though, that there are occasions when even a small degree of compressibility may produce a major effect; an example is the propagation of sound waves in water.

Incompressibility is an example of a *kinematic constraint*; it restricts the range of admissible deformations. Another example is the constraint of inextensibility in specified directions. Some highly anisotropic materials exhibit strong resistance to extension in particular directions, compared to their shear resistance and resistance to extension in other directions. Obvious examples are fibre composite materials composed of strong stiff aligned fibres reinforcing a relatively soft matrix. Materials of this kind may, approximately, be treated as inextensible in the fibre direction, and analysis of their behaviour is often greatly simplified by making this approximation. As in the case of incompressibility, some caution is needed, because slight inextensibility can produce large effects. However, the approximation is often useful, and results derived from it can be used as a basis on which to construct more accurate solutions. The mechanics of these ideal fibre-reinforced materials is described in [1].

We refer quantities to a fixed rectangular coordinate system. A typical particle has position vector $\underset{\sim}{X}$ and coordinates X_R in its reference configuration at time $t = 0$. At a subsequent time t the same particle occupies the position $\underset{\sim}{x}$ with coordinates x_i. The deformation is described by the dependence of $\underset{\sim}{x}$ on $\underset{\sim}{X}$ and t, thus

$$\underset{\sim}{x} = \underset{\sim}{x}(\underset{\sim}{X},t), \qquad \text{or} \qquad x_i = x_i(X_R,t) .$$

The deformation gradient tensor $\underset{\sim}{F}$ has components F_{iR}, where

$$F_{iR} = \partial x_i/\partial X_R .$$

We employ the finite strain tensors $\underset{\sim}{C}$ and $\underset{\sim}{B}$, with components C_{RS} and B_{ij} respectively, where

$$\underset{\sim}{C} = \underset{\sim}{F}^T\underset{\sim}{F}, \qquad \underset{\sim}{B} = \underset{\sim}{F}\underset{\sim}{F}^T,$$

$$C_{RS} = F_{iR}F_{iS} = \frac{\partial x_i}{\partial X_R}\frac{\partial x_i}{\partial X_S}, \qquad B_{ij} = F_{iR}F_{jR} = \frac{\partial x_i}{\partial X_R}\frac{\partial x_j}{\partial X_R}$$

and also the infinitesimal strain tensor $\underset{\sim}{E}$, with components E_{ij}, where

$$\underset{\sim}{E} = \tfrac{1}{2}(\underset{\sim}{F}+\underset{\sim}{F}^T) - \underset{\sim}{I}, \qquad E_{ij} = \tfrac{1}{2}\left(\frac{\partial x_i}{\partial X_j}+\frac{\partial x_j}{\partial X_i}\right) - \delta_{ij} .$$

The velocity $\underset{\sim}{v}$ is regarded as a function of $\underset{\sim}{x}$ and t. The rate-of-deformation tensor has components D_{ij}, where

$$D_{ij} = \frac{1}{2}\left(\frac{\partial v_i}{\partial x_j} + \frac{\partial v_j}{\partial x_i}\right) .$$

The condition that the material is incompressible is that, for all possible deformations,

$$\det \underset{\sim}{F} = 1 , \qquad \text{or} \qquad \det \underset{\sim}{B} = 1 , \qquad \text{or} \qquad \det \underset{\sim}{C} = 1 .$$

In terms of the rate-of deformation tensor, the incompressibility condition is

$$D_{ii} = \operatorname{tr} \underset{\sim}{D} = 0 .$$

Within the approximation of infinitesimal theory, the incompressibility condition is

$$E_{ii} = \operatorname{tr} \underset{\sim}{E} = 0 .$$

Let $\underset{\sim}{a}_0$ be a unit vector field in the reference configuration. A material line element with direction $\underset{\sim}{a}_0$ in the reference configuration has the direction of a unit vector $\underset{\sim}{a}$ in the deformed configuration, where

$$\lambda \underset{\sim}{a} = \underset{\sim}{F}\underset{\sim}{a}_0 , \qquad \text{or, in components,} \quad \lambda a_i = a_R^{(0)} F_{iR} = a_R^{(0)} \partial x_i / \partial X_R ,$$

where λ is the stretch of the line element. It follows that

$$\lambda^2 = a_R^{(0)} a_S^{(0)} \frac{\partial x_i}{\partial X_R} \frac{\partial x_i}{\partial X_S} = a_R^{(0)} a_S^{(0)} C_{RS} .$$

If the material is inextensible in the direction $\underset{\sim}{a}_0$ in the reference configuration, then $\lambda = 1$, and

$$a_i = a_R^{(0)} \partial x_i / \partial X_R , \qquad a_R^{(0)} a_S^{(0)} C_{RS} = 1 .$$

In terms of $\underset{\sim}{D}$, the inextensibility condition is

$$a_i a_j D_{ij} = 0 .$$

Within the approximation of infinitesimal theory, it is not necessary to distinguish $\underset{\sim}{a}$ and $\underset{\sim}{a}_0$, and the inextensibility condition is

$$a_i a_j E_{ij} = 0 .$$

The mechanical effect of a kinematic constraint is to produce a reaction stress. Thus in a constrained body the stress $\underset{\sim}{T}$ can be expressed as the sum of a reaction stress $\underset{\sim}{R}$ and an extra-stress $\underset{\sim}{S}$, so that

$$\underset{\sim}{T} = \underset{\sim}{S} + \underset{\sim}{R} , \qquad \text{or, in components,} \quad T_{ij} = S_{ij} + R_{ij} .$$

For an incompressible material the reaction is an arbitrary hydrostatic stress of the form

$$\underset{\sim}{R} = -p\underset{\sim}{I} , \qquad R_{ij} = -p\delta_{ij} .$$

The reaction to an inextensibility constraint is an arbitrary tension T in the inextensible direction (for convenience we shall in future call this direction a *fibre direction* and its trajectories *fibres*), so in this case

$$\underset{\sim}{R} = T\underset{\sim}{a} \otimes \underset{\sim}{a} , \qquad R_{ij} = Ta_i a_j .$$

The reaction stress does no work in a deformation which satisfies the constraint. The rate of working of the reaction stress is $R_{ij}D_{ij}$. In an incompressible material

$$R_{ij}D_{ij} = -p\delta_{ij}D_{ij} = -pD_{ii} = 0 ,$$

and in a material with an inextensible direction $\underset{\sim}{a}$

$$R_{ij}D_{ij} = Ta_i a_j D_{ij} = 0 .$$

If a material is subject to two or more constraints the total reaction stress is the sum of the reactions of each of the constraints. For example, in a material which is incompressible and inextensible in two directions defined by vectors $\underset{\sim}{a}$ and $\underset{\sim}{b}$,

$$\underset{\sim}{R} = -p\underset{\sim}{I} + T_a \underset{\sim}{a} \otimes \underset{\sim}{a} + T_b \underset{\sim}{b} \otimes \underset{\sim}{b} ,$$

where T_a and T_b represent arbitrary tensions.

The reaction pressure p and tension T are arbitrary in the sense that they are not given by constitutive equations but are determined by equilibrium or momentum equations and boundary conditions. In an incompressible material we may, without loss of generality, require that $S_{ii} = \mathrm{tr}\,\underset{\sim}{S} = 0$, and in a material with an inextensible direction $\underset{\sim}{a}$ that $a_i a_j S_{ij} = 0$. The remainder of the extra-stress $\underset{\sim}{S}$ requires a constitutive equation. Generally speaking, the constitutive equation becomes relatively less significant as the number of constraints increases. The extreme case is a rigid body, which is totally constrained and in which the total stress is an indeterminate reaction stress.

2. LINEAR ELASTICITY

In a linear elastic material the strain-energy W is a quadratic function of $\underset{\sim}{E}$. For a transversely isotropic material with preferred direction $\underset{\sim}{a}$ (whose sense is immaterial) we require

$$W(\underset{\sim}{E},\ \underset{\sim}{a}\otimes\underset{\sim}{a}) = W(\underset{\sim}{M}\underset{\sim}{E}\underset{\sim}{M}^T,\ \underset{\sim}{M}\underset{\sim}{a}\otimes\underset{\sim}{a}\underset{\sim}{M}^T)\ ,$$

for all orthogonal $\underset{\sim}{M}$. Hence, by the results of Section 8 of Chapter 8, W has the form

$$W = \tfrac{1}{2}\lambda(\mathrm{tr}\,\underset{\sim}{E})^2 + \mu_T\,\mathrm{tr}\,\underset{\sim}{E}^2 + \alpha(\underset{\sim}{a}\underset{\sim}{E}\underset{\sim}{a})\,\mathrm{tr}\,\underset{\sim}{E} + 2(\mu_L-\mu_T)\,\underset{\sim}{a}\underset{\sim}{E}^2\underset{\sim}{a} + \tfrac{1}{2}\beta(\underset{\sim}{a}\underset{\sim}{E}\underset{\sim}{a})^2\ ,$$

where λ, μ_T, μ_L, α and β are elastic constants. If the material is *incompressible*, then $\mathrm{tr}\,\underset{\sim}{E} = 0$, and

$$W = \mu_T\,\mathrm{tr}\,E^2 + 2(\mu_L-\mu_T)\,\underset{\sim}{a}\underset{\sim}{E}^2\underset{\sim}{a} + \tfrac{1}{2}\beta(\underset{\sim}{a}\underset{\sim}{E}\underset{\sim}{a})^2\ ,$$

and only three elastic constants μ_T, μ_L and β are required. If the material is inextensible in the direction $\underset{\sim}{a}$, then $\underset{\sim}{a}\underset{\sim}{E}\underset{\sim}{a} = 0$, and

$$W = \tfrac{1}{2}\lambda(\mathrm{tr}\,\underset{\sim}{E})^2 + \mu_T\,\mathrm{tr}\,\underset{\sim}{E}^2 + 2(\mu_L-\mu_T)\,\underset{\sim}{a}\underset{\sim}{E}^2\underset{\sim}{a}\ .$$

In this case the material is characterized by three elastic constants λ,

μ_L and μ_T. If the material is incompressible and inextensible in the fibre direction,

$$W = \mu_T \operatorname{tr} \underset{\sim}{E}^2 + 2(\mu_L - \mu_T) \underset{\sim}{a}\underset{\sim}{E}^2\underset{\sim}{a} ,$$

and only the shear moduli μ_T and μ_L are required to specify the material.

In each case the extra-stress S_{ij} is given by

$$S_{ij} = \partial W / \partial E_{ij} ,$$

so we have the following results.

Incompressible material

$$\underset{\sim}{S} = 2\mu_T \underset{\sim}{E} + \beta(\underset{\sim}{a}\underset{\sim}{E}\underset{\sim}{a})\underset{\sim}{a} \otimes \underset{\sim}{a} + 2(\mu_L - \mu_T)(\underset{\sim}{a} \otimes \underset{\sim}{a}\underset{\sim}{E} + \underset{\sim}{E}\underset{\sim}{a} \otimes \underset{\sim}{a}) .$$

Material which is inextensible in direction $\underset{\sim}{a}$

$$\underset{\sim}{S} = \lambda \underset{\sim}{I} \operatorname{tr} \underset{\sim}{E} + 2\mu_T \underset{\sim}{E} + 2(\mu_L - \mu_T)(\underset{\sim}{a} \otimes \underset{\sim}{a}\underset{\sim}{E} + \underset{\sim}{E}\underset{\sim}{a} \otimes \underset{\sim}{a}) .$$

Material which is incompressible and inextensible in direction $\underset{\sim}{a}$

$$\underset{\sim}{S} = 2\mu_T \underset{\sim}{E} + 2(\mu_L - \mu_T)(\underset{\sim}{a} \otimes \underset{\sim}{a}\underset{\sim}{E} + \underset{\sim}{E}\underset{\sim}{a} \otimes \underset{\sim}{a}) .$$

In each case the appropriate reaction stress must be added to give the total stress.

Similar results apply in the case of a material reinforced by two families of fibres, whose directions are specified by two unit vectors $\underset{\sim}{a}$ and $\underset{\sim}{b}$. In this case W is a function of

$$\operatorname{tr} \underset{\sim}{E} , \quad \operatorname{tr} \underset{\sim}{E}^2 , \quad \operatorname{tr} \underset{\sim}{E}^3 , \quad \underset{\sim}{a}\underset{\sim}{E}\underset{\sim}{a} , \quad \underset{\sim}{a}\underset{\sim}{E}^2\underset{\sim}{a} ,$$

$$\underset{\sim}{b}\underset{\sim}{E}\underset{\sim}{b} , \quad \underset{\sim}{b}\underset{\sim}{E}^2\underset{\sim}{b} , \quad (\underset{\sim}{a}.\underset{\sim}{b})^2 = \cos^2 2\phi , \quad \cos 2\phi \, \underset{\sim}{a}\underset{\sim}{E}\underset{\sim}{b} ,$$

where $\cos 2\phi$ is the angle between the two fibre directions. It is then straightforward to write down the most general function W which is a quadratic in $\underset{\sim}{E}$ formed from the above list. If the material is incompressible, then $\mathrm{tr}\,\underset{\sim}{E} = 0$, and if it is inextensible in the directions $\underset{\sim}{a}$ and $\underset{\sim}{b}$ then $\underset{\sim}{a}\underset{\sim}{E}\underset{\sim}{a} = 0$ and $\underset{\sim}{b}\underset{\sim}{E}\underset{\sim}{b} = 0$, and the expression for W is simplified accordingly. Orthotropic symmetry obtains if either *(a)* the fibres are orthogonal, so that $\underset{\sim}{a}.\underset{\sim}{b} = 0$, or *(b)* the families of fibres are mechanically equivalent, and then W is a symmetric function of $\underset{\sim}{a}$ and $\underset{\sim}{b}$.

3. FINITE ELASTICITY

For a finite elastic material, $W = W(C_{RS})$, and the extra-stress is given by

$$S_{ij} = \frac{\rho_0}{\rho} F_{iR} F_{jS} \left(\frac{\partial W}{\partial C_{RS}} + \frac{\partial W}{\partial C_{SR}} \right), \tag{1}$$

where ρ_0 and ρ denote densities in the reference and deformed configurations respectively.

For a transversely isotropic material, W is a function of the invariants of $\underset{\sim}{a}_0$ and $\underset{\sim}{C}$, namely

$$I_1 = \mathrm{tr}\,\underset{\sim}{C}, \qquad I_2 = \tfrac{1}{2}\{(\mathrm{tr}\,\underset{\sim}{C})^2 - \mathrm{tr}\,\underset{\sim}{C}^2\}, \qquad I_3 = \det \underset{\sim}{C} = (\rho_0/\rho)^2,$$

$$I_4 = \underset{\sim}{a}_0\underset{\sim}{C}\underset{\sim}{a}_0, \qquad I_5 = \underset{\sim}{a}_0\underset{\sim}{C}^2\underset{\sim}{a}_0 .$$

If the material is incompressible, then $I_3 = 1$. If it is inextensible in the initial direction $\underset{\sim}{a}_0$, then $I_4 = 1$. This leads to the following results. In each case $W_\alpha = \partial W/\partial I_\alpha$.

Incompressible material

$$\underset{\sim}{S} = 2\{W_1\underset{\sim}{B} - W_2\underset{\sim}{B}^{-1} + I_4W_4\underset{\sim}{a}\otimes\underset{\sim}{a} + I_4W_5(\underset{\sim}{a}\otimes\underset{\sim}{B}\underset{\sim}{a} + \underset{\sim}{a}\underset{\sim}{B}\otimes\underset{\sim}{a})\} .$$

Material inextensible in initial direction $\underset{\sim}{a}_0$

$$\underset{\sim}{S} = 2I_3^{-\frac{1}{2}}\{(I_2W_2 + I_3W_3)\underset{\sim}{I} + W_1\underset{\sim}{B} - I_3W_2\underset{\sim}{B}^{-1} + W_5(\underset{\sim}{a}\otimes\underset{\sim}{B}\underset{\sim}{a} + \underset{\sim}{a}\underset{\sim}{B}\otimes\underset{\sim}{a})\} .$$

Material incompressible and inextensible in initial direction $\underset{\sim}{a}_0$

$$\underset{\sim}{S} = 2\{W_1\underset{\sim}{B}-W_2\underset{\sim}{B}^{-1}+W_5(\underset{\sim}{a}\otimes\underset{\sim}{B}\underset{\sim}{a}+\underset{\sim}{a}\underset{\sim}{B}\otimes\underset{\sim}{a})\}\ .$$

For a material reinforced by two families of fibres, defined by unit vectors $\underset{\sim}{a}_0$ and $\underset{\sim}{b}_0$ in the reference configurations, W is a function of

$$I_1\ ,\qquad I_2\ ,\qquad I_3\ ,\qquad I_4\ ,\qquad I_5\ ,\qquad I_6 = \underset{\sim}{b}_0\underset{\sim}{C}\underset{\sim}{b}_0\ ,$$

$$I_7 = \underset{\sim}{b}_0\underset{\sim}{C}^2\underset{\sim}{b}_0\ ,\qquad I_8 = \cos 2\Phi\ \underset{\sim}{a}_0\underset{\sim}{C}\underset{\sim}{b}_0\qquad \text{and}\qquad \cos^2 2\Phi = \underset{\sim}{a}_0\cdot\underset{\sim}{b}_0\ .$$

If the material is incompressible, then $I_3 = 1$. If it is inextensible in the two fibre directions, then $I_4 = 1$ and $I_6 = 1$. If the two fibre families are mechanically equivalent, then W is a symmetric function of $\underset{\sim}{a}_0$ and $\underset{\sim}{b}_0$.

For example, for a material which is incompressible and inextensible in the two mechanically equivalent fibre directions,

$$\underset{\sim}{S} = 2\{W_1\underset{\sim}{B}-W_2\underset{\sim}{B}^{-1}+W_5(\underset{\sim}{a}\otimes\underset{\sim}{B}\underset{\sim}{a}+\underset{\sim}{a}\underset{\sim}{B}\otimes\underset{\sim}{a})+W_7(\underset{\sim}{b}\otimes\underset{\sim}{B}\underset{\sim}{b}+\underset{\sim}{b}\underset{\sim}{B}\otimes\underset{\sim}{b})\}\ .$$

In the case of an elastic material, the reaction stress may be derived from the strain-energy function by regarding p and T as Lagrangian multipliers. Thus for an incompressible material, with $I_3 = 1$, we replace W by $W-\frac{1}{2}p(I_3-1)$, and the appropriate results then follow from (1), with S_{ij} replaced by T_{ij}. Similarly, for a material which is inextensible in the direction $\underset{\sim}{a}_0$, we replace W by $W+\frac{1}{2}T(I_4-1)$ and proceed in the same way.

4. PLASTICITY - YIELD CONDITIONS

Most theories of plasticity assume the existence of a yield condition. We postulate a yield function $f(T_{ij})$, such that in admissible stress states $f \leqslant 0$, with $f = 0$ when deformation is taking place. In addition, f may depend on one or more parameters, which in turn may depend on the deformation history.

If the material is isotropic, then f can be expressed as a function of the stress invariants $\operatorname{tr} \underset{\sim}{T}$, $\operatorname{tr} \underset{\sim}{T}^2$ and $\operatorname{tr} \underset{\sim}{T}^3$. In isotropic metal plasticity, it is observed experimentally that for many materials yielding is effectively independent of a superposed hydrostatic pressure. This is incorporated in the theory by restricting f to depend on the deviatoric stress $\underset{\sim}{S}$, where

$$\underset{\sim}{S} = \underset{\sim}{T} - \tfrac{1}{3}\underset{\sim}{I} \operatorname{tr} \underset{\sim}{T} .$$

Then $\operatorname{tr} \underset{\sim}{S} = 0$ and f can be expressed as a function of $\operatorname{tr} \underset{\sim}{S}^2$ and $\operatorname{tr} \underset{\sim}{S}^3$. We note that $\underset{\sim}{S}$ is the extra-stress for an incompressible material.

For a material reinforced by a single family of fibres, with direction $\underset{\sim}{a}$, f is a function of the invariants of $\underset{\sim}{T}$ and $\underset{\sim}{a} \otimes \underset{\sim}{a}$. For a metal reinforced by inextensible fibres it is reasonable to expect that yielding is not affected by a superposed hydrostatic pressure, or by a superposed tension in the fibre direction, since such a tension produces no stress in the matrix. These conditions can be incorporated by assuming f to depend on $\underset{\sim}{T}$ only through the extra-stress $\underset{\sim}{S}$, where

$$\underset{\sim}{T} = \underset{\sim}{R} + \underset{\sim}{S} , \qquad \underset{\sim}{R} = -p\underset{\sim}{I} + T\underset{\sim}{a} \otimes \underset{\sim}{a} .$$

The indeterminacy in $\underset{\sim}{S}$ is removed by imposing the conditions

$$\operatorname{tr} \underset{\sim}{S} = 0 , \qquad \underset{\sim}{a}\underset{\sim}{S}\underset{\sim}{a} = 0 ,$$

and then

$$\underset{\sim}{S} = \underset{\sim}{T} - \tfrac{1}{2}(\operatorname{tr} \underset{\sim}{T} - \underset{\sim}{a}\underset{\sim}{T}\underset{\sim}{a})\underset{\sim}{I} + \tfrac{1}{2}(\operatorname{tr} \underset{\sim}{T} - 3\underset{\sim}{a}\underset{\sim}{T}\underset{\sim}{a})\underset{\sim}{a} \otimes \underset{\sim}{a} .$$

The invariants of $\underset{\sim}{S}$ and $\underset{\sim}{a} \otimes \underset{\sim}{a}$ are

$$J_1 = \tfrac{1}{2} \operatorname{tr} \underset{\sim}{S}^2 - \underset{\sim}{a}\underset{\sim}{S}^2\underset{\sim}{a} , \qquad J_2 = \underset{\sim}{a}\underset{\sim}{S}^2\underset{\sim}{a} , \qquad J_3 = \operatorname{tr} \underset{\sim}{S}^3 ,$$

and f is a function of J_1, J_2 and J_3. Two particular forms of f which have been found to give good agreement with experiment are

$$f = k_T^{-2} J_1 + k_L^{-2} J_2 - 1 , \tag{2}$$

and

$$f = (J_1 - k_T^2)(J_2 - k_L^2) \ . \qquad (3)$$

Here k_T and k_L represent shear yield stresses for shear in directions normal and parallel to $\underset{\sim}{a}$ respectively. The above forms are natural generalizations to transversely isotropic material of von Mises' and Tresca's yield conditions respectively. For a perfectly plastic material k_T and k_L are constants; for a hardening material they depend on the deformation history.

For a material reinforced by two families of fibres, characterized by unit vectors $\underset{\sim}{a}$ and $\underset{\sim}{b}$, the corresponding results are

$$\underset{\sim}{T} = \underset{\sim}{R} + \underset{\sim}{S} \ , \qquad \underset{\sim}{R} = -p\underset{\sim}{I} + T_a \underset{\sim}{a} \otimes \underset{\sim}{a} + T_b \underset{\sim}{b} \otimes \underset{\sim}{b} \ ,$$

with

$$\operatorname{tr} \underset{\sim}{S} = 0 \ , \qquad \underset{\sim}{a}\underset{\sim}{S}\underset{\sim}{a} = 0 \ , \qquad \underset{\sim}{b}\underset{\sim}{S}\underset{\sim}{b} = 0 \ .$$

Then

$$\underset{\sim}{S} = \underset{\sim}{T} + (1 + 3\cos^2 2\phi)^{-1} [\{\underset{\sim}{a}\underset{\sim}{T}\underset{\sim}{a} + \underset{\sim}{b}\underset{\sim}{T}\underset{\sim}{b} - (1 + \cos^2 2\phi) \operatorname{tr} \underset{\sim}{T}\}\underset{\sim}{I}$$

$$+ \{\operatorname{tr} \underset{\sim}{T} - (2\operatorname{cosec}^2 2\phi)\underset{\sim}{a}\underset{\sim}{T}\underset{\sim}{a} - (\operatorname{cosec}^2 2\phi - 3\cot^2 2\phi)\underset{\sim}{b}\underset{\sim}{T}\underset{\sim}{b}\}\underset{\sim}{a} \otimes \underset{\sim}{a}$$

$$+ \{\operatorname{tr} \underset{\sim}{T} - (2\operatorname{cosec}^2 2\phi)\underset{\sim}{b}\underset{\sim}{T}\underset{\sim}{b} - (\operatorname{cosec}^2 2\phi - 3\cot^2 2\phi)\underset{\sim}{a}\underset{\sim}{T}\underset{\sim}{a}\}\underset{\sim}{b} \otimes \underset{\sim}{b}] \ ,$$

where $\cos 2\phi = \underset{\sim}{a}.\underset{\sim}{b}$. Then f is a function of

$$J_1 \ , \qquad J_2 \ , \qquad J_3 \ , \qquad J_4 = \tfrac{1}{2} \operatorname{tr} \underset{\sim}{S}^2 - \underset{\sim}{b}\underset{\sim}{S}^2\underset{\sim}{b} \ ,$$

$$J_5 = \underset{\sim}{a}\underset{\sim}{S}\underset{\sim}{b} \cos 2\phi \ , \qquad J_6 = \underset{\sim}{a}\underset{\sim}{S}^2\underset{\sim}{b} \cos 2\phi \qquad \text{and} \qquad \cos^2 2\phi \ .$$

The most general yield function which is quadratic in the stress is (for the case in which the fibre families are mechanically equivalent)

$$f = \frac{J_2 + J_4}{c_1^2} + \frac{J_1}{c_2^2} + \frac{J_6}{c_3^2} - 1 \ ,$$

where c_1, c_2, c_3 have the dimensions of stress, are functions of $\cos^2 2\phi$ and, in general, depend on the deformation history.

5. PLASTICITY - FLOW RULES

A common procedure in plasticity theory is to assume that the yield function is a plastic potential for the plastic strain-rate $\underset{\sim}{D}_p$, so that

$$D^p_{ij} = \dot{\lambda}\partial f/\partial T_{ij}, \tag{4}$$

where $\dot{\lambda}$ is a scalar multiplier (not a material constant). In a rigid-plastic theory $\underset{\sim}{D}_p$ is the total strain-rate $\underset{\sim}{D}$; in an elastic-plastic theory the classical procedure is to decompose $\underset{\sim}{D}$ as

$$\underset{\sim}{D} = \underset{\sim}{D}_p + \underset{\sim}{D}_e,$$

where the elastic strain-rate $\underset{\sim}{D}_e$ is given by linear elasticity theory, as described in Section 2. For the yield function (2), equation (4) gives

$$\underset{\sim}{D}_p = \dot{\lambda}\left\{\frac{1}{k_T^2}\underset{\sim}{S} + \left(\frac{1}{k_L^2} - \frac{1}{k_T^2}\right)(\underset{\sim}{a}\otimes\underset{\sim}{a}\underset{\sim}{S}+\underset{\sim}{S}\underset{\sim}{a}\otimes\underset{\sim}{a})\right\} \tag{5}$$

and for the yield function (3)

$$\underset{\sim}{D}_p = \begin{cases} \dot{\mu}(\underset{\sim}{S}-\underset{\sim}{a}\otimes\underset{\sim}{a}\underset{\sim}{S}-\underset{\sim}{S}\underset{\sim}{a}\otimes\underset{\sim}{a}), & J_1 = k_T^2, \quad J_2 < k_L^2, \\ \dot{\nu}(\underset{\sim}{a}\otimes\underset{\sim}{a}\underset{\sim}{S}+\underset{\sim}{S}\underset{\sim}{a}\otimes\underset{\sim}{a}), & J_1 < k_T^2, \quad J_2 = k_L^2, \\ \dot{\mu}\underset{\sim}{S} + (\dot{\nu}-\dot{\mu})(\underset{\sim}{a}\otimes\underset{\sim}{a}\underset{\sim}{S}+\underset{\sim}{S}\underset{\sim}{a}\otimes\underset{\sim}{a}), & J_1 = k_T^2, \quad J_2 = k_L^2, \end{cases}$$

where $\dot{\mu}$ and $\dot{\nu}$ are scalar multipliers.

The $\underset{\sim}{D}_p$ obtained as above automatically satisfy the constraints of incompressibility and fibre inextensibility.

6. PLASTICITY - HARDENING RULES

In isotropic plasticity theory, the simplest hypothesis to describe strain-hardening or work-hardening is called isotropic hardening. Since this term is confusing when applied to anisotropic materials, we coin the term "proportional hardening" for the corresponding theory for anisotropic plasticity.

In proportional hardening the yield function retains its form as a function of the invariants J_α, but the parameters (such as k_L and k_T in (2) and (3)) which determine the yield surface depend on the deformation history. Hence, as the yield surface evolves, it expands (not necessarily uniformly in all directions) but successive yield surfaces are concentric (in isotropic hardening of an isotropic material, successive yield surfaces are similar as well as concentric, their scale being determined by the change of a single parameter, such as the shear yield stress).

For definiteness, we consider the yield function (2) and its associated flow rule. Other yield functions and flow rules may be treated in a similar way. In isotropic theory it is assumed that the yield stress is a function of an "equivalent strain" ε. By analogy, we postulate that k_T and k_L depend on parameters $\varepsilon_1, \varepsilon_2, \ldots$, such that $\dot{\varepsilon}_1, \dot{\varepsilon}_2, \ldots$ are scalar invariants of $\underset{\sim}{D}_p$ and $\underset{\sim}{a}$, with dimensions $(\text{time})^{-1}$. With the plastic incompressibility and inextensibility conditions

$$\operatorname{tr} \underset{\sim}{D}_p = 0 , \qquad \underset{\sim}{a}\underset{\sim}{D}_p\underset{\sim}{a} = 0 ,$$

it follows that there are three such independent parameters, which may be taken to be ε_T, ε_L and ε_3, where

$$\dot{\varepsilon}_T^2 = \tfrac{1}{2} \operatorname{tr} \underset{\sim}{D}_p^2 - \underset{\sim}{a}\underset{\sim}{D}_p^2\underset{\sim}{a} , \qquad \dot{\varepsilon}_L^2 = \underset{\sim}{a}\underset{\sim}{D}_p^2\underset{\sim}{a} , \qquad \dot{\varepsilon}_3^3 = \tfrac{1}{3}\left|\underset{\sim}{D}_p^3\right| . \tag{6}$$

We neglect dependence on ε_3, and assume that

$$k_T = k_T(\varepsilon_T, \varepsilon_L) , \qquad k_L = k_L(\varepsilon_T, \varepsilon_L) .$$

Then

$$\dot{k}_T = \frac{\partial k_T}{\partial \varepsilon_T}\dot{\varepsilon}_T + \frac{\partial k_T}{\partial \varepsilon_L}\dot{\varepsilon}_L , \qquad \dot{k}_L = \frac{\partial k_L}{\partial \varepsilon_T}\dot{\varepsilon}_T + \frac{\partial k_L}{\partial \varepsilon_L}\dot{\varepsilon}_L . \tag{7}$$

Now, from the flow rule (5), and equation (6),

$$\dot{\varepsilon}_T^2 = \dot{\lambda}^2 k_T^{-4} J_1 , \qquad \dot{\varepsilon}_L^2 = \dot{\lambda}^2 k_L^{-4} J_2 . \tag{8}$$

Hence, when $f = 0$, from (2) and (8)

$$\dot{\lambda}^2 = k_T^2\dot{\varepsilon}_T^2 + k_L^2\dot{\varepsilon}_L^2 .$$

Also, during plastic loading, when $\dot{f} = 0$, (2) gives

$$k_T^{-2}\dot{J}_1 + k_L^{-2}\dot{J}_2 - 2(k_T^{-3}\dot{k}_T J_1 + k_L^{-3}\dot{k}_L J_2) = 0 , \tag{9}$$

and it follows from (7) and (8) that

$$\dot{\lambda}\left\{J_1^{\frac{1}{2}}k_T^{-2}\left(k_T^{-3}J_1\frac{\partial k_T}{\partial \varepsilon_T} + k_L^{-3}J_2\frac{\partial k_L}{\partial \varepsilon_T}\right) + J_2^{\frac{1}{2}}k_L^{-2}\left(k_T^{-3}J_1\frac{\partial k_T}{\partial \varepsilon_L} + k_L^{-3}J_2\frac{\partial k_L}{\partial \varepsilon_L}\right)\right\}$$

$$= \tfrac{1}{2}(k_T^{-2}\dot{J}_1 + k_L^{-2}\dot{J}_2) .$$

This determines $\dot{\lambda}$ in terms of the current stress, the current values of k_T and k_L, the hardening parameters $\partial k_T/\partial\varepsilon_T$, $\partial k_T/\partial\varepsilon_L$, $\partial k_L/\partial\varepsilon_T$, $\partial k_L/\partial\varepsilon_L$ and the rate of increase of the stress invariants J_1 and J_2.

There are several possibilities for simplification. One is that k_T and k_L depend on the plastic work W_p. We have

$$\dot{W}_p = S_{ij}D_{ij}^p = 2\dot{\lambda}(k_T^{-2}J_1 + k_L^{-2}J_2) = 2\dot{\lambda} .$$

Therefore, when k_T and k_L are functions of W_p only,

$$\dot{k}_T = 2\dot{\lambda}\partial k_T/\partial W_p , \qquad \dot{k}_L = 2\dot{\lambda}\partial k_L/\partial W_p , \tag{10}$$

and it follows from (9) and (10) that

$$\dot{\lambda}(k_T^{-3}J_1\partial k_T/\partial W_p + k_L^{-3}J_2\partial k_L/\partial W_p) = \tfrac{1}{4}(k_T^{-2}\dot{J}_1 + k_L^{-2}\dot{J}_2) .$$

Another possibility is to assume that the mechanisms for hardening in shear in directions parallel and normal to the fibres are independent, and then

$$k_T = k_T(\varepsilon_T) , \qquad k_L = k_L(\varepsilon_L) .$$

Kinematic hardening. Kinematic strain-hardening is characterized by a tensor $\underset{\sim}{\alpha}$ termed the "back stress" or "shift tensor". Then, in the yield condition and flow rule, $\underset{\sim}{S}$ is replaced by $\underset{\sim}{S}-\underset{\sim}{\alpha}$, so that the yield surface is translated. The tensor $\underset{\sim}{\alpha}$ must satisfy the same constraints as $\underset{\sim}{S}$, namely, in the case of a single family of fibres,

$$\mathrm{tr}\,\underset{\sim}{\alpha} = 0\,, \qquad \underset{\sim}{a}\underset{\sim}{\alpha}\underset{\sim}{a} = 0\,.$$

Evolution equations are required for $\underset{\sim}{\alpha}$. A possible generalization to transverse isotropy of the hardening rule proposed by Prager [2] for isotropic materials is

$$\dot{\underset{\sim}{\alpha}} = g_1(\varepsilon_T,\varepsilon_L)\underset{\sim}{D}_p + g_2(\varepsilon_T,\varepsilon_L)(\underset{\sim}{a}\otimes\underset{\sim}{a}\underset{\sim}{D}_p + \underset{\sim}{D}_p\underset{\sim}{a}\otimes\underset{\sim}{a})\,,$$

where $\dot{\underset{\sim}{\alpha}}$ is an objective time-rate.

Elastic-plastic materials. In the small elastic-plastic deformation range composite materials show intense kinematic hardening. At the microstructure level this is due to interactions between fibre and matrix, leading to residual stress in both. This phenomenon has been termed "constraint hardening".

For small elastic-plastic deformations, it is no longer valid to assume that the yield function depends only on $\underset{\sim}{S}$, and it is necessary to allow dependence on all the stress components. A suitable set of invariants is

$$J_1\,, \qquad J_2\,, \qquad J_3\,, \qquad K_1 = \underset{\sim}{a}\underset{\sim}{T}\underset{\sim}{a}\,, \qquad K_2 = \mathrm{tr}\,\underset{\sim}{T} - \underset{\sim}{a}\underset{\sim}{T}\underset{\sim}{a}\,.$$

The most general quadratic yield function based on these is

$$f = \frac{J_1}{k_T^2} + \frac{J_2}{k_L^2} + \frac{K_1^2}{Y_1^2} + \frac{K_2^2}{Y_2^2} + \frac{K_1K_2}{Y_3^2} - 1\,.$$

Now Y_1 and Y_2 are yield stresses in tension in the fibre direction, and equibiaxial tension in planes normal to $\underset{\sim}{a}$, respectively. In practice they are much larger than the shear yield stresses k_T and k_L. To admit constraint hardening f is modified to take the form

$$f = \frac{J_1}{k_T^2} + \frac{J_2}{k_L^2} + \frac{(K_1-\alpha_1)^2}{Y_1^2} + \frac{(K_2-\alpha_2)^2}{Y_2^2} + \frac{(K_1-\alpha_1)(K_2-\alpha_2)}{Y_3^2} - 1\,.$$

The parameters α_1 and α_2 represent a back-stress and may be expected to depend on the history of $\underset{\sim}{a}\underset{\sim}{E}_p\underset{\sim}{a}$ and $\mathrm{tr}\,\underset{\sim}{E}_p$. The very high rates of hardening associated with constraint hardening mean that the fibre strain and the dilatation are restricted to small magnitudes which may be neglected in the large deformation theory.

Further discussion and applications of plasticity theory for fibre-reinforced materials, with references, is given in [1, 3, 4].

REFERENCES

1. SPENCER, A.J.M., *Deformations of Fibre-reinforced Materials*, Oxford University Press, 1972

2. PRAGER, W., A new method of analyzing stresses and strains in work-hardening plastic solids, *J.Appl.Mech.* 23 (1956): 493-496

3. SPENCER, A.J.M., The formulation of constitutive equations for anisotropic solids, in *Mechanical Behaviour of Anisotropic Solids* (ed. J. P. Boehler), Editions du CNRS, Paris and M. Nijhoff, The Hague, 1982: 2-26

4. SPENCER, A.J.M., Yield conditions and hardening rules for fibre-reinforced materials with plastic response, in *Failure Criteria of Structured Media* (ed. J. P. Boehler), A. A. Balkema, 1986

Chapter 11

INVARIANTS OF FOURTH-ORDER TENSORS

J. Betten
Technical University Aachen, F.R. Germany

1. INTRODUCTION

In solid mechanics representing scalar-valued tensor functions or second-order tensor-valued tensor functions is of major concern. For instance, the plastic potential is scalar-valued, whereas constitutive equations are tensor-valued.

Many scientists, for instance, PIPKIN, RIVLIN, SMITH, SPENCER, WINEMAN, to name just a few, have had an ongoing about the response of anisotropic materials. This discussion was based upon constitutive expressions invariant under the group of transformations which defines the symmetry of material.

In this course, however, anisotropic effects are characterized by additional material tensors of rank two and four. Therefore, scalar-valued and tensor-valued functions involving both second-order and fourth-order argument tensors are represented. Under any orthogonal transformation these functions have to fulfil the conditions of invariance and form-invariance, respectively.

Many mathematicians have studied the theory of algebraic invariants in detail. The results have ben published, for instance, by GRACE and YOUNG [1], GUREVICH [2], RIVLIN [3], SCHUR [4], WEITZENBÖCK [5], WEYL [6]. Very extensive contributions to algebraic invariant theory with respect to its application in modern continuum mechanics have been presented, for example, by SPENCER [7], TRUESDELL and NOLL [8].

Emploing the theory of invariants to the mechanics of isotropic and anisotropic materials has proven to be of practical worth [9,10,11]. In the theory of algebraic invariants the central problem is: For a given set of tensors which are not necessarily of the same order, and a given group of transformations, one must find an integrity basis, whose elements are algebraic invariants. An integrity basis is a set of polynomials, each invariant under the group of transformations, so that any polynomial function invariant under the group is expressible as a polynomial in elements of the integrity basis [3,4,6,7,12].

2. INTEGRITY BASIS FOR A SECOND-ORDER TENSOR

Let $F = F(\underset{\sim}{\sigma})$ be a scalar-valued function of a second-order tensor, for instance CAUCHY's stress tensor $\underset{\sim}{\sigma}$. This function is said to be isotropic if the condition of invariance

$$F(a_{ip}a_{jq}\sigma_{pq}) \equiv F(\sigma_{ij}) \tag{1}$$

is fulfilled under any orthogonal transformation ($a_{ik}a_{jk} = \delta_{ij}$), where the summation convention is utilized, and $\underset{\sim}{\delta}$ represents KRONECKER's tensor.

For all purposes the function F in (1) could be the plastic potential. Then, from the theory of isotropic tensor functions [7,9,13], it is evident that for an isotropic medium the plastic potential F might be expressed as a single-valued function of the irreducible basic invariants

$$S_\nu \equiv \operatorname{tr} \underset{\sim}{\sigma}^\nu \quad ; \qquad \nu = 1,2,3 \tag{2}$$

or, alternatively, of the irreducible principal invariants

$$J_1 \equiv \sigma_{ii} , \quad J_2 \equiv -\sigma_{i[i]}\sigma_{j[j]} , \quad J_3 \equiv \sigma_{i[i]}\sigma_{j[j]}\sigma_{k[k]} \qquad (3a,b,c)$$

of the stress tensor $\underset{\sim}{\sigma}$, that is,

$$F = F[S_\nu(\underset{\sim}{\sigma})] \quad \text{or} \quad F = F[J_\nu(\underset{\sim}{\sigma})]; \quad \nu = 1,2,3 , \qquad (4a,b)$$

respectively. In (3), i.e., $(-1)^{3-\nu} J_\nu \equiv \sigma_{i_1[i_1]} \cdots \sigma_{i_\nu[i_\nu]}$, the operation of alternation is used. This process is indicated by placing square brackets around those indices to which it applies, that is, the ν bracketed indices i...k are permutated in all possible ways, whereas indices which are excluded from the alternation are not bracketed. They keep their position. Thus, we obtain $\nu!$ terms. The terms corresponding to even permutations are given a plus sign, and those which correspond to odd permutations a minus sign, and then they are added and divided by $\nu!$. Comparing (2) and (3), we find the following relations:

$$J_1 = S_1 , \quad J_2 = (S_2 - S_1^2)/2, \quad J_3 = (2S_3 - 3S_2S_1 + S_1^3)/6 . \qquad (5a,b,c)$$

The irreducible (basic or principal) invariants in (2) or, alternatively, in (3) are the elements of the integrity basis for the stress tensor $\underset{\sim}{\sigma}$ under the proper orthogonal group, i.e. $|a_{ij}| = +1$, and this integrity basis also forms a functional basis. One can say, that a set of invariants constitutes a functional basis, if any invariant can be expressed as a function of them [7]. It has been shown by PIPKIN and RIVLIN [14] and PIPKIN and WINEMANN [15] that an integrity basis will also form a functional basis which may not necessarily be irreducible; then all invariants can be expressed as functions of the invariants of an integrity basis.

It is evident, that the principal values σ_I, σ_{II}, σ_{III} of the stress tensor $\underset{\sim}{\sigma}$ are independent of the choice of the coordinate system, i.e., the principal values can be regarded as invariants of the tensor. For isotropic behaviour the plastic potential F must be a symmetric scalar-valued function of the principal values, and an irreducible representation is given in terms of the elementary symmetric functions

$$\sigma_I + \sigma_{II} + \sigma_{III} = J_1, \qquad \sigma_I\sigma_{II} + \sigma_{II}\sigma_{III} + \sigma_{III}\sigma_I = -J_2 \tag{6a,b}$$

$$\sigma_I\sigma_{II}\sigma_{III} = J_3 , \tag{6c}$$

which are identical to the principal invariants (3a,b,c). Thus, the representations (4a,b) imply isotropy.

The elementary symmetric functions (6a,b,c) are the coefficients of the characteristic polynomial $P_3(\lambda)$ of the indeterminate λ,

$$P_3(\lambda) = (\lambda - \sigma_I)(\lambda - \sigma_{II})(\lambda - \sigma_{III}) \equiv \lambda^3 - J_1\lambda^2 - J_2\lambda - J_3 , \tag{7}$$

whose roots are the principal values σ_I, σ_{II}, σ_{III}. All symmetric functions can be expressed in terms of the elementary symmetric functions (6a,b,c). In other words: the elementary symmetric functions form a functional basis for the symmetric functions. For instance, the symmetric function $S_4 = \sigma_I^4 + \sigma_{II}^4 + \sigma_{III}^4$ can be expressed in the following way:

$$S_4 = 4J_1J_3 + 4J_1^2J_2 + 2J_2^2 + J_1^4 \tag{8}$$

that is, in terms of the elementary symmetric functions [16].

As we have already seen, the representations (4a,b) imply isotropy. In the anisotropic case the restriction (1) on the plastic potential F is less severe. Then, the function F is invariant under the group of transformations ($s_{ik}s_{jk} = \delta_{ij}$) associated with the symmetry properties of the material [13], where $\underset{\sim}{s}$ is a subgroup of the orthogonal group $\underset{\sim}{a}$. In other words, the symmetry properties of the material impose restrictions upon the manner in which the function F depends on the stress components [17].

For a particular crystal class [18] the plastic potential F in (1) may be represented as a polynomial which is invariant under the subgroup $\underset{\sim}{s}$ of transformations associated with the symmetry properties of the crystal class considered. The function F is then expressible as a polynomial in these invariants, which form a functional basis.

3. SIMPLIFIED CHARACTERISTIC POLYNOMIAL

Instead of the plastic potential being represented by an integrity basis under a subgroup it can be expressed in the following way:

$$F = F(\sigma_{ij}, A_{ijkl}) \tag{9}$$

where A_{ijkl} are the components of a fourth-order constitutive tensor characterizing the anisotropic properties of the material. Then, by analogy to (1), we have the invariance condition

$$F(a_{ip}a_{jq}\sigma_{pq};\ a_{ip}a_{jq}a_{kr}a_{ls}A_{pqrs}) \equiv F(\sigma_{ij};\ A_{ijkl}). \tag{10}$$

In this case the central problem is: to construct an irreducible integrity basis for the tensors $\underset{\sim}{\sigma}$ and $\underset{\sim}{A}$. Together with the invariants of the single argument tensors, like (2) or (3), we have to consider the system of simultaneous or joint invariants.

Let us first construct a set of irreducible principal invariants of the fourth-order tensor A_{ijkl}, which may be a linear operator, i.e.,

$$Y_{ij} = A_{ijkl}X_{kl} \qquad \text{or} \qquad Y_{\alpha} = A_{\alpha\beta}X_{\beta}\ , \tag{11a,b}$$

where i,j,k,l = 1,2,3 or α,β = 1,2,...,9, respectively. In (11) two possible indicial notations are used. The transformation laws of the Cartesian tensor components can be written as

$$Y^{*}_{ij} = a_{ip}a_{jq}Y_{pq} \qquad \text{or} \qquad Y^{*}_{\alpha} = a_{\alpha\beta}Y_{\beta} \tag{12a,b}$$

and

$$A^*_{ijkl} = a_{ip}a_{jq}a_{kr}a_{ls}A_{pqrs} \quad \text{or:} \quad A^*_{\alpha\beta} = a_{\alpha\xi}a_{\beta\eta}A_{\xi\eta} \,, \tag{13a,b}$$

with the orthonormal conditions:

$$a_{it}a_{jt} = \delta_{ij} \quad \text{or:} \quad a_{\alpha\zeta}a_{\beta\zeta} = \delta_{\alpha\beta} \tag{14a,b}$$

where:

$$\left.\begin{array}{l} a_{\alpha\zeta} \mathrel{\hat{=}} a_{ip}a_{jq} \\ a_{\beta\zeta} \mathrel{\hat{=}} a_{kp}a_{lq} \end{array}\right\} \Rightarrow a_{\alpha\zeta}a_{\beta\zeta} \mathrel{\hat{=}} a_{ip}a_{jq}a_{kp}a_{lq} = \delta_{ik}\delta_{jl} \mathrel{\hat{=}} \delta_{\alpha\beta} \,. \tag{15}$$

In (12b) the operator $\underset{\sim}{A}$ defines a linear transformation on a 9-dimensional vector space V_9. This transformation is a correspondence which assigns to every vector $\underset{\sim}{X}$ in V_9 a vector $\underset{\sim}{A}\underset{\sim}{X}$ in V_9, in such a way that

$$\underset{\sim}{A}(a_1\underset{\sim}{X}_1 + a_2\underset{\sim}{X}_2) = a_1\underset{\sim}{A}\underset{\sim}{X}_1 + a_2\underset{\sim}{A}\underset{\sim}{X}_2$$

identically in the vectors $\underset{\sim}{X}_1$, $\underset{\sim}{X}_2$ and the scalars a_1, a_2.

Let X_α; $\alpha = 1,2,\ldots,9$; be the components of an arbitrary vector of unit magnitude which we shall call a direction vector or simply, a direction. We then ask: For what direction $\underset{\sim}{X}$ does the linear transformation $\underset{\sim}{A}$ yield a vector $\underset{\sim}{Y}$ according to (11) which is in the same direction as $\underset{\sim}{X}$? That is,

$$(A_{ijkl} - \lambda A^{(0)}_{ijkl})X_{kl} = 0_{ij} \quad \text{or} \quad (A_{\alpha\beta} - \lambda A^{(0)}_{\alpha\beta})X_\beta = 0_\alpha \,, \tag{16a,b}$$

where λ is a real scalar to be determined, and $A^{(0)}_{ijkl} \equiv \delta_{ik}\delta_{jl}$ or $A^{(0)}_{\alpha\beta} \equiv \equiv \delta_{\alpha\beta}$ are the components of the unit tensor $\underset{\sim}{A}^0$, whereas 0_{ij} or 0_α are the components of the zero tensor $\underset{\sim}{0}$.

For a nontrivial solution of (16a,b) we must have

$$\det(A_{ijkl} - \lambda A^{(0)}_{ijkl}) = 0 \quad \text{or} \quad \det(A_{\alpha\beta} - \lambda A^{(0)}_{\alpha\beta} = 0 \tag{17a,b}$$

in order to determine the principal or proper values of the linear transformation $\underset{\sim}{A}$.

In order to construct a set of principal invariants of a fourth-order tensor $\underset{\sim}{A}$, we shall note that the determinant (17) is an invariant, and we therefore consider the characteristic polynomial

$$P_n(\lambda) \equiv \det(A_{ijkl} - \lambda A^{(0)}_{ijkl}) = \sum_{\nu=0}^{n} J_\nu(\underset{\sim}{A})\lambda^{n-\nu} \tag{18}$$

in which, as we can see from (17a,b), the first index pair $(ij) \equiv \alpha$ characterizes the rows and the second one $(kl) \equiv \beta$ the columns of a $n \times n$ matrix $\underset{\sim}{A}$, generally $n = 9$.

Analogous to (3), the principal invariants J_ν in (18) can be determined by performing the operation of alternation [19]:

$$(-1)^{n-\nu} J_\nu \equiv A_{\alpha_1[\alpha_1]} A_{\alpha_2[\alpha_2]} \cdots A_{\alpha_\nu[\alpha_\nu]} \tag{19}$$

where $(-1)^n J_0 \equiv 1$. The right hand side in (19) is equal to the sum of all $\binom{n}{\nu} = \frac{n!}{\nu!(n-\nu)!}$ principal minors of order $\nu \leqq n$, where $\nu = 1$ and $\nu = n$ lead to $\mathrm{tr}\underset{\sim}{A}$ and $\det\underset{\sim}{A}$, respectively.

Assuming we have the usual symmetry conditions

$$A_{ijkl} = A_{jikl} = A_{ijlk} = A_{klij} \tag{20a}$$

or alternatively written as

$$A_{\alpha\beta} = A_{\beta\alpha}\,, \qquad \alpha,\beta = 1,2,\dots,6 \tag{20b}$$

then we can express the zero power tensor of fourth-order in (18) in the following way:

$$A^{(0)}_{ijkl} \equiv A_{ijpq}A^{(-1)}_{pqkl} = (\delta_{ik}\delta_{jl} + \delta_{il}\delta_{jk})/2 = A^{(-1)}_{ijpq}A_{pqkl}\,, \tag{21}$$

as we can see from (11) for $\underset{\sim}{Y} \Rightarrow \underset{\sim}{X}$.

If we expand (19) with n = 6, then we can find a set of irreducible principal invariants of a fourth-order tensor:

$$J_1 \equiv - S_1, \quad J_2 \equiv (S_1^2 - S_2)/2!, \quad J_3 \equiv - (S_1^3 - 3S_1S_2 + 2S_3)/3! ,$$

$$J_4 \equiv (S_1^4 + 8S_1S_3 - 6S_2S_1^2 + 3S_2^2 - 6S_4)/4! ,$$

$$J_5 \equiv - (S_1^5 - 30S_1S_4 + 15 S_1S_2^2 - 20 S_2S_3 - 10 S_2S_1^3 + 20 S_3S_1^2 + 25 S_5)/5!,$$

$$J_6 \equiv (S_1^6 + 144S_1S_5 - 120S_1S_2S_3 - 15S_2S_1^4 + 90S_2S_4 + 40 S_3S_1^3 - 15S_2^3 -$$
$$- 90S_4S_1^2 + 40 S_3^2 + 45S_2^2S_1^2 - 120S_6)/6! \tag{22}$$

where

$$S_\nu \equiv \mathrm{tr}\underset{\sim}{A}^\nu \equiv A_{i_1j_1i_2j_2} A_{i_2j_2i_3j_3} \cdots A_{i_\nu j_\nu i_1j_1} \tag{23}$$

are irreducible basic invariants found by forming irreducible traces of tensor powers. We see, that the principal invariants (22) of a fourth-order tensor can be determined uniquely by polynomial relations from the irreducible basic invariants (23).

The set of the six quantities (22) or, alternatively, the six quantities (23) are irreducible invariants of the fourth-order symmetric tensor (20) under the orthogonal group. However, the systems (22) and (23) are not complete, because some irreducible invariants like A_{iijj}, $A_{iipq}A_{pqjj}$, $A_{ijip}A_{pjqr}A_{rsqs}$ etc. cannot be expressed through (23). Thus, the characteristic polynomial (18) must be generalized. Proposals are made in sections 7 and 8.

4. THE HAMILTON-CAYLEY THEOREM

In the theory of invariants the HAMILTON-CAYLEY theorem plays an important role [7,12,20]. It states that

$$A_{ij}^{(3)} - J_1 A_{ij}^{(2)} - J_2 A_{ij} - J_3 \delta_{ij} \equiv 0_{ij} , \tag{24}$$

where (A_{ij}) is any 3×3 matrix. This theorem can be applied to the fourth-order tensor (20):

$$\sum_{\nu=0}^{n} J_\nu A_{\beta\gamma}^{(n-\nu)} \equiv 0_{\beta\gamma} \quad ; \qquad \beta,\gamma = 1,2,\ldots,n \tag{25}$$

as has been pointed out in detail in [21]. Thus, a second-order tensor A_{ij} or a fourth-order tensor A_{ijkl} satisfies its own characteristic equation (7) or (18), respectively. From (25) we find, that a tensor of power n and all higher powers can be expressed in terms of powers lower than n:

$$A_{\beta\gamma}^{(p)} = \sum_{\nu=1}^{n} {}^{p}Q_{n-\nu} A_{\beta\gamma}^{(n-\nu)} \quad ; \qquad p \geqq n , \tag{26}$$

where n = 6, if $\underset{\sim}{A}$ is a symmetric fourth-order tensor (20), or where n = 3, if $\underset{\sim}{A}$ is a symmetric second-order tensor. The coefficients ${}^{p}Q_{n-\nu}$ in (26) are scalar polynomials of degree $p - n + \nu$ in the principal invariants $J_1,\ldots,J_\nu$. For the symmetric fourth-order tensor (20) with n = 6 we have listed the polynomials ${}^{p}Q_{6-\nu}$ in table 1.

Table 1 Scalar Polynomials ${}^{p}Q_{n-\nu}$ in (26) for a symmetric fourth-order tensor (n = 6)

ν \ p	6	7	8
1	$-J_1$	$-J_2 + J_1^2$	$-J_3 + 2J_1J_2 - J_1^3$
2	$-J_2$	$-J_3 + J_1J_2$	$-J_4 + J_1J_3 + J_2(J_2 - J_1^2)$
3	$-J_3$	$-J_4 + J_1J_3$	$-J_5 + J_1J_4 + J_3(J_2 - J_1^2)$
4	$-J_4$	$-J_5 + J_1J_4$	$-J_6 + J_1J_5 + J_4(J_2 - J_1^2)$
5	$-J_5$	$-J_6 + J_1J_5$	$J_1J_6 + J_5(J_2 - J_1^2)$
6	$-J_6$	J_1J_6	$J_6(J_2 - J_1^2)$

From (22), (26) and table 1, we see, that all invariants formed by traces of tensor powers can be expressed by scalar polynomials in the irreducible basic invariants $S_1,\ldots,S_6$ defined in (23).

The scalar coefficients in (26) can be determined by using the recursion relation [22,23]

$$ {}^{p}Q_{n-\nu} = (-1)^{n-1}\left(J_{p-n+\nu} + \sum_{\mu=1}^{p-n} {}^{p-\mu}Q_{n-\nu}J_{\mu}\right) , \tag{27}$$

which is valid for the arbitrary dimension n. For the symmetric tensor (20) of rank four (n = 6) we can use (27) to find the scalar polynomials listed in table 1, and for a symmetric second-order tensor (n = 3) we can calculate polynomials listed in [16].

5. CONSTRUCTION OF SIMULTANEOUS INVARIANTS

In the cases where we have several different tensor variables (9), besides the irreducible invariants (3), (19), (22), we must take simultaneous or joint invariants into consideration. In order to construct a set of simultaneous invariants of the stress tensor $\underset{\sim}{\sigma}$ and the fourth-order constitutive tensor $\underset{\sim}{A}$ we could begin by applying the following theorem:

A scalar-valued function $f(\underset{\sim}{v},\underset{\sim}{T})$ of one n-dimensional vector $\underset{\sim}{v}$ and one symmetric second-order tensor $\underset{\sim}{T}$ is an orthogonal invariant, i.e. invariant under the orthogonal group, if and only if it can be expressed as a function of the 2n special invariants

$$ J_1(\underset{\sim}{T}),\ldots,J_n(\underset{\sim}{T}),\; v^2,\; \underset{\sim}{v}\cdot\underset{\sim}{T}\underset{\sim}{v},\ldots,\underset{\sim}{v}\cdot\underset{\sim}{T}^{n-1}\underset{\sim}{v} \; . \tag{28}$$

This theorem is valid for the arbitrary dimension n [8]. By using the HAMILTON-CAYLEY theorem (25), which states that $\underset{\sim}{T}^n$ and all higher powers $\underset{\sim}{T}^{n+h}$ can be expressed in terms of $\underset{\sim}{\delta},\underset{\sim}{T},\underset{\sim}{T}^2,\ldots,\underset{\sim}{T}^{n-1}$ according to (26), and by assuming the symmetry conditions (20a,b), we can find the set of 15 simultaneous invariants [19,24] from (28):

$$\left.\begin{aligned} &\Pi_2^{[\nu]} \equiv \sigma_{ij}A_{ijkl}^{(\nu)}\sigma_{kl}, \quad \Pi_3^{[\nu]} \equiv \sigma_{ij}A_{ijkl}^{(\nu)}\sigma_{kl}^{(2)}, \\ &\Pi_4^{[\nu]} \equiv \sigma_{ij}^{(2)}A_{ijkl}^{(\nu)}\sigma_{kl}^{(2)}, \quad \nu = 1,2,\ldots,5, \end{aligned}\right\} \tag{29}$$

where ν in a square bracket denotes ν's several invariants, whereas ν in parenthesis is an exponent.

Some of these simultaneous invariants can be constructed in the following way. If we take the second-order tensor $B_{pq} \equiv \sigma_{ij}A_{ijpq}$ into account, then, for instance, the basic invariant $B_{pq}B_{qp}$ is identical to the simultaneous invariant $\Pi_2^{[2]}$. In a different example we can consider the second-order tensor $C_{pq} \equiv \sigma_{ij}^{(2)}A_{ijpq}$ and find, for instance, the invariant $C_{pq}C_{qp} \equiv \Pi_4^{[2]}$. Further examples could be:

$$B_{pq}\sigma_{qp} \equiv \Pi_2^{[1]}, \quad B_{pq}\sigma_{qp}^{(2)} = C_{pq}\sigma_{qp} \equiv \Pi_3^{[1]}, \quad C_{pq}\sigma_{qp}^{(2)} \equiv \Pi_4^{[1]}.$$

The isotropic special case can be expressed by the isotropic constitutive tensor

$$A_{ijkl}^{(\nu)} = a_\nu\delta_{ij}\delta_{kl} + b_\nu(\delta_{ik}\delta_{jl} + \delta_{il}\delta_{jk}) \tag{30}$$

of power $\nu = 1,2,\ldots$. Then the simultaneous invariants $\Pi_2^{[\nu]}$ in (29) are equal to the principal invariant $J_2(\underset{\sim}{\sigma})$ for $a_\nu = -1/2$, $b_\nu = 1/4$ and equal to the basic invariant $S_2(\underset{\sim}{\sigma})$ for $a_\nu = 0$, $b_\nu = 1/2$. Similarly, the invariants $\Pi_3^{[\nu]}$ in (29) are equal to $J_3(\underset{\sim}{\sigma}) - J_1^3(\underset{\sim}{\sigma})/6$ for $a_\nu = -1/6$, $b_\nu = 1/6$ and equal to the cubic basic invariant $S_3(\underset{\sim}{\sigma})$ for $a_\nu = 0$, $b_\nu = 1/2$. Furthermore, the invariants $\Pi_1^{[\nu]} \equiv \delta_{ij}A_{ijkl}^{(\nu)}\sigma_{kl}$ or $\Pi_{1*}^{[\nu]} \equiv \delta_{ij}A_{ijkl}^{(\nu)}\sigma_{kl}^{(2)}$ are equal to $S_1(\underset{\sim}{\sigma})$ or $S_2(\underset{\sim}{\sigma})$, respectively, if $3a_\nu + 2b_\nu = 1$.

It must be mentioned, that the irreducible invariants $\Pi_1^{[\nu]}$ and $\Pi_{1*}^{[\nu]}$ are not contained in (29). Furthermore, a vector $\underset{\sim}{v}$ of arbitrary dimension n has only one invariant, namely v^2 in (28). However, in contrast to $\underset{\sim}{v}$, a second-order tensor in three dimension has three invariants (3a, b,c). Thus, the system (22) of irreducible principal invariants or the irreducible set (29) of simultaneous invariants cannot be complete. This fact must be emphazised and to taken into account, if we use the notation (20b) for the fourth-order tensor (20a) or the notation σ_α, $\alpha = 1,2,\ldots,6$, for the symmetric second-order tensor $\sigma_{ij} = \sigma_{ji}$ with $i,j = 1,2,3$.

6. CONSTRUCTION OF INVARIANTS BY THE POLARIZATION PROCESS

In the case of further tensor variables in the scarlar function (9), for instance $F = F(\sigma_{ij}, B_{ij}, A_{ijkl})$, we can find new invariants from the systems (2), (23) or (29) by using the polarization process. Some examples are listed below:

$$B_{ij}\partial\Pi_3^{[\nu]}/\partial\sigma_{ij} = B_{ij}A_{ijpq}^{(\nu)}\sigma_{pq}^{(2)} + 2B_{ij}\sigma_{jk}A_{kipq}^{(\nu)}\sigma_{pq} \equiv \operatorname{tr}\underset{\sim\sim}{BC} + 2\operatorname{tr}\underset{\sim\sim\sim}{B\sigma C}\ ;$$

$$A_{ijkl}\frac{\partial^2 S_2(\underset{\sim}{\sigma})}{\partial\sigma_{ij}\partial\sigma_{kl}} = 2A_{ijij} \equiv 2J_1(\underset{\sim}{A})\ ;$$

$$A_{ijkl}^{(\nu)}\sigma_{kl}\partial S_2(\underset{\sim}{\sigma})/\partial\sigma_{ij} = 2\sigma_{ij}A_{ijkl}^{(\nu)}\sigma_{kl} \equiv 2\Pi_2^{[\nu]};$$

$$A_{ijkl}^{(\nu)}\sigma_{kl}\partial S_3(\underset{\sim}{\sigma})/\partial\sigma_{ij} = 3\sigma_{ij}^{(2)}A_{ijkl}^{(\nu)}\sigma_{kl} \equiv 3\Pi_3^{[\nu]};$$

$$B_{ij}\frac{\partial\Pi_3^{[\nu]}}{\partial\sigma_{ij}} = B_{ij}A_{ijpq}^{(\nu)}\sigma_{pq}^{(2)} + 2B_{ij}\sigma_{jk}A_{kipq}^{(\nu)}\sigma_{pq} \equiv \operatorname{tr}\underset{\sim\sim}{BC} + 2\operatorname{tr}\underset{\sim\sim\sim}{B\sigma D}\ ,$$

where $C_{ij} \equiv A_{ijpq}^{(\nu)}\sigma_{pq}^{(2)}$ and $D_{ij} \equiv A_{ijpq}^{(\nu)}\sigma_{pq}$;

$$B_{ij}\sigma_{kl}\partial S_2(\underset{\sim}{A})/\partial A_{ijkl} = 2B_{ij}A_{ijkl}\sigma_{kl} \equiv 2\operatorname{tr}\underset{\sim\sim}{B\tau}\ ,$$

where $\tau_{ij} \equiv A_{ijkl}\sigma_{kl}$, .

$$B_{ij}B_{kl}\partial\Pi_3^{[1]}/\partial A_{ijkl} = B_{ij}\sigma_{ji}B_{kl}\sigma_{lk}^{(2)} \equiv (\mathrm{tr}\underset{\sim\sim}{B\sigma})(\mathrm{tr}\underset{\sim\sim}{B\sigma^2}) ;$$

$$B_{ij}\sigma_{kl}\partial\Pi_3^{[1]}/\partial A_{ijkl} = B_{ij}\sigma_{ji}\sigma_{kk}^{(3)} \equiv (\mathrm{tr}\underset{\sim\sim}{B\sigma})(\mathrm{tr}\underset{\sim}{\sigma}^3) .$$

Some applications of the theory of invariants to the formulation of yield criteria and failure criteria are discussed in [25,26], where theoretical results have been compared with experimental data for sintered powder materials and several polymers.

7. EXTENDED CHARACTERISTIC POLYNOMIAL

The characteristic polynomial (18) in section 3 should be admitted as a simplified one. An extended form of the characteristic polynomial (18) can be achieved by using the isotropic fourth-order tensor

$$I_{ijkl} := \lambda\delta_{ij}\delta_{kl} + \mu(\delta_{ik}\delta_{jl} + \delta_{il}\delta_{jk}) \tag{31}$$

instead of the spherical tensor $\lambda A_{ijkl}^{(0)}$, where $A_{ijkl}^{(0)}$ is the zero power tensor defined in (21). Thus we take into consideration the characteristic polynomial

$$P(\lambda,\mu) = \det(A_{ijkl} - I_{ijkl}) = 0 , \tag{32}$$

i.e., we formulate the eigenvalue problem

$$A_{ijkl}X_{kl} = I_{ijkl}X_{kl} \tag{33}$$

instead of (16a,b). The isotropic tensor (31) yield an image dyad

$$Y_{ij} = I_{ijkl}X_{kl} = \lambda X_{rr}\delta_{ij} + 2\mu X_{ij} , \tag{34}$$

which is coaxial with the dyad $\underset{\sim}{X}$.
In continuum mechanics equation (34) is known as the constitutive equation for an isotropic linear elastic solid. Such a material is cha-

racterized by the two elastic constants λ and μ, which are called the LAMÉ constants. Using the notation $Y_\alpha = I_{\alpha\beta} X_\beta$, $\alpha,\beta = 1,2,\dots,6$, we see that the isotropic tensor $\underset{\sim}{I}$ in (34) can be represented in the matrix form

$$I_{ijkl} \equiv \begin{pmatrix} 2\mu+\lambda & \lambda & \lambda & 0 & 0 & 0 \\ \lambda & 2\mu+\lambda & \lambda & 0 & 0 & 0 \\ \lambda & \lambda & 2\mu+\lambda & 0 & 0 & 0 \\ 0 & 0 & 0 & 2\mu & 0 & 0 \\ 0 & 0 & 0 & 0 & 2\mu & 0 \\ 0 & 0 & 0 & 0 & 0 & 2\mu \end{pmatrix} \tag{35}$$

which is quasi-diagonal of the structure $\{3,1,1,1\}$. The inverse tensor of (31) can be found from $X_{ij} = I^{(-1)}_{ijkl} Y_{kl}$ in connection with (34):

$$I^{(-1)}_{ijkl} = \frac{1}{4\mu}(\delta_{ik}\delta_{jl} + \delta_{il}\delta_{jk}) - \frac{\lambda}{2\mu(3\lambda+2\mu)}\delta_{ij}\delta_{kl} \; . \tag{36}$$

In order to control this result we prove:

$$I^{(-1)}_{ijpq} I_{pqkl} = I_{ijpq} I^{(-1)}_{pqkl} \equiv A^{(0)}_{ijkl} \; , \tag{37}$$

where the zero power tensor has the diagonal form

$$A^{(0)}_{ijkl} = \{1,1,1,1,1,1\} \; . \tag{38}$$

We see, in the special case ($\lambda = 0$, $\mu = 1/2$) the isotropic tensor (35) tends to the unit tensor (38). Then we find from (32) the simplified characteristic polynomial (18). The determinant of (35) is

$$\det(I_{ijkl}) = 32(2\mu + 3\lambda)\mu^5 \; ,$$

which tends to $\det(A^{(0)}_{ijkl}) = 1$ in the special case ($\lambda = 0$, $\mu = 1/2$).

In the following we should calculate irreducible invariants as coefficients in the characteristic polynomial (32). Because of (11a,b), (20a,b) and (35) the characteristic polynomial (32) can be written as:

$$\begin{vmatrix} A_{1111}-\Omega & A_{1122}-\lambda & A_{1133}-\lambda & 2A_{1112} & 2A_{1123} & 2A_{1131} \\ A_{2211}-\lambda & A_{2222}-\Omega & A_{2233}-\lambda & 2A_{2212} & 2A_{2223} & 2A_{2231} \\ A_{3311}-\lambda & A_{3322}-\lambda & A_{3333}-\Omega & 2A_{3312} & 2A_{3323} & 2A_{3331} \\ A_{1211} & A_{1222} & A_{1233} & 2A_{1212}-2\mu & 2A_{1223} & 2A_{1231} \\ A_{2311} & A_{2322} & A_{2333} & 2A_{2312} & 2A_{2323}-2\mu & 2A_{2331} \\ A_{3111} & A_{3122} & A_{3133} & 2A_{3112} & 2A_{3123} & 2A_{3131}-2\mu \end{vmatrix} = 0 \tag{39}$$

where the abbreviation $\Omega \equiv 2\mu + \lambda$ is used. The <u>non-symmetric</u> determinant in (39) is a polynomial of degree <u>3</u> in λ and degree <u>6</u> in μ:

$$P(\lambda,\mu) = \sum_{p,q} C_{(p,q)}\lambda^p\mu^q, \qquad p = 0,1,2,3;\ q = 0,1,\ldots,6, \quad p + q \leqq 6 \tag{40}$$

with 22 coefficients $C_{(p,q)}$. However, the characteristic equation $P(\lambda,\mu) = 0$ can we devided by the coefficient $C_{(0,6)}$, so that we could find no more than <u>21</u> invariants. Now, the main problem is: to expand the determinant (39) and to find out if all 21 coefficients are <u>irreducible</u> invariants [27].

In order to avoid the lengthy expansion of the determinant (39) we propose the following way. If we consider a second-order tensor we can start from the diagonal form

$$\sigma_{ij} = \mathrm{diag}\{\sigma_I, \sigma_{II}, \sigma_{III}\} . \tag{41}$$

Then, the determinant det $(\sigma_{ij} - \lambda\delta_{ij})$ is identical to the characteristic polynomial (7). Similarly, we could consider the <u>orthopropic</u> case with

$$A_{1112} = A_{1123} = A_{1131} = A_{2212} = A_{2223} = A_{2231} = A_{3312} = A_{3323} = A_{3331} = 0.$$

Then, the dyads $\underset{\sim}{X}$ and $\underset{\sim}{Y}$ in (11a) are <u>coaxial</u> and, instead of (39), the following determinant is to expand:

$$\begin{vmatrix} A_I-(2\mu+\lambda) & B_I-\lambda & B_{II}-\lambda & 0 & 0 & 0 \\ B_I-\lambda & A_{II}-(2\mu+\lambda) & B_{III}-\lambda & 0 & 0 & 0 \\ B_{II}-\lambda & B_{III}-\lambda & A_{III}-(2\mu+\lambda) & 0 & 0 & 0 \\ 0 & 0 & 0 & A_{IV}-2\mu & 0 & 0 \\ 0 & 0 & 0 & 0 & A_V-2\mu & 0 \\ 0 & 0 & 0 & 0 & 0 & A_{VI}-2\mu \end{vmatrix} = 0 \qquad (42)$$

From (42) we find the characteristic polynomial:

$$P(\lambda,\mu) = 64\mu^6 - 32I_1\mu^5 + 16I_2\mu^4 - 8I_3\mu^3 + 4I_4\mu^2 - 2I_5\mu + 3I_6\lambda^2 + I_7\lambda +$$
$$+ I_8\lambda\mu + I_9\lambda\mu^2 + I_{10}\lambda\mu^3 + I_{11}\lambda\mu^4 + 96\lambda\mu^5 - 6I_{12}\lambda^2\mu + 12I_{13}\lambda^2\mu^2 -$$
$$- 24I_{14}\lambda^2\mu^3 + I_{15}\,, \qquad (43)$$

where the <u>15</u> invariants $I_1,\ldots,I_{15}$ are listed below:

$$I_1 \equiv K_1 + M_1\,, \qquad I_2 \equiv K_2 + M_2 + K_1M_1 - L_1^2 + 2L_2\,,$$

$$I_3 \equiv D + M_3 + K_1M_2 + M_1(K_2 - L_1^2 + 2L_2)\,,$$

$$I_4 \equiv M_1D + K_1M_3 + M_2(K_2 - L_1^2 + 2L_2)\,,$$

$$I_5 \equiv M_2D + M_3(K_2 - L_1^2 + 2L_2)\,, \qquad I_6 \equiv K_1M_3\,,$$

$$I_7 \equiv M_3(L_1^2 - 4L_2 - K_2 + 2N)\,,$$

$$I_8 \equiv 4M_3(K_1 - L_1) - 2M_2(L_1^2 - 4L_2 - K_2 + 2N)\,,$$

$$I_9 \equiv 4M_1(L_1^2 - 4L_2 - K_2 + 2N) - 8M_2(K_1 - L_1) - 12M_3\,,$$

$$I_{10} \equiv 8\left[2M_1(K_1 - L_1) - L_1^2 + 4L_2 + K_2 - 2N + 3M_2\right] ,$$

$$I_{11} \equiv - 16\left[2(K_1 - L_1) + 3M_1\right] , \quad I_{12} \equiv K_1M_2 ,$$

$$I_{13} \equiv K_1M_1 , \quad I_{14} \equiv K_1 , \quad I_{15} \equiv M_3D . \tag{44}$$

In the system (44) the following abbreviations are used:

$$\left.\begin{aligned}
&K_1 \equiv A_I + A_{II} + A_{III} , \qquad M_1 \equiv A_{IV} + A_V + A_{VI} ,\\
&K_2 \equiv A_IA_{II} + A_{II}A_{III} + A_{III}A_I , \quad M_2 \equiv A_{IV}A_V + A_VA_{VI} + A_{VI}A_{IV} ,\\
&K_3 \equiv A_IA_{II}A_{III} , \qquad M_3 \equiv A_{IV}A_VA_{VI} ,\\
&L_1 \equiv B_I + B_{II} + B_{III} ,\\
&L_2 = B_IB_{II} + B_{II}B_{III} + B_{III}B_I ,\\
&L_3 = B_IB_{II}B_{III} ,\\
&N \equiv A_IB_{III} + A_{II}B_{II} + A_{III}B_I ,\\
&D \equiv K_3 + 2L_3 - (A_IB_{III}^2 + A_{II}B_{II}^2 + A_{III}B_I^2) .
\end{aligned}\right\} \tag{45}$$

We see, in the system (44) the invariants I_1 and I_{15} are the trace and determinant of the fourth-order tensor, respectively.

In the special case ($\lambda = 0$) only the six invariants $I_1,\ldots,I_5$, I_{15} are relevant, and they are identical to the six invariants in (22), if furthermore $B_I = B_{II} = B_{III} = 0$.

In the two dimensional case ($i,j,k,l = 1,2$) the linear transformation (11a) can be represented in the following matrix notation

$$\begin{pmatrix} Y_{11} \\ Y_{22} \\ Y_{12} \end{pmatrix} = \begin{pmatrix} A_{1111} & A_{1122} & 2A_{1112} \\ A_{2211} & A_{2222} & 2A_{2212} \\ A_{1211} & A_{1222} & 2A_{1212} \end{pmatrix} \begin{pmatrix} X_{11} \\ X_{22} \\ X_{12} \end{pmatrix} \tag{46a}$$

$$\underset{\sim}{Y} = \mathfrak{A} \underset{\sim}{X} \tag{46b}$$

where the matrix $\mathfrak{A}$ is non-symmetric.
The isotropic case (34) can now be written as

$$\begin{pmatrix} Y_{11} \\ Y_{22} \\ Y_{12} \end{pmatrix} = \begin{pmatrix} \lambda + 2\mu & \lambda & 0 \\ \lambda & \lambda + 2\mu & 0 \\ 0 & 0 & 2\mu \end{pmatrix} \begin{pmatrix} X_{11} \\ X_{22} \\ X_{12} \end{pmatrix} . \tag{47}$$

Considering (46a,b) and (47), the eigenvalue problem (33) leads to the characteristic polynomial

$$P(\lambda, \mu) = \begin{vmatrix} A_{1111} - (\lambda + 2\mu) & A_{1122} - \lambda & 2A_{1112} \\ A_{2211} - \lambda & A_{2222} - (\lambda + 2\mu) & 2A_{2212} \\ A_{1211} & A_{1222} & 2A_{1212} - 2\mu \end{vmatrix} \tag{48}$$

from which we find the characteristic equation $P(\lambda,\mu) = 0$:

$$8\mu^3 - 4J_1(\mathfrak{A})\mu^2 - 2J_2(\mathfrak{A})\mu - J_3(\mathfrak{A}) - 2K_1\lambda - 2K_2\lambda\mu + 8\lambda\mu^2 = 0 \tag{49}$$

with the following <u>5</u> irreducible invariants:

$$\left.\begin{aligned}
J_1(\mathfrak{A}) &\equiv A_{1111} + A_{2222} + 2A_{1212} \equiv \operatorname{tr} \mathfrak{A} , \\
J_2(\mathfrak{A}) &\equiv [\operatorname{tr} \mathfrak{A}^2 - (\operatorname{tr} \mathfrak{A})^2]/2 , \\
J_3(\mathfrak{A}) &\equiv \det(\mathfrak{A}) , \\
K_1 &\equiv (A_{1112} - A_{2211})^2 - A_{1212}(A_{1111} + A_{2222} - 2A_{1122}) , \\
K_2 &\equiv J_1(\mathfrak{A}) + 2(A_{1212} - A_{1122}) .
\end{aligned}\right\} \qquad (50)$$

The orthotropic special case is characterized by $A_{1112} = A_{1222} = 0$. Then, the matrix $\mathfrak{A}$ in (46a,b) is symmetric, and the invariants in (50) are reduced to simpler forms. However, all 5 invariants are relevant in this special case, too. But the invariants K_1 and K_2 in (49) are not relevant, if we take a simplified characteristic equation with $\lambda = 0$ into consideration. Then, the number of 5 invariants in (50) is reduced to 3.

8. THE LAGRANGE MULTIPLIER METHOD

In order to find the characteristic equation of a second-order tensor one can utilize the LAGRANGE multiplier method, i.e., we vary the orthogonal transformation $\underset{\sim}{a}$ in the tensor law

$$A^*_{ij} = a_{ip} a_{jq} A_{pq} \qquad (51)$$

in such a way that the tensor components A^*_{11}, A^*_{22}, A^*_{33} will be extremal, where the "auxiliary" or "subsidiary" condition

$$M_{ij} \equiv a_{ip} a_{jp} - \delta_{ij} = 0_{ij} \qquad (52)$$

is to fulfil. This is the condition, which follows from the property $a^{(-1)}_{ij} = a_{ji}$ of an orthogonal matrix. From (51) and (52) we form the tensor

$$\Phi_{ij} = a_{ip}a_{jq}A_{pq} - \lambda(a_{ip}a_{jp} - \delta_{ij}) \ , \tag{53}$$

where the undetermined factor λ is called a LAGRANGE multiplier. Now, the tensor components A^*_{11}, A^*_{22}, A^*_{33} are extremal, if we require

$$\partial\Phi_{(rr)}/\partial a_{(r)k} \overset{!}{=} 0_{kr} \ . \tag{54}$$

From (54) we find

$$(A_{ij} - \lambda_{(r)}\delta_{ij})a_{(r)j} = 0_{ri} \tag{55}$$

with 3 eigenvalues $\lambda_{(1)} \equiv \lambda_{I},\ldots,\lambda_{(3)} \equiv \lambda_{III}$ and 3 eigenvectors $n_i^{(1)} \equiv$ $\equiv n_i^{I},\ldots,n_i^{(3)} \equiv n_i^{III}$, where

$$n_i^{(r)} = (a_{r1}, a_{r2}, a_{r3}) \tag{56}$$

is the r-th row of the orthogonal matrix $\underset{\sim}{a}$. Note, that there is no sum on the bracketed index r in (54) and (55). The condition for (55) to have non-trivial solutions (56) yields the characteristic equation $\det(A_{ij} - \lambda\delta_{ij}) = 0$.

We can also start from the quadratic form

$$F = A_{ij}x_ix_j \tag{57}$$

with the "auxiliary" or "secondary" condition

$$M = \delta_{ij}x_ix_j - x^2 \tag{58}$$

where the invariant $x \equiv \sqrt{x_ix_i}$ is the magnitude of the vector $\underset{\sim}{x}$. Now, the LAGRANGE multiplier method requires

$$\partial\Phi/\partial x_i \equiv \partial(F - \lambda M)/\partial x_i \overset{!}{=} 0_i \tag{59}$$

and yields

$$(A_{ij} - \lambda\delta_{ij})x_j = 0_i \ . \tag{60}$$

From (60) we again read the characteristic equation.

In order to find the characteristic equation of a fourth-order tensor we can also apply the LAGRANGE multiplier method, and, in extending (51), (52), (53), we have the transformation law

$$A^*_{ijkl} = a_{ip}a_{jq}a_{kr}a_{ls}A_{pqrs} \ , \tag{61}$$

the $\underline{2}$ "auxiliary" conditions

$$L_{ijkl} = a_{ip}a_{jp}a_{kq}a_{lq} - \delta_{ij}\delta_{kl} = 0_{ijkl} \ , \tag{62a}$$

$$M_{ijkl} = a_{ip}a_{kp}a_{jq}a_{lq} + a_{ip}a_{lp}a_{jq}a_{kq} - \delta_{ik}\delta_{jl} - \delta_{il}\delta_{jk} = 0_{ijkl} \ , \tag{62b}$$

and the fourth-order "modified" tensor

$$\Phi_{ijkl} = A^*_{ijkl} - \lambda L_{ijkl} - \mu M_{ijkl} \ , \tag{63}$$

respectively. Analogously with (54), we require

$$\partial\Phi_{(iiii)}/\partial a_{(i)u} \stackrel{!}{=} 0_{iu} \quad \text{or:} \quad \partial\Phi_{(ijij)}/\partial a_{(i)u} \stackrel{!}{=} 0_{iju} \tag{64}$$

(no sum on the bracketed indices i or j). Going on in this way, we find more details in [27].

The other possibility to apply the LAGRANGE multiplier method is illustrated in the following. Similarly to (57), we start from the scalar-valued function

$$F = A_{ijkl}X_{ij}X_{kl} \tag{65}$$

which may be an analogous function with the elastic potential (strain-energy function). Because the second-order tensor $\underset{\sim}{X}$ in (65) has $\underline{3}$ irreducible invariants $S_1 \equiv \mathrm{tr}\,\underset{\sim}{X}, \ldots, S_3 \equiv \mathrm{tr}\,\underset{\sim}{X}^3$, being the elements of the integrity basis, in contrast to the vector $\underset{\sim}{x}$ in (57) having only one invariant, namely its magnitude x in (58), we take into consideration the following $\underline{3}$ "auxiliary" conditions:

$$L = \delta_{ij}\delta_{kl}X_{ij}X_{kl} - S_1^2 = 0 \,, \tag{66a}$$

$$M = (\delta_{ik}\delta_{jl} + \delta_{il}\delta_{jk})X_{ij}X_{kl} - 2S_2 \,, \tag{66b}$$

$$N = (\delta_{jk}\delta_{pl} + \delta_{jl}\delta_{pk})X_{ip}X_{ij}X_{kl} - 2S_3 \,. \tag{66c}$$

Thus, the quadratic form (65) of the dyad $\underset{\sim}{X}$ is modified to the form

$$\Phi = F - \lambda L - \mu M - \nu N \tag{67}$$

which is to be made stationary, i.e., similarly to (59), we require

$$\partial\Phi/\partial X_{rs} \overset{!}{=} 0_{rs} \,. \tag{68}$$

From (68) we find the system of non-linear equations in the dyad $\underset{\sim}{X}$,

$$(A_{ijkl} - I_{ijkl})X_{kl} = 3\nu X_{ij}^{(2)} \,, \tag{69}$$

where I_{ijkl} are the components of the fourth-order isotropic tensor defined in (31). In the special case when $\nu = 0$ we find from (69) the eigenvalue problem (33) which yields the characteristic equation (32). Some more details concerning the system (69) can be found in [27]. Furthermore, applications of the LAGRANGE multiplier method to tensors of order six are discussed in [27], too.

REFERENCES

1. GRACE, J.H. and YOUNG, A., The Algebra of Invariants, Cambridge Univ. Press. London and New York 1903.

2. GUREVICH, G.B., Foundations of the Theory of Algebraic Invariants, P. Noordhoff, Groningen 1964.

3. RIVLIN, R.S., An Introduction to Non-linear Continuum Mechanics, in: Non-linear Continuum Theories in Mechanics and Physics and their applications (Ed. R.S. Rivlin), Edizioni Cremonese, Rome 1970.

4. SCHUR, I., Vorlesungen über Invariantentheorie, bearbeitet und herausgegeben von H. Grunsky, Die Grundlehren der mathematischen Wissenschaften in Einzeldarstellungen, Bd. 143, Springer-Verlag, Berlin/Heidelberg/New York 1968.

5. WEITZENBÖCK, R., Invariantentheorie, P. Noordhoff, Groningen 1923.

6. WEYL, H., The Classical Groups, Their Invariants and Representation, Princeton Univ. Press, Princeton and New Jersey 1946.

7. SPENCER, A.J.M., Theory of Invariants, in: Continuum Physics (ed. A.C. Eringen), Academic Press, New York 1971, 239-353.

8. TRUESDELL, C. and NOLL, W., The Non-Linear Field Theories of Mechanics, in: Handbuch der Physik (Ed. S. Flügge), Vol. III/3, Springer-Verlag, Berlin/Heidelberg/New York 1965.

9. BETTEN; J., Ein Beitrag zur Invariantentheorie in der Plastomechanik anisotroper Stoffe, Z. Angew. Math. Mech. (ZAMM), 56 (1976), 557-559.

10. BETTEN, J., Theory of Invariants in Creep Mechanics of Anisotropic Solids (Ed. J.P. Boehler), Martinus Nijhoff Publishers, The Hague/Boston/London 1982, 65-80.

11. BETTEN, J., Creep Theory of Anisotropic Solids, Journal of Rheology, 25 (1981), 565-581.

12. SPENCER, A.J.M. and RIVLIN, R.S., The Theory of Matrix Polynomials and its Application to the Mechanics of Isotropic Continua, Arch. Rational Mech. Anal. 2 (1958/59), 309-336.

13. SMITH, G.F., On the Yield Condition for Anisotropic Materials, Quart. Appl. Math. 20 (1962), 241-247.

14. PIPKIN, A.C. and RIVLIN, R.S., The Formulation of Constitutive Equations in Continuum Physics, Arch. Rational Mech. Anal. 4 (1959), 129-144.

15. PIPKIN, A.C. and WINEMAN, A.S., Material Symmetry Restrictions on Non-Polynomial Constitutive Equations, Arch. Rational Mech. Anal. 12 (1963), 420-426.

16. BETTEN, J., Elementare Tensorrechnung für Ingenieure, Vieweg-Verlag, Braunschweig 1977.

17. WINEMAN, A.S. and PIPKIN, A.C., Material Symmetry Restrictions on Constitutive Equations, Arch. Rational Mech. Anal. 17 (1964), 184-214.

18. SMITH, G.F., SMITH, M.M. and RIVLIN, R.S., Integrity Bases for a Symmetric Tensor and a Vector - The Crystal Classes, Arch. Rational Mech. Anal. 12 (1963), 93-133.

19. BETTEN, J., Integrity Basis for a Second-Order and a Fourth-Order Tensor, Internat. J. Math. & Math. Sci. 5 (1982), 87-96.

20. RIVLIN, R.S. and SMITH, G.F., Orthogonal Integrity Basis for N Symmetric Matrices, in: Contributions to Mechanics (Ed. D. Abir), Pergamon Press, Oxford/.../Braunschweig 1969, 121-141.

21. BETTEN, J., Zur Anstellung von Stoffgleichungen in der Kriechmechanik anisotroper Körper, Rheol. Acta 20 (1981), 527-535.

22. BETTEN, J., Elastizitäts- und Plastizitätslehre, Vieweg-Verlag, Braunschweig/Wiesbaden 1985.

23. BETTEN, J., Irreducible Invariants of Fourth-Order Tensors, Fifth International Conference on Mathematical Modelling, University of California, Berkeley 1985, to be published in the proceedings: Mathematical Modelling in Science and Technology (eds. X.J.R. Avula, G. Leitmann, C.D. Mote, Jr. and E.Y. Rodin), Pergamon Press, New York 1986.

24. BETTEN, J., Representation of Constitutive Equations in Creep Mechanics of Isotropic and Anisotropic Materials, in: Creep in Structures (Ed. A.R.S. Ponter and D.R. Hayhurst), Springer-Verlag, Berlin/Heidelberg/New York 1981, 179-201.

25. BETTEN, J., Pressure-dependent Yield Behaviour of Isotropic and Anisotropic Materials, IUTAM-Symposium on Deformation and Failure of Granular Materials, Delft 1982, published in the proceedings (eds. P.A. Vermeer and H.J. Luger), A.A. Balkema, Rotterdam 1982, 81-89.

26. BETTEN, J., Formulation of Failure Criteria for Anisotropic Materials under Multi-Axial States of Stress, Colloque International du CNRS n° 351 on Failure Criteria of Structured Media, Grenoble 1983, to be published in the proceedings (ed. J.P. Boehler).

27. BETTEN, J., The Eigenvalue Problem of a Fourth-Order Tensor, in preparation.

Chapter 12

FORMULATION OF ANISOTROPIC CONSTITUTIVE EQUATIONS

J. Betten
Technical University Aachen, F.R. Germany

1. INTRODUCTION

This chapter is concerned with the formulation of constitutive expressions of the form

$$d_{ij} = f_{ij}(\sigma_{pq}, \omega_{pq}, A_{pqrs}) , \tag{1}$$

where $d_{ij} = (v_{i,j} + v_{j,i})/2$ are the cartesian components of the so called "rate-of-deformation tensor" $\underset{\sim}{d}$. Other common names are the "rate-of-strain" or "strain-rate-tensor". Note that d_{ij} is linear in the velocity components v_i, and that this linearity is exact and no approximation has been made in deriving it. Furthermore, the tensor $\underset{\sim}{d}$ is not to confuse with the "material time derivative" $\dot{\underset{\sim}{\varepsilon}}$ of the infinitesimal strain tensor $\varepsilon_{ij} = (u_{i,j} + u_{j,i})/2$, because we have [1]:

$$\dot{\varepsilon}_{ij} = d_{ij} - (u_{i,p}v_{p,j} + u_{j,p}v_{p,i})/2 .$$

Only in the case of small displacement gradients ($u_{i,j} \equiv \partial u_i/\partial x_j$) and small velocity gradient tensors ($v_{i,j} \equiv \partial v_i/\partial x_j$) we have: $\varepsilon_{ij} \approx d_{ij}$.

The symmetric strain-rate tensor ($d_{ij} = d_{ji}$) is represented as a second-order tensor-valued function (1) of two symmetric argument tensors of rank two and one symmetric argument tensor of rank four, being the CAUCHY stress tensor $\underset{\sim}{\sigma}$, the damage tensor $\underset{\sim}{\omega}$, and a fourth-order constitutive tensor $\underset{\sim}{A}$ characterizing the anisotropy from, perhaps rolling, i.e. the anisotropy of the material in its undamaged state. Constitutive equations of the form (1) are applied in creep mechanics [1,2,3,4,5], because the creep behaviour of isotropic and anisotropic materials in the tertiary stage is characterized by a damage tensor.

The general representation of (1) is given through a linear combination

$$\underset{\sim}{d} = \sum_{\alpha} \varphi_{\alpha} \underset{\sim}{G}_{\alpha} , \tag{2}$$

where the $\underset{\sim}{G}$'s are symmetric tensor generators of rank two involving the argument tensors $\underset{\sim}{\sigma}$, $\underset{\sim}{\omega}$, $\underset{\sim}{A}$. The coefficients φ_{α} in (2) are scalar-valued functions of the integrity basis associated with the representation (2). They must also contain experimental data measured in uni-axial creep tests [5]. The main problems are: to construct an irreducible set of tensor generators and to determine the scalar coefficients involving experimental data.

A further aim is to represent the constitutive equations (2) in canonical form

$$d_{ij} = {}^{0}H_{ijkl}\delta_{kl} + {}^{1}H_{ijkl}\sigma_{kl} + {}^{2}H_{ijkl}\sigma^{(2)}_{kl} , \tag{3}$$

where ${}^{0}H_{ijkl}$, ${}^{1}H_{ijkl}$, ${}^{2}H_{ijkl}$ are the components of fourth-order tensor-valued functions ${}^{0}\underset{\sim}{H}$, ${}^{1}\underset{\sim}{H}$, ${}^{2}\underset{\sim}{H}$, depending on the damage tensor $\underset{\sim}{\omega}$ and the anisotropy tensor $\underset{\sim}{A}$. The canonical form (3) is a representation in three terms, which are the contributions of zero, first, and second orders in the stress tensor $\underset{\sim}{\sigma}$, influenced by the functions ${}^{0}\underset{\sim}{H}$, ${}^{1}\underset{\sim}{H}$, and ${}^{2}\underset{\sim}{H}$, respectively.

2. THE DAMAGE STATE IN A CONTINUUM

Before formulating constitutive equations of the form (1), we should describe the damage state in a continuum.

In a uni-axial tension specimen material deterioration can be described by introducing an additional variable ω or, alternatively, $\psi \equiv 1 - \omega$ into constitutive equations, i.e. the strain rate d can be expressed in the form $d = f(\sigma,\omega)$, or $d = f(\sigma,\psi)$, where σ is the uni-axial stress [6,7]. The material parameters ω and ψ describe the current damage state and the "continuity" of the material, respectively. The parameter of "continuity", ψ, represents that fraction of the cross-sectional area which is not occupied by either voids or internal fissures. The net stress acting over the cross-section of the uni-axial tension specimen is then $\hat{\sigma} = \sigma/\psi$. When $\psi = 1$ the material is in its virgin undamaged state, and when $\psi = 0$ the material can no longer sustain any load. In the latter case the constitutive equation would be required in order to approach an infinite strain rate. Furthermore, it is assumed that the damage rate $\dot{\omega}$ or, alternatively, the rate of continuity change $\dot{\psi}$ is also governed by the uni-axial stress and by the current state of continuity, i.e. $\dot{\omega} = g(\sigma,\omega)$, or $\dot{\psi} = - g(\sigma,\psi)$.

For example, the functions f and g may have the forms

$$d/d_0 = (\sigma/\sigma_0)^n/(1-\omega)^m \ , \qquad \dot{\omega}/\dot{\omega}_0 = (\sigma/\sigma_0)^\nu/(1-\omega)^\mu \ , \tag{4a,b}$$

where n, m, ν,μ , d_0, $\dot{\omega}_0$, and σ_0 are constants. Then, the undamaged case ($\omega = 0$) leads to NORTON's power law, which is assumed to be valid for the secondary creep stage, while the strain rate d approaches infinity as ω approaches 1.

Because of its microscopic nature, damage generally has an anisotropic character even if the material was originally isotropic. The fissure orientation and length cause anisotropic macroscopic behaviour. Therefore, damage in an isotropic or anisotropic material, which is in a state of multi-axial stress, can only be described in tensorial form.

When generalizing the above uni-axial concept, constitutive equations and anisotropic growth equations are expressed as tensor-valued functions

$$d_{ij} = f_{ij}(\underset{\sim}{\sigma},\underset{\sim}{\omega}) \ , \qquad \overset{\circ}{\omega}_{ij} = g_{ij}(\underset{\sim}{\sigma},\underset{\sim}{\omega}) \ , \tag{5a,b}$$

respectively, where $(^{o})$ denotes the JAUMANN derivative and $\underset{\sim}{\omega}$ is an appropiately defined damage tensor.

Damage tensors are constructed, for instance, in [8]. Furthermore, we refer to the work of MURAKAMI and OHNO [9]. They assumed that damage accumulating in the process of creep can be expressed through a symmetric tensor of rank two.

RABOTNOV [10] also introduced a symmetric second-order tensor of damage and defined a symmetric net stress tensor $\underset{\sim}{\hat{\sigma}}$ by way of a linear transformation

$$\sigma_{ij} = \Omega_{ijkl}\hat{\sigma}_{kl} \ , \tag{6}$$

where the fourth-order tensor $\underset{\sim}{\Omega}$ is assumed to be symmetric.

However, in [11] is has been pointed out that the fourth-order tensor in (6) is only symmetric when corresponding to the first index pair ij, but not to the second one kl. Thus, the net stress tensor $\underset{\sim}{\hat{\sigma}}$ is not symmetric in the anisotropic damage case. The net stress tensor can be decomposed into a symmetric part and into an antisymmetric one, where only the symmetric part is equal to the net stress tensor introduced by RABOTNOV [10], as shown in [11].

A tensor of continuity and an anisotropic damage tensor can be constructed in the following way.

In three-dimensional space a parallelogram formed by the vectors $\underset{\sim}{v}$ and $\underset{\sim}{w}$ can be represented by

$$S_i = \varepsilon_{ijk}v_j w_k \equiv (\underset{\sim}{v} \times \underset{\sim}{w})_i \tag{7a}$$

or in the dual form

$$S_{ij} = \varepsilon_{ijk} S_k \iff S_i = \varepsilon_{ijk} S_{jk}/2 \ , \tag{7b}$$

where ε_{ijk} is the third-order permutation tensor. From the relations (7a,b) we immediately find the decomposition:

$$S_{ij} = 2!\ v_{[i}w_{j]} \equiv \begin{vmatrix} v_i & v_j \\ w_i & w_j \end{vmatrix} \ . \tag{8}$$

Such an alternating product of two vectors $\underset{\sim}{v}$ and $\underset{\sim}{w}$ is called a simple bi-vector $\underset{\sim}{S}$, which has the three nonvanishing essential components S_{12}, S_{23}, S_{31}. In rectilinear components in three-dimensional space, we see that the absolute values of the components S_{12}, S_{23}, S_{31} are projections of the area of the parallelogram (7a) on the coordinate planes. Thus S_{ij}, according to (8), represents an area vector in three-dimensional space and has an orientation fixed by the cross-product (7a).

Now we consider at a point "0" in an undamaged continuum a differential tetrahedron, as shown in Fig. 1a.

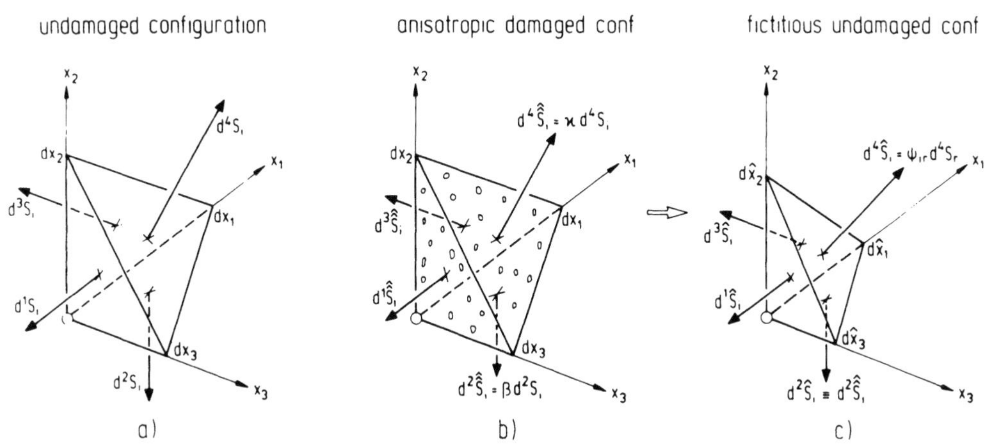

Fig. 1 Anisotropic Tensor of Continuity

Such a tetrahedron can be characterized by a system of bi-vectors:

$$\left.\begin{aligned} d^1S_i &= -\varepsilon_{ijk}(dx_2)_j\,(dx_3)_k/2\ , \\ d^2S_i &= -\varepsilon_{ijk}(dx_3)_j\,(dx_1)_k/2\ , \\ d^3S_i &= -\varepsilon_{ijk}(dx_1)_j\,(dx_2)_k/2\ , \\ d^4S_i &= \varepsilon_{ijk}[(dx_1)_j - (dx_3)_j]\,[(dx_2)_k - (dx_3)_k]/2\ , \end{aligned}\right\} \tag{9}$$

where the sum is equal to the zero vector:

$$d^1S_i + d^2S_i + d^3S_i + d^4S_i = 0_i\ . \tag{10}$$

The same tetrahedron in a damaged state (Fig. 1b) can be characterized by the following system of bi-vectors:

$$\left.\begin{aligned} d^1\hat{S}_i &= -\alpha_{ijk}(dx_2)_j\,(dx_3)_k/2 \equiv \alpha d^1S_i\ , \\ d^2\hat{S}_i &= -\beta_{ijk}(dx_3)_j\,(dx_1)_k/2 \equiv \beta d^2S_i\ , \\ d^3\hat{S}_i &= -\gamma_{ijk}(dx_1)_j\,(dx_2)_k/2 \equiv \gamma d^3S_i\ , \\ d^4\hat{S}_i &= \varkappa_{ijk}[(dx_1)_j - (dx_3)_j]\,[(dx_2)_k - (dx_3)_k]/2 \equiv \varkappa d^4S_i\ , \end{aligned}\right\} \tag{11}$$

where $\alpha_{ijk} \equiv \alpha\,\varepsilon_{ijk}$, $\beta_{ijk} \equiv \beta\,\varepsilon_{ijk}$, etc. are total skew-symmetric tensor of rank three, which have the essential components $\alpha_{123} \equiv \alpha$, $\beta_{123} \equiv \beta$, etc., respectively. We see, the vectors $d^1\hat{S}_i$, ..., $d^4\hat{S}_i$ in Fig. 1b differ from the corresponding vectors of the undamaged configuration (Fig. 1a) only in lengths. It should also be noted that the vector sum is not equal to the zero vector:

$$d^1\hat{S}_i + d^2\hat{S}_i + d^3\hat{S}_i + d^4\hat{S}_i \neq 0_i\ , \tag{12}$$

unless there is no damage ($\alpha = \beta = \gamma = \varkappa = 1$) according to (10) or the damage is only isotropic ($\alpha = \beta = \gamma = \varkappa$).

Next we should consider a fictitious undamaged configuration (Fig. 1c) characterized by the following system of bi-vectors:

$$\left.\begin{aligned}
d^1\hat{S}_i &= -\varepsilon_{ijk}(d\hat{x}_2)_j\,(d\hat{x}_3)_k/2 \equiv d^1\hat{\hat{S}}_i\,,\\
d^2\hat{S}_i &= -\varepsilon_{ijk}(d\hat{x}_3)_j\,(d\hat{x}_1)_k/2 \equiv d^2\hat{\hat{S}}_i\,,\\
d^3\hat{S}_i &= -\varepsilon_{ijk}(d\hat{x}_1)_j\,(d\hat{x}_2)_k/2 \equiv d^3\hat{\hat{S}}_i\,,\\
d^4\hat{S}_i &= \varepsilon_{ijk}\left[(d\hat{x}_1)_j - (d\hat{x}_3)_j\right]\left[(d\hat{x}_2)_k - (d\hat{x}_3)_k\right]/2\,,
\end{aligned}\right\} \tag{13}$$

where, by analogy to (10), the vector sum is equal to the zero vector:

$$d^1\hat{S}_i + d^2\hat{S}_i + d^3\hat{S}_i + d^4\hat{S}_i = 0_i\,. \tag{14}$$

The three area vectors $d^1\hat{S}_i, \ldots, d^3\hat{S}_i$ in (13) are identical to the corresponding vectors of the damaged configuration. The fourth vector $d^4\hat{S}_i$, having the same magnitude as $d^4\hat{\hat{S}}_i$, differs from d^4S_i not only in length, but also in its direction. Therefore, the vectors $d^4\hat{S}_i$ and d^4S_i are connected by a linear operator $\underset{\sim}{\psi}$ of rank two (second-order tensor):

$$d^4\hat{S}_i = \psi_{ir}\, d^4S_r\,. \tag{15}$$

By comparing the systems of bi-vectors (9), (11), (13) and by using the equations (10), (14) in connection with the transformation (15), we find the following relation:

$$\begin{aligned}
&\psi_{ir}\varepsilon_{rjk}\left[(dx_2)_j(dx_3)_k + (dx_3)_j(dx_1)_k + (dx_1)_j(dx_2)_k\right] = \\
&\quad = \alpha_{ijk}(dx_2)_j(dx_3)_k + \beta_{ijk}(dx_3)_j(dx_1)_k + \gamma_{ijk}(dx_1)_j(dx_2)_k]\,,
\end{aligned} \tag{16}$$

where the transvection $\psi_{ir}\varepsilon_{rjk}$ leads to a third-order tensor of continuity:

$$\psi_{ir}\varepsilon_{rjk} \equiv \psi_{ijk} = \psi_{i[jk]} , \tag{17}$$

which is skew-symmetric with respect to the bracketed index pair $[jk]$.

Because of $\alpha_{ijk} \equiv \alpha\,\varepsilon_{ijk}$, etc. the terms on the right-hand side of (16) are vectors with magnitudes of $|d^1\hat{S}_1| = \alpha_{1jk}(dx_2)_j(dx_3)_k/2$, etc. and with the same directions as the base vectors ${}^1e_i$, ${}^2e_i$, ${}^3e_i$ belonging to the Cartesian coordinate system. Therefore, in connection with (17), the relation (16) can be written in the form of:

$$\psi_{ijk}[(dx_2)_j(dx_3)_k + \ldots] = {}^1e_i\alpha_{1jk}(dx_2)_j(dx_3)_k + \ldots , \tag{18}$$

in which we can immediately see the decomposition:

$$\psi_{ijk} = {}^1e_i\alpha_{1jk} + {}^2e_i\beta_{2jk} + {}^3e_i\gamma_{3jk} , \tag{19}$$

or because of $\alpha_{1jk} \equiv \alpha\,\varepsilon_{1jk}$, etc., we can write:

$$\psi_{ijk} = \alpha\,{}^1e_i\varepsilon_{1jk} + \beta\,{}^2e_i\varepsilon_{2jk} + \gamma\,{}^3e_i\varepsilon_{3jk} . \tag{20}$$

From (17) we can find the dual relation

$$\psi_{ijk} \equiv \psi_{i[jk]} = \varepsilon_{jkr}\psi_{ir} \;\Longleftrightarrow\; \psi_{ir} = \varepsilon_{rjk}\psi_{ijk}/2 , \tag{21}$$

which is similar to that in (7b). Finally, we can use (21) to deduce the diagonal form:

$$\psi_{ij} = \psi_{ipq}\varepsilon_{jpq}/2 = \text{diag}\,\{\alpha,\ \beta,\ \gamma\} . \tag{22a}$$

Inserting (20) into (22a) and replacing $\delta_{1j} \equiv {}^1e_j$, etc. we see, that the second rank tensor of continuity can be decomposed in terms of dyadics formed from the base vectors:

$$\psi_{ij} = \alpha({}^1\underset{\sim}{e} \otimes {}^1\underset{\sim}{e})_{ij} + \beta({}^2\underset{\sim}{e} \otimes {}^2\underset{\sim}{e})_{ij} + \gamma({}^3\underset{\sim}{e} \otimes {}^3\underset{\sim}{e})_{ij} . \tag{22b}$$

The relations (21) and (22a) are illustrated in Fig. 2.

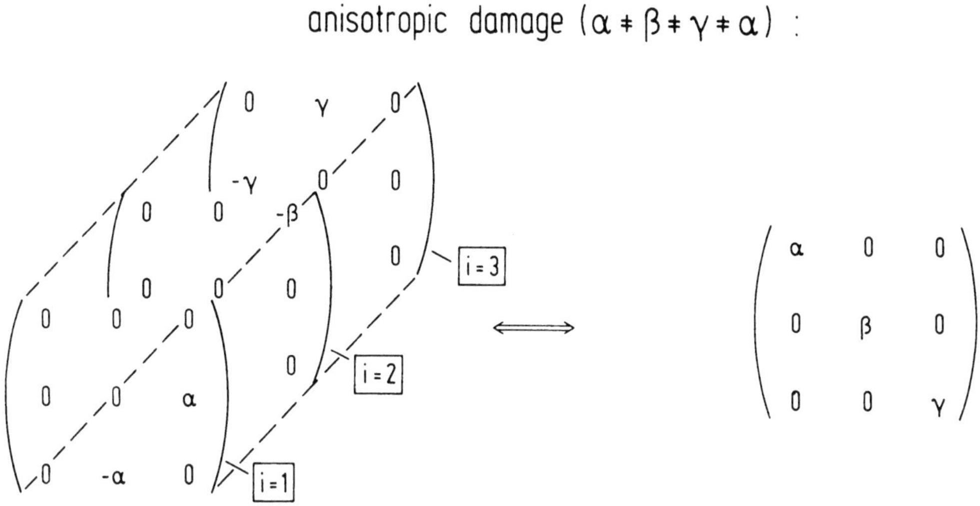

Fig. 2 Dual Tensor of Continuity

Especially, from Fig. 2 we can see the skew-symmetric character of the third order tensor of continuity indicated in (17) and its three essential components α, β, γ. These values are fractions which represent the net cross-section elements perpendicular to the coordinate axes x_1, x_2, x_3 (Fig. 1b) and which can be measured in tests on specimens cut along three mutually perpendicular directions x_1, x_2, x_3.

The damage may sometimes develop isotropically, as observed by JOHNSON [12] for R.R. 59 aluminium alloy. In this special case ($\alpha = \beta = \gamma \equiv \psi$) the second rank tensor of continuity (21) is a spherical tensor:

$$\psi_{ijk} = \psi\, \varepsilon_{jkr}\delta_{ir} = \psi\, \varepsilon_{ijk} \quad <=> \psi_{ir} = \psi\, \varepsilon_{rjk}\varepsilon_{ijk}/2 = \psi\, \delta_{ir} \tag{23}$$

and the third-order tensor is now total skew-symmetric ($\psi_{ijk} \equiv \psi_{[ijk]}$), in contrast to (21).

Instead of the continuity tensor $\underset{\sim}{\psi}$ to be found in (21) we can use the damage tensor $\underset{\sim}{\omega}$ defined as:

$$\omega_{ijk} \equiv \varepsilon_{ijk} - \psi_{ijk}\,, \qquad \omega_{ij} \equiv \delta_{ij} - \psi_{ij} \tag{24a,b}$$

and characterized by the dual relation:

$$\omega_{ijk} = \omega_{i[jk]} = \varepsilon_{jkr}\omega_{ir} \quad <=> \quad \omega_{ir} = \varepsilon_{rjk}\omega_{ijk}/2 \ . \tag{25}$$

3. STRESSES IN A DAMAGED CONTINUUM

In the undamaged continuum (Fig. 3a) CAUCHY's formula

$$p_i = \sigma_{ji}n_j \tag{26}$$

was derived from the principle of equilibrium, where p_i, n_i are the components of the stress vector $\underset{\sim}{p}$ and the unit vector $\underset{\sim}{n}$, respectively.

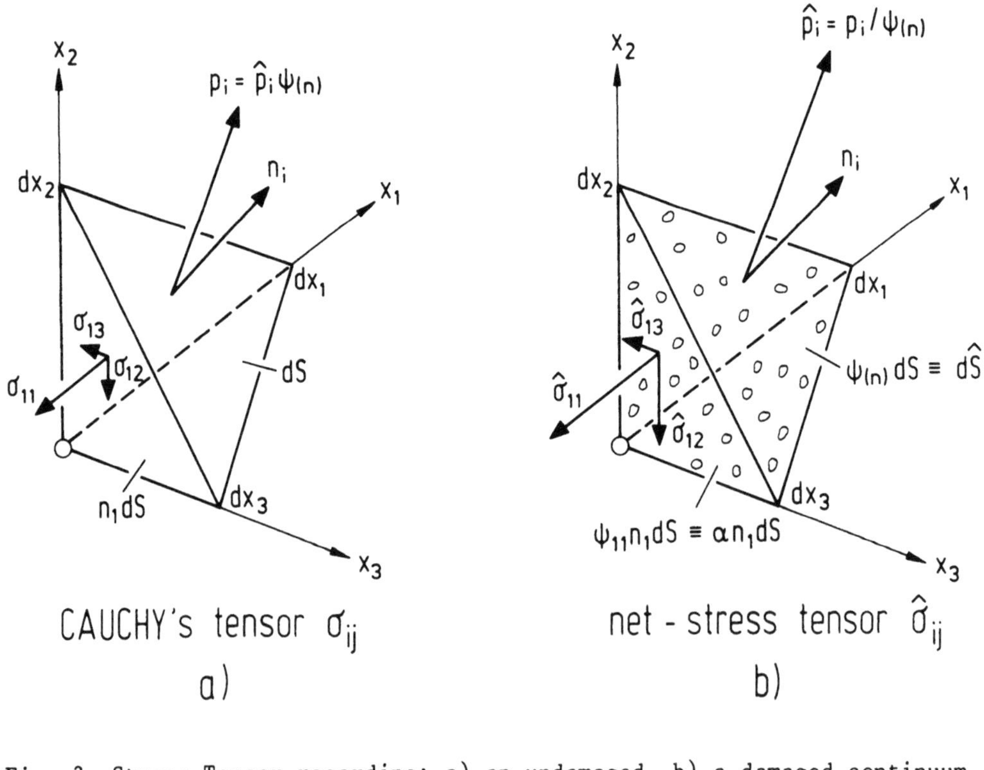

Fig. 3 Stress Tensor regarding: a) an undamaged, b) a damaged continuum

In the same way we can also find the corresponding relation for a damaged continuum:

$$\hat{p}_i \, \psi_{(n)} = \psi_{jk} \, \hat{\sigma}_{ki} \, n_j \,, \tag{27}$$

where ψ_{jk} are the components of the continuity tensor $\underset{\sim}{\psi}$ according to (22a,b).

The surface elements dS and $d\hat{S}$ in Fig. 3 are subjected to the same force vector:

$$dP_i = p_i dS \equiv \hat{p}_i d\hat{S} = d\hat{P}_i \,. \tag{28}$$

Thus, after comparing (26) to (27) we can finally find the actual net-stress tensor $\underset{\sim}{\hat{\sigma}}$, which is simply a transformation from CAUCHY's tensor:

$$\sigma_{ij} = \psi_{ir} \hat{\sigma}_{rj} = \sigma_{ji} \quad \Longleftrightarrow \quad \hat{\sigma}_{ij} = \psi_{ir}^{(-1)} \sigma_{rj} \neq \hat{\sigma}_{ji} \,. \tag{29a,b}$$

By suitable transvections we find $\sigma_{ij}\sigma_{jk}^{(-1)} = \psi_{ik}$ and $\hat{\sigma}_{ij}\sigma_{jk}^{(-1)} = \psi_{ik}^{(-1)}$. As indicated in (29a,b) the actual net-stress tensor $\underset{\sim}{\hat{\sigma}}$ is non-symmetric, unless we have isotropic damage expressed as in (23).

Because of the symmetry $\sigma_{ij} = (\sigma_{ij} + \sigma_{ji})/2$ of CAUCHY's stress tensor $\underset{\sim}{\sigma}$ we can write the following representations:

$$\sigma_{ij} = (\psi_{ip}\delta_{jq} + \delta_{iq}\psi_{jp})\hat{\sigma}_{pq}/2 \equiv \varphi_{ijpq}\hat{\sigma}_{pq} \,, \tag{29c}$$

$$\hat{\sigma}_{ij} = (\psi_{ip}^{(-1)}\delta_{jq} + \psi_{iq}^{(-1)}\delta_{jp})\sigma_{pq}/2 \equiv \Phi_{ijpq}\sigma_{pq} \,, \tag{29d}$$

from (29a,b). We see, the fourth-order tensors $\underset{\sim}{\varphi}$ and $\underset{\sim}{\Phi}$ defined in (29c,d) are only symmetric with respect to two indices:

$$\varphi_{ijpq} = \varphi_{jipq} \,, \qquad \Phi_{ijpq} = \Phi_{ijqp} \,, \tag{30a,b}$$

so that, for instance, the tensor $\underset{\sim}{\varphi}$ can be decomposed into a symmetric and an antisymmetric part [11]:

$$\varphi_{ijpq} = (\psi_{ip}\delta_{jq} + \delta_{iq}\psi_{jp} + \psi_{iq}\delta_{jp} + \delta_{ip}\psi_{jq})/4 +$$

$$+ (\psi_{ip}\delta_{jq} + \delta_{iq}\psi_{jp} - \psi_{iq}\delta_{jp} - \delta_{ip}\psi_{jq})/4 , \qquad (31)$$

where the symmetric part is identical to the tensor $\underset{\sim}{\Omega}$ in (6) introduced by RABOTNOV [10].

Furthermore, from (29) and (30) one can see, that the actual net-stress tensor $\underset{\sim}{\hat{\sigma}}$ is non-symmetric in the anisotropic damage case:

$$\hat{\sigma}_{12}/\hat{\sigma}_{21} = \beta/\alpha , \quad \hat{\sigma}_{23}/\hat{\sigma}_{32} = \gamma/\beta , \quad \hat{\sigma}_{31}/\hat{\sigma}_{13} = \alpha/\gamma . \qquad (32)$$

This fact is disadvantageous, and therefore it is awkward to use the actual net-stress tensor $\underset{\sim}{\hat{\sigma}}$ in constitutive equations with a symmetric strain rate tensor $\underset{\sim}{d}$. A transformed net-stress tensor $\underset{\sim}{t}$ is introduced in [8], which is symmetric, and thusly more advantageous. This tensor is defined through the following operation:

$$t_{ij} = (\hat{\sigma}_{ik}\psi_{kj}^{(-1)} + \psi_{ki}^{(-1)}\hat{\sigma}_{jk})/2 . \qquad (33)$$

Inserting (29d) into (33), one can find:

$$t_{ij} = C_{ijpq}^{(-1)}\sigma_{pq} \qquad \Rightarrow \qquad \sigma_{ij} = C_{ijpq}t_{pq} , \qquad (34)$$

where

$$C_{ijpq} = (\psi_{ip}\psi_{jq} + \psi_{iq}\psi_{jp})/2 \text{ with } C_{ijpq}^{(-1)} = (\psi_{ip}^{(-1)}\psi_{jq}^{(-1)} + \psi_{iq}^{(-1)}\psi_{jp}^{(-1)})/2 \qquad (35)$$

is a symmetric fourth-order tensor of continuity having the following properties:

$$C_{ijpq} = C_{jipq} = C_{ijqp} = C_{pqij} . \qquad (36)$$

Because of (22a,b) and (36) the fourth-order tensor of continuity in

(35) can be represented by a 6 x 6 matrix, which has the diagonal form:

$$C_{ijkl} = \mathrm{diag}\,\{C_{1111},\ C_{2222},\ C_{3333},\ C_{1212},\ C_{2323},\ C_{3131}\}\ , \tag{37a}$$

$$C_{ijkl} = \mathrm{diag}\,\{\alpha^2,\ \beta^2,\ \gamma^2,\ \alpha\beta/2,\ \beta\gamma/2,\ \gamma\alpha/2\}\ , \tag{37b}$$

that is, the components of the "pseudo-net-stress" tensor $\underset{\sim}{t}$ are given as

$$\left.\begin{array}{lll} t_{11} = \sigma_{11}/\alpha^2\ , & t_{12} = \sigma_{12}/(\alpha\beta)\ , & t_{13} = \sigma_{13}/(\alpha\beta)\ , \\ t_{21} = t_{12}\ , & t_{22} = \sigma_{22}/\beta^2\ , & t_{23} = \sigma_{23}/(\beta\gamma)\ , \\ t_{31} = t_{13}\ , & t_{32} = t_{23}\ , & t_{33} = \sigma_{33}/\gamma^2\ . \end{array}\right\} \tag{38}$$

An interpretation of the introduced "pseudo-net-stress" tensor in (33) can be given in the following way. An alternative form of CAUCHY's formula (26) is:

$$dP_i = \sigma_{ji}\ dS_j\ , \tag{39}$$

where dP_i is the actual force vector (28). According to (15) we can write:

$$dP_i = \sigma_{ji}\psi^{(-1)}_{jr}\,d\hat{S}_r\ , \tag{40a}$$

or substituting $\sigma_{ji} = C_{jipq}t_{pq}$ we find the relation:

$$dP_i = \psi_{ip}t_{pr}d\hat{S}_r \tag{40b}$$

which can be multiplied by $\psi^{(-1)}_{ki}$, so that we have $\psi^{(-1)}_{ki}dP_i = t_{kr}d\hat{S}_r$, or after changing the indices:

$$\psi^{(-1)}_{ik}dP_k \equiv d\tilde{P}_i = t_{ji}d\hat{S}_j\ . \tag{41}$$

Comparing (39) to (41) we see, that (41) can be interpreted as CAUCHY's formula for the fictitious undamaged configuration (Fig. 1c), which is subjected to the pseudo-force $d\tilde{P}_i \equiv \psi_{ik}^{(-1)} dP_k$ instead of the actual force dP_i. The "pseudo-force" vector $d\tilde{P}_i$ is equal to the actual force vector dP_i, if the continuum is in an undamaged state ($\psi_{ij} = \delta_{ij}$). In the total damaged state ($\psi_{ij} = 0_{ij}$), however, $d\tilde{P}_i$ is approaching infinity.

4. CONSTITUTIVE EQUATIONS INVOLVING DAMAGE AND INITIAL ANISOTROPY

One must consider the fact that the symmetric tensor-valued functions (5a,b) are valid for an isotropic material in an anisotropic damaged state. Furthermore, one must differentiate between the anisotropic damage growth and the initial anisotropy resulting from a forming process, for instance rolling. Then constitutive equations can be found in expressions such as those in (1).

In formulating constitutive equations like (1) one has to take the following into account: the undamaged case ($\underset{\sim}{\omega} \rightarrow \underset{\sim}{0}$) immediately leads to (5a), while the strain rate tensor $\underset{\sim}{d}$ approaches infinity as $\underset{\sim}{\omega}$ approaches $\underset{\sim}{\delta}$. With respect to polynomial representations of constitutive equations it is convenient to use the tensor +)

$$D_{ij} \equiv (\delta_{ij} - \omega_{ij})^{(-1)} \equiv \psi_{ij}^{(-1)} \qquad (42)$$

as an argument tensor instead of the tensorial damage variable $\underset{\sim}{\omega}$. Thus, expressions such as

$$d_{ij} = f_{ij}\,(\sigma_{pq},\, D_{pq},\, A_{pqrs}) \qquad (43)$$

must be taken into consideration, i.e., where $f_{ij} = f_{ji}$ are the components of a second-order tensor-valued function of two symmetric tensors of rank two and one symmetric tensor of rank four. The general represen-

---

+) Then, the pseudo-force vector (41) is defined as $d\tilde{P}_i \equiv D_{ij} P_j$, for instance.

tation is given through a linear combination (2), where the $\underset{\sim}{G}$'s are symmetric tensor generators of rank two involving the argument tensors $\underset{\sim}{\sigma}$, $\underset{\sim}{D}$, $\underset{\sim}{A}$. Such generators, for instance, are given by matrix products of types [14]:

$$\{\underset{\sim}{\sigma}^{\lambda}\underset{\sim}{D}^{\nu} + \underset{\sim}{D}^{\nu}\underset{\sim}{\sigma}^{\lambda}\}_{ij} \equiv \sigma_{ik}^{(\lambda)} D_{kj}^{(\nu)} + D_{ik}^{(\nu)}\sigma_{kj}^{(\lambda)} \tag{44}$$

or [8]:

$$\{\underset{\sim}{\sigma}^{\lambda}\underset{\sim}{A}^{\mu} \cdot \underset{\sim}{D}^{\nu} + \underset{\sim}{A}^{\mu} \cdot \underset{\sim}{D}^{\nu}\underset{\sim}{\sigma}^{\lambda}\}_{ij} \equiv \sigma_{ik}^{(\lambda)} A_{kjpq}^{(\mu)} D_{pq}^{(\nu)} + A_{ikpq}^{(\mu)} D_{pq}^{(\nu)}\sigma_{kj}^{(\lambda)} \ . \tag{45}$$

Then, the condition of form-invariance

$$a_{ik}a_{jl}f_{kl}(\sigma_{pq},\ldots,A_{pqrs}) \equiv f_{ij}(a_{pt}a_{qu}\sigma_{tu},\ldots,a_{pt}a_{qu}a_{rv}a_{sw}A_{tuvw}) \tag{46}$$

is fulfilled under the transformation of the orthogonal group $\underset{\sim}{a}$.

In finding an irreducible set of tensor generators we note, that, according to the HAMILTON-CAYLEY theorem [14] none of the exponents λ,ν of the second-order tensors $\underset{\sim}{\sigma}$, $\underset{\sim}{D}$ need be larger than 2, and note that a symmetric fourth-order tensor of power six and all higher powers can be expressed in terms of lower powers, that is $\mu \leqq 5$, as shown in [1,15, 16].

The coefficients φ_{α} in the linear combination (2) are scalar-valued functions of the integrity basis the elements of which are the irreducible invariants. In finding the integrity basis associated with the representation (2) of the constitutive equations (43) we must take the theorem into account [17], which states that the trace of any product formed from matrices and their transposes is a polynomial invariant under the full or proper orthogonal group. Thus we form the following traces:

$$\left.\begin{aligned} &\mathrm{tr}\ \underset{\sim}{d}\ \underset{\sim}{\sigma} = \mathrm{tr} \sum_{\alpha} \varphi_{\alpha}\ \underset{\sim}{G}_{\alpha}\ \underset{\sim}{\sigma} \equiv \sum_{\alpha} \varphi_{\alpha}\ \mathrm{tr}\ \underset{\sim}{G}_{\alpha}\ \underset{\sim}{\sigma}\ , \\ &\mathrm{tr}\ \underset{\sim}{d}\ \underset{\sim}{D}\ , \quad \mathrm{tr}\ \underset{\sim}{d}\ \underset{\sim}{A}\cdot\underset{\sim}{D}\ , \quad \mathrm{tr}\ \underset{\sim}{d}\ \underset{\sim}{A}\cdot\underset{\sim}{D} \equiv d_{ij}A_{ijpq}D_{pq}\ . \end{aligned}\right\} \tag{47}$$

Alternatively, we can form tr $\underset{\sim}{d}$, tr $\underset{\sim}{d}^2$, and tr $\underset{\sim}{d}^3$ from the linear combination (2). The sets of invariants so formed contain reducible and equivalent elements, i.e. redundant elements. In writing down the possible irreducible invariants, we can use some results from the theory of invariants, for instance the following LEMMAs [17]:

- If $\underset{\sim}{G}$ is a reducible generator, then tr $\underset{\sim}{X}\underset{\sim}{G}$ and tr $\underset{\sim}{G}\underset{\sim}{X}$ are reducible, where $\underset{\sim}{X}$ is an arbitrary tensor.
- The trace of a matrix product is unaltered by cyclic permutations of its factors.
- The trace of a matrix product is equal to the trace of the transpose of the product.
- The transpose of a matrix product π is the matrix product formed by writing down the transposes of the factors of π in reverse order, for instance, tr $\underset{\sim}{a}\underset{\sim}{b}\underset{\sim}{c}$ = tr $\underset{\sim}{c}'\underset{\sim}{b}'\underset{\sim}{a}'$.

Further deliberations concerning reducibility and equivalence and concerning the relevant consequences of the HAMILTON-CAYLEY theorem are given in [13,14,17,18].

In the following some examples of constitutive equations (43) represented through a linear combination (2) or in a canonical form (3) are discussed.

4.1 Representation by Superposition

First, we assume that a representation of the tensor-valued function (43) in the three tensor variables $\underset{\sim}{\sigma}$, $\underset{\sim}{D}$, $\underset{\sim}{A}$ of rank two, two, four, respectively, is given by "superposition":

$$\mathfrak{M}(\underset{\sim}{\sigma}, \underset{\sim}{D}, \underset{\sim}{A}) = \mathfrak{M}_1(\underset{\sim}{\sigma}, \underset{\sim}{D}) + \mathfrak{M}_2(\underset{\sim}{\sigma}, \underset{\sim}{A}) + \mathfrak{M}_3(\underset{\sim}{D}, \underset{\sim}{A}) \ . \tag{48}$$

Then a tensor polynomial in $\underset{\sim}{\sigma}$, $\underset{\sim}{D}$, $\underset{\sim}{A}$ is made up by a linear combination (2) which includes the tensor generator

$${}^1G_{ij}^{[\lambda,\nu]} \equiv (L_{ij}^{[\lambda,\nu]} + L_{ji}^{[\lambda,\nu]})/2, \quad {}^2G_{ij}^{[\lambda,\mu,\nu]} \equiv (M_{ij}^{[\lambda,\mu,\nu]} + M_{ji}^{[\lambda,\mu,\nu]})/2, \tag{49a,b}$$

$$ {}^{3}G_{ij}^{[\lambda,\mu,\nu]} \equiv (N_{ij}^{[\lambda,\mu,\nu]} + N_{ji}^{[\lambda,\mu,\nu]})/2 \; ; \; \lambda,\nu = 0,1,2; \; \mu = 0,1,\dots,5, \quad (49c) $$

where the following matrix products are introduced:

$$ \{\underset{\sim}{L}^{[\lambda,\nu]}\}_{ij} \equiv L_{ij}^{[\lambda,\nu]} \equiv \sigma_{ik}^{(\lambda)} D_{kj}^{(\nu)} \equiv \{\underset{\sim}{\sigma}^{\lambda} \underset{\sim}{D}^{\nu}\}_{ij} , \quad (50a) $$

$$ M_{ij}^{[\lambda,\mu,\nu]} \equiv \sigma_{ik}^{(\lambda)} A_{kjpq}^{(\mu)} \sigma_{pq}^{(\nu)} , \qquad N_{ij}^{[\lambda,\mu,\nu]} \equiv D_{ik}^{(\lambda)} A_{kjpq}^{(\mu)} D_{pq}^{(\nu)} . \quad (50b,c) $$

The representation of (43) is then given in the form:

$$ f_{ij} = \sum_{\lambda,\nu} {}^{1}\varphi_{(\lambda,\nu)} \; {}^{1}G_{ij}^{[\lambda,\nu]} + \dots + \sum_{\lambda,\mu,\nu} {}^{3}\varphi_{(\lambda,\mu,\nu)} \; {}^{3}G_{ij}^{[\lambda,\mu,\nu]} , \quad (51) $$

which contains 99 irreducible tensor generators: 9 from the set (49a) and 45 generators each from the sets (49b,c). Note, that in (50a,b,c) those values of λ,μ,ν which are bracketed by parentheses are exponents, whereas a square bracket gives any set λ,ν or λ,μ,ν the character of a label which indicates several matrix products.

In the undamaged case ($\underset{\sim}{\omega} = \underset{\sim}{0}$) the tensor $\underset{\sim}{D}$ defined in (42) is identical to KRONECKER's tensor $\underset{\sim}{\delta}$, and the expressions (49a,c) are then the generators of the single argument tensors $\underset{\sim}{\sigma}$, $\underset{\sim}{A}$, only. This special case leads to constitutive equations appropiate for the description of the response of an anisotropic material in its undamaged state.

In the special case of an undamaged ($\underset{\sim}{D} \to \underset{\sim}{\delta}$) and isotropic ($\underset{\sim}{A}^{\mu} \to \underset{\sim}{A}^{0}$) continuum we find from (49a,b,c) that ${}^{1}G_{ij}^{[\lambda,\nu]} \to \sigma_{ij}^{(\lambda)}$, ${}^{2}G_{ij}^{[\lambda,\mu,\nu]} \to \sigma_{ij}^{(\lambda+\nu)}$ and ${}^{3}G_{ij}^{[\lambda,\mu,\nu]} \to \delta_{ij}$, i.e., the representation (51) contains the constitutive equation

$$ d_{ij} = \varphi_0 \delta_{ij} + \varphi_1 \sigma_{ij} + \varphi_2 \sigma_{ij}^{(2)} \quad (52) $$

as a special case.

The representation (51) may be extented by using matrix products of the types:

$$P_{ij}^{[\lambda,\mu,\nu]} \equiv \sigma_{ik}^{(\lambda)} A_{kjpq}^{(\mu)} D_{pq}^{(\nu)} , \qquad Q_{ij}^{[\lambda,\mu,\nu]} \equiv D_{ik}^{(\nu)} A_{kjpq}^{(\mu)} \sigma_{pq}^{(\lambda)} , \tag{53a,b}$$

that is by considering further tensor generators:

$$^{4}G_{ij}^{[\lambda,\mu,\nu]} \equiv P_{(ij)}^{[\lambda,\mu,\nu]} , \qquad ^{5}G_{ij}^{[\lambda,\mu,\nu]} \equiv Q_{(ij)}^{[\lambda,\mu,\nu]} , \tag{54a,b}$$

each of which are 45 irreducible generators. Instead of (51) we then find a representation with 139 irreducible tensor generators.

The scalar-valued functions $^{1}\varphi_{(\lambda,\nu)}$, $^{2}\varphi_{(\lambda,\mu,\nu)}$, etc. in (51) are functions in the elements of the integrity basis, which can be found by forming the following transvections:

$$f_{ij}\sigma_{ji}, \; f_{ij}D_{ji}, \; f_{ij}A_{jipq}\sigma_{pq}, \; f_{ij}A_{jipq}\sigma_{pq}^{(2)}, \; f_{ij}A_{jipq}D_{pq}, \; f_{ij}A_{jipq}D_{pq}^{(2)}, \tag{55}$$

or, alternatively, by considering the basis invariants f_{ii}, $f_{ij}f_{ji}$, $f_{ij}f_{jk}f_{ki}$ of the tensorial expression (51), as was pointed out in detail after (47).

We see, that the representation (51) with 99 irreducible generators or its extended form with 139 irreducible elements are too complicated. Therefore, we have to search constitutive equations for more practical use.

4.2 Simplified Representations

The effect of anisotropy characterized by a fourth-order tensor $\underset{\sim}{A}$ can be expressed through a symmetric "image" tensor $\underset{\sim}{\tau}$ of rank two, i.e. through a linear transformation:

$$\tau_{ij} = A_{ijkl}\sigma_{kl} = \tau_{ji} . \tag{56}$$

Then we have to consider the constitutive expression:

$$d_{ij} = f_{ij}(\sigma_{pq}, D_{pq}, \tau_{pq}) \; , \tag{57}$$

being a tensor-valued function of three symmetric tensor variables $\underset{\sim}{\sigma}$, $\underset{\sim}{D}$, $\underset{\sim}{\tau}$. As was pointed out by SPENCER and RIVLIN [13,18] in detail, the representation of the tensor function (57) is characterized by 46 irreducible tensor generators and an integrity basis with 28 irreducible invariants. The representation of a non-polynomial function (57), which was discussed by WANG[21], SMITH[19]and BOEHLER[22], involves 16 tensor generators and only 22 invariants. In[11] the author of this chapter has considered also 19 tensor generators, but the associated integrity basis consits of 28 irreducible elements. It is remarkable that these 28 elements are identical to those given by RIVLIN and SPENCER [13,14,17,18], although these authors used a more extended representation than the representation discussed in [11]. On the one hand, the former with 46 tensor generators and, on the other hand, the latter with 19. Furthermore, in [11] a canonical form (3) for (57) is found by constructing fourth-order tensor-valued functions ${}^{0}\underset{\sim}{H}$, ${}^{1}\underset{\sim}{H}$, ${}^{2}\underset{\sim}{H}$ of the argument tensors $\underset{\sim}{D}$ and $\underset{\sim}{\tau}$.

A further simplification in representing constitutive equations is shown in [11] by considering the tensor-valued function

$$d_{ij} = f_{ij}(\hat{\sigma}_{pq}, \tau_{pq}) \tag{58}$$

in two second-rank tensor variables each of which is a linear transformation (29b) and (56) from CAUCHY's stress tensor $\underset{\sim}{\sigma}$, where the linear operators $\underset{\sim}{\Phi}$ and $\underset{\sim}{A}$ of rank four are determined from the effect of damage ($\underset{\sim}{\psi} = \underset{\sim}{\delta} - \underset{\sim}{\omega}$) and from the anisotropy of the material in its undamaged state.

The only inconvenience in finding a set of irreducible tensor generators for (58) is the fact, that the net-stress tensor $\hat{\underset{\sim}{\sigma}}$ is non-symmetric. The list of tensor generators given in [11] contains 13 symmetric tensor generators, and the associated integrity basis has 20 irreducible invariants. A canonical form for (58) can also be found in [11].

Another way of simplifying the form (43) is the tensor function

$$d_{ij} = f_{ij}(t_{pq}, A_{pqrs}) \tag{59}$$

in the symmetric "pseudo-net-stress" tensor $\underset{\sim}{t}$ defined in (34) and the symmetric anisotropy tensor $\underset{\sim}{A}$ of rank four. The representation (2) for (59) is characterized by 48 tensor generators, i.e. we have:

$$d_{ij} = \frac{1}{2} \sum_{\lambda,\mu,\nu} \varphi_{(\lambda,\mu,\nu)} (t_{ik}^{(\lambda)} A_{kjpq}^{(\mu)} t_{pq}^{(\nu)} + t_{jk}^{(\lambda)} A_{kipq}^{(\mu)} t_{pq}^{(\nu)}) \tag{60}$$

$(\lambda,\nu = 0,1,2;\ \mu = 0,1,\ldots,5)$, where the φ's are scalar-valued functions of the integrity basis.

Finally, instead of (58), we should consider the function

$$d_{ij} = f_{ij}(t_{pq}, \tau_{pq}) \tag{61}$$

in the two symmetric second-rank tensors each of which is a linear transformation (34) and (56) from CAUCHY's stress tensor $\underset{\sim}{\sigma}$, where the symmetric linear operators $\underset{\sim}{C}^{-1}$ and $\underset{\sim}{A}$ of rank four are determined by the effect of damage and by the anisotropy of the material in its undamaged state.

It is well known [14], that the representation (2) of the tensor function (61) involves nine generators, i.e., we have:

$$d_{ij} = \frac{1}{2} \sum_{\lambda,\nu=0}^{2} \varphi_{(\lambda,\nu)} (t_{ik}^{(\lambda)} \tau_{kj}^{(\nu)} + \tau_{ik}^{(\nu)} t_{kj}^{(\lambda)}) , \tag{62}$$

where $\varphi_{(\lambda,\nu)}$ are scalar-valued functions of the integrity basis associated with the representation (62), which can be found by forming tr $\underset{\sim}{d}\underset{\sim}{t}$ and tr $\underset{\sim}{d}\underset{\sim}{\tau}$. In this way we find the invariants

$$t_{ij}^{(\lambda+1)} \tau_{ji} \equiv \text{tr}\ \underset{\sim}{t}^{\lambda+1} \underset{\sim}{\tau}^{\nu} \quad \text{and} \quad t_{ij}^{(\lambda)} \tau_{ji}^{(\nu+1)} \equiv \text{tr}\ \underset{\sim}{t}^{\lambda} \underset{\sim}{\tau}^{\nu+1} , \tag{63a,b}$$

some of which are reducible or equivalent. In seperating the redundant elements, as was pointed out in detail after (47), we find from (63a,b) the integrity basis:

$$\left.\begin{array}{l} \mathrm{tr}\, \underset{\sim}{t},\ \mathrm{tr}\, \underset{\sim}{t}^2,\ \mathrm{tr}\, \underset{\sim}{t}^3;\ \mathrm{tr}\, \underset{\sim}{\tau},\ \mathrm{tr}\, \underset{\sim}{\tau}^2,\ \mathrm{tr}\, \underset{\sim}{\tau}^3; \\ \mathrm{tr}\, \underset{\sim}{t}\underset{\sim}{\tau},\ \mathrm{tr}\, \underset{\sim}{t}\underset{\sim}{\tau}^2,\ \mathrm{tr}\, \underset{\sim}{t}^2\underset{\sim}{\tau},\ \mathrm{tr}\, \underset{\sim}{t}^2\underset{\sim}{\tau}^2, \end{array}\right\} \tag{64}$$

which is associated with the representation (62).

We can write the constitutive equation (62), for instance, in a canonical form, like (3), if we introduce [20] the following identities for tensor generators:

Let $\underset{\sim}{X}$ and $\underset{\sim}{Y}$ be two symmetric second-order tensors, i.e. $X_{ij} = X_{ji}$ and $Y_{ij} = Y_{ji}$, respectively. Then, the identities

$$\{\tfrac{1}{2}(\underset{\sim}{X}^{\lambda}\underset{\sim}{Y}^{\nu} + \underset{\sim}{Y}^{\nu}\underset{\sim}{X}^{\lambda})\}_{ij} \equiv \eta^{[\nu]}_{ijkl} X^{(\lambda)}_{kl} \equiv \xi^{[\lambda]}_{ijkl} Y^{(\nu)}_{kl} \tag{65}$$

are valid for arbitrary values λ and ν, where the symbols $\xi^{[\lambda]}_{ijkl}$ and $\eta^{[\nu]}_{ijkl}$ are λ and ν several fourth-order tensors defined in the following way:

$$\xi^{[\lambda]}_{ijkl} \equiv (X^{(\lambda)}_{ik}\delta_{jl} + X^{(\lambda)}_{il}\delta_{jk} + \delta_{ik}X^{(\lambda)}_{jl} + \delta_{il}X^{(\lambda)}_{jk})/4\ , \tag{66a}$$

$$\eta^{[\nu]}_{ijkl} \equiv (Y^{(\nu)}_{ik}\delta_{jl} + Y^{(\nu)}_{il}\delta_{jk} + \delta_{ik}Y^{(\nu)}_{jl} + \delta_{il}Y^{(\nu)}_{jk})/4\ . \tag{66b}$$

Similar identities for non-symmetric tensors were introduced in [11].

4.3 Separation of Tensor Variables

In the following a simple method will be suggested [3] where one can find representations for tensor-valued functions by separating tensor variables. For instance, the function (61) may be represented by separating the two variables $\underset{\sim}{t}$ and $\underset{\sim}{\tau}$ in the following way:

$$d_{ij} = f_{ij}(\underset{\sim}{t},\underset{\sim}{\tau}) = [\delta_{ik}\, {}^0X_{kj}(\underset{\sim}{\tau}) + {}^0X_{ik}(\underset{\sim}{\tau})\delta_{kj}]/2 +$$

$$+ [t_{ik}\, {}^1X_{kj}(\underset{\sim}{\tau}) + {}^1X_{ik}(\underset{\sim}{\tau})t_{kj}]/2 +$$

$$+ [t_{ik}^{(2)}\, {}^2X_{kj}(\underset{\sim}{\tau}) + {}^2X_{ik}(\underset{\sim}{\tau})t_{kj}^{(2)}]/2 \,, \tag{67a}$$

$$d_{ij} = \frac{1}{2}\sum_{\lambda=0}^{2} [t_{ik}^{(\lambda)}\, {}^\lambda X_{kj}(\underset{\sim}{\tau}) + {}^\lambda X_{ik}(\underset{\sim}{\tau})t_{kj}^{(\lambda)}] \,. \tag{67b}$$

This is a representation consisting of three parts ($\lambda = 0,1,2$), which may be interpreted as the contributions of the zero, first, and second orders found in the variable $\underset{\sim}{t}$. The three functions ${}^0\underset{\sim}{X}$, ${}^1\underset{\sim}{X}$, ${}^2\underset{\sim}{X}$ are isotropic tensor functions of the other variable $\underset{\sim}{\tau}$:

$${}^\lambda X_{ij}(\underset{\sim}{\tau}) = \sum_{\nu=0}^{2} \varphi_{(\lambda,\nu)}\, \tau_{ij}^{(\nu)} \,. \tag{68}$$

Inserting (68) in (67b), the usual polynomial (62) is obtained.

In a similar way one can represent the symmetric tensor-valued function (57) of three symmetric argument tensors:

$$\left.\begin{aligned} d_{ij} = {} & \frac{1}{2}\sum_{\lambda=0}^{2} [\sigma_{ik}^{(\lambda)}\, {}^\lambda X_{kj}(\underset{\sim}{D},\underset{\sim}{\tau}) + {}^\lambda X_{ik}(\underset{\sim}{D},\underset{\sim}{\tau})\sigma_{kj}^{(\lambda)}] + \\ & + \frac{1}{2}\sum_{\lambda=0}^{2} [D_{ik}^{(\lambda)}\, {}^\lambda Y_{kj}(\underset{\sim}{\tau},\underset{\sim}{\sigma}) + {}^\lambda Y_{ik}(\underset{\sim}{\tau},\underset{\sim}{\sigma})D_{kj}^{(\lambda)}] + \\ & + \frac{1}{2}\sum_{\lambda=0}^{2} [\tau_{ik}^{(\lambda)}\, {}^\lambda Z_{kj}(\underset{\sim}{\sigma},\underset{\sim}{D}) + {}^\lambda Z_{ik}(\underset{\sim}{\sigma},\underset{\sim}{D})\tau_{kj}^{(\lambda)}] \,, \end{aligned}\right\} \tag{69}$$

where nine isotropic tensor functions each containing two variables are introduced:

$${}^\lambda X_{ij}(\underset{\sim}{D},\underset{\sim}{\tau}) = \frac{1}{2}\sum_{\mu,\nu=0}^{2} \varphi_{(\lambda,\mu,\nu)}[D_{ik}^{(\mu)}\tau_{kj}^{(\nu)} + \tau_{ik}^{(\nu)}D_{kj}^{(\mu)}] \;;\; \lambda = 0,1,2. \tag{70}$$

The other functions ${}^\lambda\underset{\sim}{Y}$, ${}^\lambda\underset{\sim}{Z}$ in (69) have been defined in a similar manner.

Thus, the representation (69) consists of 81 terms, where some of them are equivalent. After selecting the equivalent elements, one can find a set of 43 irreducible terms [1].

The obove method of "separation of tensor variables" can be used in order to find tensor-valued functions for several argument tensors.

REFERENCES

1. BETTEN, J., Elasticitäts- und Plastizitätslehre, Vieweg-Verlag, Braunschweig/Wiesbaden 1985.

2. BETTEN, J., Zur Aufstellung von Stoffgleichungen in der Kriechmechanik anisotroper Körper, Rheol. Acta 20 (1981): 527-535.

3. BETTEN, J., Constitutive Equations of Isotropic and Anisotropic Materials in the Secondary and Tertiary Creep Stage, in: Creep and Fracture of Engineering Materials and Structures, Part II, (eds. B. Wilshire and D.R.J. Owen), Pineridge Press, Swansea 1984, 1291-1305.

4. BETTEN, J., Materialgleichungen zur Beschreibung des sekundären und tertiären Kriechverhaltens anisotroper Stoffe, Z. Angew. Math. Mech. (ZAMM) 64 (1984): 211-220.

5. BETTEN, J., Applications of Tensor Functions to the Formulation of Constitutive Equations involving Damage and Initial Anisotropy, IUTAM-Symposium on Mechanics of Damage and Fatigue, Haifa and Tel Aviv 1985, to be published in the proceedings (eds. S.R. Bodner and Z. Hashin).

6. KACHANOV, L.M., On the Time to Failure under Creep Conditions (in Russian), Izv. Ak. Nauk USSR Otdel. Tekh. Nauk, 8 (1958): 26-31.

7. RABOTNOV, Y.N., Creep Problems in Structural Members, North-Holland 1969.

8. BETTEN, J., Damage Tensors in Continuum Mechanics, Euromech Colloquium 147 on "Damage Mechanics", Paris VI, Cachan 1981, published in Journal de Mécanique theórique et appliquée 2 (1983): 13-32.

9. MURAKAMI, S. and OHNO, N., A Continuum Theory of Creep and Creep Damage, in: Creep in Structures (eds. A.R.S. Ponter and D.R. Hayhurst), Springer-Verlag, Berlin/Heidelberg/New York 1981, 422-444.

10. RABOTNOV, Yu. N., Creep Rupture, in: Applied Mechanics Conference (eds. M. Hetenyi and H. Vincenti), Stanford University 1968, 342-349.

11. BETTEN, J., Net-Stress Analysis in Creep Mechanics, Ingenieur-Archiv 52 (1982): 405-419.

12. JOHNSON, A.E., Complex-Stress Creep of Metals, Metallurgical Reviews 5 (1960): 447-506.

13. SPENCER, A.J.M. and Rivlin, R.S., Finite Integrity Bases for Five or Fewer Symmetric 3 x 3 Matrices, Arch. Rational Mech. Anal. 2 (1958/59): 435-446.

14. SPENCER, A.J.M., Theory of Invariants, in: Continuum Physics. Vol. I, (ed. A.C. Eringen), Academic Press, New York 1971, 239-353.

15. BETTEN, J., Integrity Basis for a Second-Order and a Fourth-Order Tensor, International J. Math. & Math. Sci. 5 (1982): 87-96.

16. BETTEN, J., Irreducible Invariants of Fourth-Order Tensors, Fifth International Conference on Mathematical Modelling, University of California, Berkeley 1985, to be published in the proceedings: Mathematical Modelling in Science and Technology (eds. X.J.R. Avula, G. Leitmann, C.D. Mote, Jr. and E.Y. Rodin), Pergamon Press, New York 1986.

17. RIVLIN, R.S. and SMITH, G.F., Orthogonal Integrity Basis for N Symmetric Matrices, in: Contributions to Mechanic (ed. D. Abir), Pergamon Press, Oxford/.../Braunschweig 1969, 121-141.

18. SPENCER, A.J.M. and RIVLIN, R.S., The Theory of Matrix Polynomials and its Application to the Mechanics of Isotropic Continua, Arch. Rational Mech. Anal. 2 (1958/59): 309-336.

19. SMITH, G.F., On Isotropic Functions of Symmetric Tensors, Skew-Symmetric Tensors, and Vectors, Int. J. Engng. Sc. 9 (1971): 899-916.

20. BETTEN, J., On the Representation of the Plastic Potential of Anisotropic Solids, Colloque International du CNRS n° 319, Comportement plastique des solides anisotropes, Grenoble 1981, published in the the proceedings: Plastic Behavior of Anisotropic Solids (ed. J.P. Boehler), CNRS, Paris 1985, 213-228.

21. WANG, C.C., On Representations for Isotropic Functions, Part I, Arch. Rational Mech. Anal. 33 (1969): 249-267.

22. BOEHLER, J. P., On Irreducible Representations for Isotropic Scalar Functions, Z. Angew. Math. Mech. (ZAMM) 57 (1977) : 323-327.

Chapter 13

INTERPOLATION METHOS FOR TENSOR FUNCTIONS

J. Betten
Technical University Aachen, F.R. Germany

1. INTRODUCTION

In this chapter polynomial interpolation is considered and extended to tensor-valued functions of one and two argument tensors. Let x_α $(\alpha = 1,2,\ldots,n)$ be distinct points and y_α corresponding values. The polynomial of degree n-1

$$y = f(x) = \sum_{\alpha=1}^{n} L_\alpha(x) y_\alpha + R_n \tag{1}$$

is called "LAGRANGE interpolation formula", where the polynomials

$$L_\alpha(x) = \prod_{\substack{\beta=1 \\ \beta\neq\alpha}}^{n} (x - x_\beta)/(x_\alpha - x_\beta), \qquad \alpha = 1,2,\ldots,n \tag{2}$$

are introduced. It is clear that $L_\alpha(x_\beta)$ is equal to one for $\alpha = \beta$ and equal to zero for $\alpha \neq \beta$. The remainder in (1) is given by

$$R_{n-1} = \frac{1}{n!} (x - x_1) \ldots (x - x_n) \frac{d^n f(\xi)}{dx^n} \tag{3}$$

where min $(x, x_1, \ldots, x_n) < \xi <$ max $(x, x_1, \ldots, x_n)$.

Since the above interpolation has a unique solution, all other representations must coincide with the LAGRANGE polynomial (1). But the LAGRANGE formula is disadvantageous, if we pass from a space of dimension n to a space of one higher dimension. Then the "NEWTON formula"

$$f(x) = a_0 + a_1(x - x_1) + a_2(x - x_1)(x - x_2) + \ldots \tag{4}$$

is to prefer, where $a_0 \equiv f(x_1)$. The coefficients $a_1, a_2, \ldots, a_n$ are called the "devided differences" of the first, second, ..., n-th order, respectively.

2. TENSOR FUNCTION OF ONE VARIABLE

In extending the LAGRANGE interpolation method to a tensor-valued function

$$Y_{ij} = f_{ij}(\underset{\sim}{X}) = Y_{ji} , \qquad i,j = 1,2,3 \tag{5}$$

of one symmetric argument tensor $(X_{ij} = X_{ji})$, we consider the principal values $\lambda_I \equiv X_I$, $\lambda_{II} \equiv X_{II}$, $\lambda_{III} \equiv X_{III}$ of the tensor $\underset{\sim}{X}$ as interpolating points and find, by analogy to (1) and (2), the tensorial representation [1]

$$Y_{ij} = f_{ij}(\underset{\sim}{X}) = \sum_{\alpha=I}^{III} f_\alpha {}^{\alpha}L_{ij}(\underset{\sim}{X}) + R_{ij}(\underset{\sim}{X}) , \tag{6}$$

where the values $f_\alpha \equiv Y_\alpha$, $\alpha = I,II,III$, correspond to the principal values X_α. The tensors ${}^{\alpha}\underset{\sim}{L}$, $\alpha = I,II,III$, are defined by

$${}^{\alpha}L_{ij} \equiv \frac{(X_{ik} - \lambda_{(\alpha+I)}\delta_{ik})(X_{kj} - \lambda_{(\alpha+II)}\delta_{kj})}{(\lambda_{(\alpha)} - \lambda_{(\alpha+I)})(\lambda_{(\alpha)} - \lambda_{(\alpha+II)})} , \tag{7}$$

where the summation convention is utilized, and $\underset{\sim}{\delta}$ represents the KRONECKER tensor. The tensor-valued remainder in (6) is given by

$$R_{ij} = \frac{1}{3!}(X_{ik} - \lambda_I\delta_{ik})(X_{kl} - \lambda_{II}\delta_{kl})(X_{lj} - \lambda_{III}\delta_{lj})\,\frac{d^3f(\underset{\sim}{X})}{d\underset{\sim}{X}^3} \tag{8a}$$

and can be expressed by

$$R_{ij} = \frac{1}{3!}(X^{(3)}_{ij} - J_1X^{(2)}_{ij} - J_2X_{ij} - J_3\delta_{ij})\,\frac{d^3f(\underset{\sim}{X})}{d\underset{\sim}{X}^3} \tag{8b}$$

with the principal invariants (elementary symmetric functions)

$$J_1 = \lambda_I + \lambda_{II} + \lambda_{III}\ , \tag{9a}$$

$$J_2 = -(\lambda_{II}\lambda_{III} + \lambda_{III}\lambda_I + \lambda_I\lambda_{II})\ , \tag{9b}$$

$$J_3 = \lambda_I\lambda_{II}\lambda_{III}\ , \tag{9c}$$

which are the elements of the integrity basis.
In (8b) we see the HAMILTON-CAYLEY equation

$$X^{(3)}_{ij} - J_1X^{(2)}_{ij} - J_2X_{ij} - J_3\delta_{ij} = 0_{ij}\ , \tag{10}$$

i.e., the tensor-valued remainder (8a,b) is always equal to the zero tensor.

The tensors (7) can be written in the form

$$^{\alpha}L_{ij} = {}^{\alpha}\varphi_0\delta_{ij} + {}^{\alpha}\varphi_1X_{ij} + {}^{\alpha}\varphi_2X^{(2)}_{ij}\ , \tag{11}$$

if we introduce the abbreviations

$$^{\alpha}\varphi_0 \equiv \lambda_{(\alpha+I)}\lambda_{(\alpha+II)}/N_{(\alpha)}\ , \tag{12a}$$

$$^{\alpha}\varphi_1 \equiv -(\lambda_{(\alpha+I)} + \lambda_{(\alpha+II)})/N_{(\alpha)}\ , \tag{12b}$$

$$ {}^{\alpha}\varphi_2 \equiv 1/N_{(\alpha)} \tag{12c} $$

with

$$ N_{(\alpha)} \equiv (\lambda_{(\alpha)} - \lambda_{(\alpha+I)})(\lambda_{(\alpha)} - \lambda_{(\alpha+II)}) \; . \tag{13} $$

Thus, we can represent the tensorial interpolation formula (6) as

$$ Y_{ij} = \Phi_0 \delta_{ij} + \Phi_1 X_{ij} + \Phi_2 X_{ij}^{(2)} \; , \tag{14} $$

where the scalars Φ_0, Φ_1, Φ_2 are given by

$$ \Phi_\nu \equiv \sum_{\alpha=I}^{III} {}^{\alpha}\varphi_\nu f_\alpha \; , \qquad \nu = 0,1,2 \; . \tag{15} $$

The coefficients (15) can also be found by solving the system of linear equations

$$ \left. \begin{aligned} Y_I &= \Phi_0 + \Phi_1 X_I + \Phi_2 X_I^2 \\ Y_{II} &= \Phi_0 + \Phi_1 X_{II} + \Phi_2 X_{II}^2 \\ Y_{III} &= \Phi_0 + \Phi_1 X_{III} + \Phi_2 X_{III}^2 \; , \end{aligned} \right\} \tag{16} $$

which is obtained by inserting the distinct principal values $\lambda_I \equiv X_I$ etc. into equation (14).
The tensor-valued function (14) is an "isotropic tensor function" because the relation

$$ a_{ip} a_{jq} Y_{pq} = f_{ij}(a_{ip} a_{jq} X_{pq}) \tag{17} $$

is satisfied under any orthogonal transformation ($a_{ik} a_{jk} = a_{ki} a_{kj} = \delta_{ij}$). According to the definition (17), we can say: A tensor-valued function (5) of one tensor variable $\underset{\sim}{X}$ is isotropic if and only if it has a representation of the form (14), where the Φ_ν are invariants of $\underset{\sim}{X}$ and hence can be expressed as functions of the principal invariants of $\underset{\sim}{X}$. The determination of the Φ_ν is given by (15).

As an alternate approach, we find, by extending the NEWTON formula (4), the tensorial representation

$$Y_{ij} = f_{ij}(\underset{\sim}{X}) = a_0\delta_{ij} + a_1(X_{ij} - \lambda_I\delta_{ij}) + a_2(X_{ik} - \lambda_I\delta_{ik})(X_{kj} - \lambda_{II}\delta_{kj}) +$$

$$+ a_3(X_{ik} - \lambda_I\delta_{ik})(X_{kl} - \lambda_{II}\delta_{kl})(X_{lj} - \lambda_{III}\delta_{lj}) \,. \tag{18}$$

Further terms are not possible, because the second rank tensor $\underset{\sim}{X}$ has only three principal values. Moreover, the last term in (18) vanishes because of the HAMILTON-CAYLEY theorem. Thus we have the formula

$$Y_{ij} = a_0\delta_{ij} + a_1(X_{ij} - \lambda_I\delta_{ij}) + a_2(X_{ik} - \lambda_I\delta_{ik})(X_{kj} - \lambda_{II}\delta_{kj}) \,, \tag{19}$$

where the coefficients can be found by inserting the principal values:

$$a_0 = Y_I \,, \quad a_1 = (Y_I - Y_{II})/(\lambda_I - \lambda_{II}) \,, \tag{20a,b}$$

$$a_2 = \frac{a_1 - (Y_{III} - Y_I)/(\lambda_{III} - \lambda_I)}{\lambda_{II} - \lambda_{III}} \,. \tag{20c}$$

The interpolation formula (19) can be written as

$$Y_{ij} = \psi_0\delta_{ij} + \psi_1 X_{ij} + \psi_2 X_{ij}^{(2)} \,, \tag{21}$$

where the scalars are defined by

$$\psi_0 \equiv a_0 - a_1\lambda_I + a_2\lambda_I\lambda_{II} \,, \tag{22a}$$

$$\psi_1 \equiv a_1 - a_2(\lambda_I + \lambda_{II}) \,, \quad \psi_2 \equiv a_2 \,. \tag{22b,c}$$

The tensor-valued function (21) is an isotropic tensor function and must coincide with (14), i.e., the scalar coefficients (15) and (22a,b,c) are identical:

$$\Phi_\nu \equiv \psi_\nu \qquad \nu = 0,1,2 \, , \tag{23}$$

as we can easily prove.

3. TENSOR FUNCTION OF TWO VARIABLES

The problem, to construct a scalar polynomial

$$z = f(x,y) = \sum_{\alpha,\rho} a_{\alpha\rho} x^\alpha y^\rho \tag{24}$$

of two scalar variables sastisfying the mn interpolation conditions

$$f(x_\alpha, y_\rho) = z_{\alpha\rho} \qquad \begin{matrix} \alpha = 1,2,\ldots,m \\ \rho = 1,2,\ldots,n \end{matrix} \tag{25}$$

is solved by the interpolation formula

$$z = \sum_{\alpha=1}^{m} \sum_{\rho=1}^{n} L_\alpha(x) L_\rho(y) f(x_\alpha, y_\rho) + R(x,y) \tag{26}$$

which may be regarded as the generalization of the LAGRANGE formula (1) to two variables. Thus the polynomials $L_\alpha(x)$ and $L_\rho(y)$ are given in the form (2), while the remainder R(x,y) can be expressed by

$$R(x,y) = R_{m-1}(x) + R_{n-1}(y) - M(xy) \, , \tag{27}$$

where the first and second term can be read from

$$R_{m-1} = \frac{(x - x_1) \ldots (x - x_m)}{m!} \, \frac{\partial^m f(\xi_1, y)}{\partial x^m} \tag{28a}$$

$$R_{n-1} = \frac{(y - y_1) \ldots (y - y_n)}{n!} \, \frac{\partial^n f(x, \eta_1)}{\partial y^n} \tag{28b}$$

and the third term in (27) is given by

$$M(xy) = \frac{(x - x_1) \ldots (x - x_m)(y - y_1) \ldots (y - y_n)}{m!\; n!} \frac{\partial^{(m+n)} f(\xi_2, \eta_2)}{\partial x^m \partial y^n} \tag{28c}$$

The points (ξ_1, η_1) and (ξ_2, η_2) are lying in the interpolation region. Extensions to any number of variables will follow in a similar fashion.

An extension to a tensor-valued function

$$Z_{ij} = \left[f_{ij}(\underset{\sim}{X}, \underset{\sim}{Y}) + f_{ji}(\underset{\sim}{X}, \underset{\sim}{Y}) \right] / 2 = Z_{ji} \tag{29}$$

of two symmetric tensor variables ($X_{ij} = X_{ji}$, $Y_{ij} = Y_{ji}$) is achieved by the formula

$$f_{ij}(\underset{\sim}{X}, \underset{\sim}{Y}) = \sum_{\alpha,\rho=I}^{III} {}^{\alpha}L_{ik}(\underset{\sim}{X})\, {}^{\rho}L_{kj}(\underset{\sim}{Y})\, f_{(\alpha,\rho)} + R_{ij}(\underset{\sim}{X}, \underset{\sim}{Y}) \; , \tag{30}$$

using the following tensor polynomials:

$${}^{\alpha}L_{ik}(\underset{\sim}{X}) \equiv \frac{(X_{ip} - \lambda_{(\alpha+I)}\delta_{ip})(X_{pk} - \lambda_{(\alpha+II)}\delta_{pk})}{(\lambda_{(\alpha)} - \lambda_{(\alpha+I)})(\lambda_{(\alpha)} - \lambda_{(\alpha+II)})} \tag{31a}$$

$${}^{\rho}L_{kj}(\underset{\sim}{Y}) \equiv \frac{(Y_{kq} - \mu_{(\rho+I)}\delta_{kq})(Y_{qj} - \mu_{(\rho+II)}\delta_{qj})}{(\mu_{(\rho)} - \mu_{(\rho+I)})(\mu_{(\rho)} - \mu_{(\rho+II)})} \tag{31b}$$

where $\lambda_{(\alpha)}$, $\alpha = I, II, III$, and $\mu_{(\rho)}$, $\rho = I, II, III$, are the principal values of the argument tensors $\underset{\sim}{X}$ and $\underset{\sim}{Y}$, respectively. The remainder in (30) is given by

$$R_{ij}(\underset{\sim}{X}, \underset{\sim}{Y}) = R_{ij}(\underset{\sim}{X}) + R_{ij}(\underset{\sim}{Y}) - M_{ij}(\underset{\sim}{X}\underset{\sim}{Y}) \; , \tag{32}$$

and is a tensorial extension of (27). Because of the HAMILTON-CAYLEY equation (10), the tensor-valued remainder (32) vanishes like (8a,b).

The generalization of the formula (19) to two tensor variables can be expressed by

$$f_{ij}(\underset{\sim}{X},\underset{\sim}{Y}) = a_0\delta_{ij} + a_1\,{}^1A_{ij} + a_2\,{}^2A_{ij} + b_1\,{}^1B_{ij} + b_2\,{}^2B_{ij} + c_1\,{}^1A_{ik}\,{}^1B_{kj} + $$
$$+ c_2\,{}^1A_{ik}\,{}^2B_{kj} + c_3\,{}^2A_{ik}\,{}^1B_{kj} + c_4\,{}^2A_{ik}\,{}^2B_{kj} \tag{33}$$

where the abbreviations

$${}^1A_{ij} \equiv (X_{ij} - \lambda_I\delta_{ij}) , \quad {}^1B_{ij} \equiv (Y_{ij} - \mu_I\delta_{ij}) , \tag{34a,b}$$

$${}^2A_{ij} \equiv (X_{ik} - \lambda_I\delta_{ik})(X_{kj} - \lambda_{II}\delta_{kj}) , \tag{35a}$$

$${}^2B_{ij} \equiv (Y_{ik} - \mu_I\delta_{ik})(Y_{kj} - \mu_{II}\delta_{kj}) \tag{35b}$$

are used. The nine coefficients a_0, a_1, ..., c_4 in formula (33) are determined by the nine interpolation conditions

$$f_{11}(X_\alpha,\ Y_\rho) \equiv f_{(\alpha,\rho)} , \qquad \alpha,\rho = I,II,III . \tag{36}$$

For example, we find a_0 by substituting $X_{11} \equiv X_I = \lambda_I$, $Y_{11} \equiv Y_I = \mu_I$ and $f_{11} \equiv f_{(I,I)}$. The coefficient c_2 can be determined by substituting $X_{11} \equiv X_{II} = \lambda_{II}$, $Y_{11} \equiv Y_{III} = \mu_{III}$ and $f_{11}(X_{II},\ Y_{III}) \equiv f_{(II,III)}$. In this way we can express the coefficients in (33) by the given values (36):

$$a_0 \equiv f_{(I,I)}$$

$$a_1 \equiv \frac{f_{(I,I)} - f_{(II,I)}}{\lambda_I - \lambda_{II}} , \qquad b_1 \equiv \frac{f_{(I,I)} - f_{(I,II)}}{\mu_I - \mu_{II}}$$

$$a_2 \equiv \sum_{\alpha=I}^{III} f_{(\alpha,I)}/M_{(\alpha)} , \qquad b_2 \equiv \sum_{\rho=I}^{III} f_{(I,\rho)}/N_{(\rho)}$$

$$c_1 \equiv \frac{f_{(I,I)} - f_{(I,II)} + f_{(II,II)} - f_{(II,I)}}{(\lambda_I - \lambda_{II})(\mu_I - \mu_{II})}$$

$$c_2 \equiv \sum_{\rho=I}^{III} (f_{(I,\rho)} - f_{(II,\rho)})/[(\lambda_I - \lambda_{II})N_{(\rho)}]$$

$$c_3 \equiv \sum_{\alpha=I}^{III} (f_{(\alpha,I)} - f_{(\alpha,II)})/[(\mu_I - \mu_{II})M_{(\alpha)}]$$

$$c_4 \equiv \sum_{\alpha,\rho=I}^{III} f_{(\alpha,\rho)}(\lambda_{(\alpha+I)} - \lambda_{(\alpha+II)})(\mu_{(\rho+I)} - \mu_{(\rho+II)})/$$

$$/\prod_{\beta=I}^{III} (\lambda_{(\beta)} - \lambda_{(\beta+I)})(\mu_{(\beta)} - \mu_{(\beta+II)}) ,$$

where the abbreviations

$$M_{(\alpha)} \equiv (\lambda_{(\alpha)} - \lambda_{(\alpha+I)})(\lambda_{(\alpha)} - \lambda_{(\alpha+II)}) ,$$

$$N_{(\rho)} \equiv (\mu_{(\rho)} - \mu_{(\rho+I)})(\mu_{(\rho)} - \mu_{(\rho+II)})$$

are used.

The tensor-valued function (30) or, alternatively, the approach (33) can be represented by the same minimum polynomial

$$f_{ij}(\underset{\sim}{X},\underset{\sim}{Y}) = \sum_{\varkappa,\nu=0}^{2} \psi_{(\varkappa,\nu)} X_{ik}^{(\varkappa)} Y_{kj}^{(\nu)} , \qquad (37)$$

where the system $\psi_{(\varkappa,\nu)}$ of scalar coefficients is given by:

$$\psi_{(0,0)} \equiv a_0 - a_1\lambda_I + a_2\lambda_I\lambda_{II} - b_1\mu_I + b_2\mu_I\mu_{II} + c_1\lambda_I\mu_I - c_2\lambda_I\mu_I\mu_{II} -$$

$$- c_3\mu_I\lambda_I\lambda_{II} + c_4\lambda_I\lambda_{II}\mu_I\mu_{II}$$

$$\psi_{(0,1)} \equiv b_1 - b_2(\mu_I + \mu_{II}) - c_1\lambda_I + c_2\lambda_I(\mu_I + \mu_{II}) + c_2\lambda_I\lambda_{II} -$$

$$- c_4\lambda_I\lambda_{II}(\mu_I + \mu_{II})$$

$$\psi_{(0,2)} \equiv b_2 - c_2\lambda_I + c_4\lambda_I\lambda_{II}$$

$$\psi_{(1,0)} \equiv a_1 - a_2(\lambda_I + \lambda_{II}) - c_1\mu_I + c_2\mu_I\mu_{II} + c_3\mu_I(\lambda_I + \lambda_{II}) -$$

$$- c_4\mu_I\mu_{II}(\lambda_I + \lambda_{II})$$

$$\psi_{(1,1)} \equiv c_1 - c_2(\mu_I + \mu_{II}) - c_3(\lambda_I + \lambda_{II}) + c_4(\lambda_I + \lambda_{II})(\mu_I + \mu_{II})$$

$$\psi_{(1,2)} \equiv c_2 - c_4(\lambda_I + \lambda_{II})$$

$$\psi_{(2,0)} \equiv a_2 - c_3\mu_I + c_4\mu_I\mu_{II}$$

$$\psi_{(2,1)} \equiv c_3 - c_4(\mu_I + \mu_{II}), \; \psi_{(2,2)} \equiv c_4 \; .$$

By analogy to (14) the polynomial (37) of two tensor variables is an isotropic tensor function because, in extension of (17), the relation

$$a_{ip}a_{jq}f_{pq}(\underset{\sim}{X},\underset{\sim}{Y}) = f_{ij}(a_{ip}a_{jq}X_{pq},\; a_{ip}a_{jq}Y_{pq}) \tag{38}$$

is satisfied under any orthogonal transformation.
The polynomial (37) with the above determined scalar coefficients $\psi_{(\varkappa,\nu)}$ may be regarded as the generalization of the simple case (14) to two argument tensors. Extensions to any number of tensor variables can follow in a similar fashion.

4. INTERPOLATION AT COINCIDENT POINTS

In formulating the problem of polynomial interpolation for tensor functions, we have assumed that the principal values of the argument tensors, i.e. the interpolating points, are distinct. In the case of coincident points, we need the derivatives of the tensor function (5), which can be obtained from (19) or (21):

$$f'_{ij} \equiv \partial Y_{ip}/\partial X_{pj} = \psi_1 \delta_{ij} + 2\psi_2 X_{ij} \ , \qquad (39)$$

$$f''_{ij} \equiv \partial f'_{iq}/\partial X_{qj} = 2\psi_2 \delta_{ij} \ . \qquad (40)$$

For example, in the case of $\lambda_I = \lambda_{III} \neq \lambda_{II}$, we find from (19) and (39):

$$a_0 = Y_I \ , \qquad a_1 = (Y_I - Y_{II})/(\lambda_I - \lambda_{II}) \ , \qquad (41a,b)$$

$$a_2 = (f'_I - a_1)/(\lambda_I - \lambda_{II}) \ , \qquad (41c)$$

if we substitute $Y_I \equiv f_{11}(X_{11} \equiv X_I = \lambda_I)$, $Y_{II} \equiv f_{22}(X_{22} \equiv X_{II} = \lambda_{II})$, and $f'_{11}(X_{11} \equiv \lambda_I) \equiv f'_I$. In a similar way we obtain

$$a_0 = Y_I \ , \qquad a_1 = f'_I \ , \qquad (42a,b)$$

$$a_2 = \left[f'_I - (Y_I - Y_{III})/(\lambda_I - \lambda_{III})\right]/(\lambda_I - \lambda_{III}) \qquad (42c)$$

for the case $\lambda_I = \lambda_{II} \neq \lambda_{III}$ and

$$a_0 = Y_I \ , \qquad a_1 = (Y_I - Y_{II})/(\lambda_I - \lambda_{II}) \ , \qquad (43a,b)$$

$$a_2 = (a_1 - f'_{II})/(\lambda_I - \lambda_{II}) \ , \qquad (43c)$$

if the case of $\lambda_I \neq \lambda_{II} = \lambda_{III}$ is considered.
Finally, if all principal values coincide, $\lambda_I = \lambda_{II} = \lambda_{III}$, we determine

$$a_0 = f_I \ , \qquad a_1 = f'_I \ , \qquad a_2 = f''_I/2 \ . \qquad (44a,b,c)$$

But in this special case the argument tensor is a spherical one, $X_{ij} = \lambda_I \delta_{ij}$, and therefore the formula (19) reduces to the trivial result:

$f_{ij} = f_I \delta_{ij}$. Note, that the interpolation formulas for scalar functions $y = f(x)$ approaches the TAYLOR expansion for $f(x)$ at x_0, if we make x_α, $\alpha = 1,2,\ldots,n$, coincide at x_0.

In extending the interpolation at coincident points to tensor functions of two variables we need derivatives of the function (29) with respect to both the variable $\underset{\sim}{X}$ and $\underset{\sim}{Y}$. For example, we have to use the derivatives

$$Z'_{ij} \equiv (\partial Z_{ip}/\partial X_{pj} + \partial Z_{jp}/\partial X_{pi})/2 \ , \tag{45a}$$

$$\dot{Z}_{ij} \equiv (\partial Z_{ip}/\partial Y_{pj} + \partial Z_{jp}/\partial Y_{pi})/2 \ , \tag{45b}$$

$$\dot{Z}' \equiv \frac{1}{4}\left(\frac{\partial Z'_{ip}}{\partial Y_{pj}} + \frac{\partial Z'_{jp}}{\partial Y_{pi}} + \frac{\partial \dot{Z}_{ip}}{\partial X_{pj}} + \frac{\partial \dot{Z}_{jp}}{\partial X_{pi}}\right) \ , \tag{45c}$$

and so on.

5. POLYNOMIALS OF SECOND-ORDER AND FOURTH-ORDER TENSORS

In the following we consider a tensor-valued function

$$Z_{ij} = f_{ij}(X_{pq}, Y_{pq}, A_{pqrs}) \tag{46}$$

of two second-order and one fourth-order tensor the general representation of which is given by a linear combination

$$\underset{\sim}{Z} = \sum_\alpha \varphi_\alpha \underset{\sim}{G}_\alpha \ , \tag{47}$$

where the $\underset{\sim}{G}$'s are symmetric tensor generators of rank two involving the argument tensors $\underset{\sim}{X}$, $\underset{\sim}{Y}$, $\underset{\sim}{A}$. Such generators, for instance, are given by matrix products of types

$$\{\underset{\sim}{X}^{\varkappa}\underset{\sim}{Y}^{\nu} + \underset{\sim}{Y}^{\nu}\underset{\sim}{X}^{\varkappa}\}_{ij} \equiv X_{ik}^{(\varkappa)}Y_{kj}^{(\nu)} + Y_{ik}^{(\nu)}X_{kj}^{(\varkappa)}$$

$$\{\underset{\sim}{X}^{\varkappa}\underset{\sim}{A}^{\rho}\cdot\underset{\sim}{Y}^{\nu} + \underset{\sim}{A}^{\rho}\cdot\underset{\sim}{Y}^{\nu}\underset{\sim}{X}^{\varkappa}\}_{ij} \equiv X_{ik}^{(\varkappa)}A_{kjpq}^{(\rho)}Y_{pq}^{(\nu)} + A_{ikpq}^{(\rho)}Y_{pq}^{(\nu)}X_{kj}^{(\varkappa)} .$$

In finding an irreducible set of tensor generators we note, that, according to the HAMILTON-CAYLEY theorem, none of the exponents $\varkappa,\nu$ of the second-rank tensors $\underset{\sim}{X},\underset{\sim}{Y}$ needs be larger than two. In generalizing the HAMILTON-CAYLEY theorem it is pointed out in detail in [2] that a symmetric fourth-order tensor of power six and all higher powers can be expressed in terms of lower powers, that is $\rho \leqq 5$. The coefficients φ_α in the linear combination (47) are scalar-valued functions of the integrity basis the elements of which are the irreducible invariants. Together with the invariants of the single argument tensors we have to consider the set of simultaneous or joint invariants.

Tensor valued functions like (47) are used in [3,4,5] in order to represent constitutive equations of creep involving CAUCHY's stress tensor $\underset{\sim}{X}$, a damage tensor $\underset{\sim}{Y}$ of rank two, and a fourth-order tensor $\underset{\sim}{A}$ characterizing the initial anisotropy or the anisotropy of the material in its undamaged state.

6. SIMPLE EXAMPLES

Given the second-rank tensor

$$X_{ij} = \begin{pmatrix} 3 & 0 & 1 \\ 0 & 2 & 0 \\ 1 & 0 & 3 \end{pmatrix} , \tag{48}$$

we will find the third root

$$Y_{ij} = X_{ij}^{(1/3)} . \tag{49}$$

The principal values of (49) are $\lambda_I = 4$, $\lambda_{II} = \lambda_{III} = 2$. From (43a,b,c) we calculate:

$$a_0 = 4^{1/3} , \qquad a_1 = (4^{1/3} - 2^{1/3})/2 ,$$

$$a_2 = (a_1 - 2^{-2/3}/3)/2 ,$$

and find the coefficients (22a,b,c) of the polynomial (21):

$$\psi_0 = 4^{1/3} - \frac{4}{3} 2^{-2/3} \tag{50a}$$

$$\psi_1 = 2^{1/3} - 4^{1/3} + 2^{-2/3} \tag{50b}$$

$$\psi_2 = \frac{1}{4} (4^{1/3} - 2^{1/3} - \frac{2}{3} 2^{-2/3}) . \tag{50c}$$

Thus, the third root (49) of the tensor (48) is given by

$$X_{ij}^{(1/3)} = \begin{pmatrix} \frac{2^{1/3} + 1}{2^{2/3}} & 0 & \frac{2^{1/3} - 1}{2^{2/3}} \\ 0 & 2^{1/3} & 0 \\ \frac{2^{1/3} - 1}{2^{2/3}} & 0 & \frac{2^{1/3} + 1}{2^{2/3}} \end{pmatrix} \tag{51}$$

This result was obtained by BETTEN [6] in another way.

In the theory of finite deformation the tensorial HENCKY measure of strain and strain rate plays an important role [5,7] because it can be decomposed into a sum of an isochoric distortion and a volume change [5, 8,9,10,11,12]. The problem to represent the logarithmic function

$$\underset{\sim}{Y} = \ln \underset{\sim}{X} \tag{52}$$

as an isotropic tensor function (14) is solved by determining the scalar functions in (15):

$$\Phi_\nu \equiv {}^{I}\varphi_\nu \ln\lambda_I + {}^{II}\varphi_\nu \ln\lambda_{II} + {}^{III}\varphi_\nu \ln\lambda_{III} \tag{53}$$

Especially for the given tensor (48) we find the result

$$\ln \underset{\sim}{X} = \ln 2 \begin{pmatrix} 3/2 & 0 & 1 \\ 0 & 1 & 0 \\ 1 & 0 & 3/2 \end{pmatrix} \tag{54}$$

by considering the relations (21), (22a,b,c), and (43a,b,c).
Other examples are $\underset{\sim}{Y} = \exp \underset{\sim}{X}$, $\underset{\sim}{Y} = \sin \underset{\sim}{X}$ etc. which can be treated in the same way.

Finally, the elastic-plastic behaviour of solids loaded under uni-axial stress σ may be expressed by the stress-strain relations [13]

$$\sigma/\sigma_F = [\tanh(E\varepsilon/\sigma_F)^n]^{1/n} , \tag{55a}$$

$$\sigma/\sigma_F = (E\varepsilon/\sigma_F)/[1 + (E\varepsilon/\sigma_F)^n]^{1/n} , \tag{55b}$$

where E is YOUNG's modulus, and σ_F represents the yield stress determined in an uni-axial tension test. The exponent n regulates the elastic-plastic transition. For instance, an elastic-perfectly plastic behaviour is characterized by $n \to \infty$. For engineering applications it is important to generalize the relations (55a,b) to the multi-axial state of stress. This can be achieved by an isotropic tensor function (21), where the scalars (22a,b,c) are to express by the empirical relations (55a,b).

In the next paragraph a similar example is discussed in detail.

7. TENSORIAL GENERALIZATION OF NORTON's CREEP LAW

The NORTON creep law

$$d/d_0 = (\sigma/\sigma_0)^n \quad \text{or} \quad d = K\sigma^n , \tag{56a,b}$$

which is often used in order to describe the secondary creep behaviour under uni-axial states of stress (σ) should be generalized to multi-axial states of stress [14], i.e., we consider an isotropic tensor function

$$d_{ij} = f_{ij}(\underset{\sim}{\sigma}) = \varphi_0^*\delta_{ij} + \varphi_1^*\sigma_{ij} + \varphi_2^*\sigma_{ij}^{(2)}, \tag{57}$$

where d_{ij} are the cartesian components of the "rate-of-deformation tensor". The scalar coefficients φ_0^*, φ_1^*, φ_2^* in (57) should be determined as functions of the experimental data (K,n) in (56a,b).

Alternatively, we can represent the constitutive equation in the form

$$d_{ij} = f_{ij}(\underset{\sim}{\sigma}') = \varphi_0\delta_{ij} + \varphi_1\sigma'_{ij} + \varphi_2\sigma'^{(2)}_{ij}, \tag{58}$$

where $\sigma'_{ij} = \sigma_{ij} - \sigma_{kk}\delta_{ij}/3$ are the components of the stress deviator $\underset{\sim}{\sigma}'$. For the special case of incompressible behaviour ($d_{kk} = 0$) we find from (58):

$$3\varphi_0 + \varphi_2\sigma'^{(2)}_{kk} = 0 \Rightarrow \varphi_0 = -2\varphi_2 J'_2/3 \tag{59}$$

with $J'_2 \equiv \sigma'_{ik}\sigma'_{ki}/2$, so that the constitutive equation (58) is reduced to the form

$$d_{ij} = \varphi_1\sigma'_{ij} + \varphi_2\sigma''_{ij} \tag{60}$$

containing the traceless tensors

$$\sigma'_{ij} \equiv \partial J'_2/\partial\sigma_{ij} \quad \text{and} \quad \sigma''_{ij} \equiv \partial J'_3/\partial\sigma_{ij} \tag{61a,b}$$

with $J'_3 \equiv \sigma'_{ij}\sigma'_{jk}\sigma'_{ki}/3$.

The uni-axial equivalent state of stress (index V) is characterized through the tensor variables

$$(\sigma_{ij})_V = \mathrm{diag}\{\sigma,0,0\}, \ (\sigma'_{ij})_V = \mathrm{diag}\{\tfrac{2}{3}\sigma, -\tfrac{1}{3}\sigma, -\tfrac{1}{3}\sigma\} , \quad (62a,b)$$

$$(d_{ij})_V = \mathrm{diag}\{d, -\nu d, -\nu d\} , \quad (62c)$$

where ν is the transverse contraction ratio.

In the following the diagonal elements in (62a,b,c) are considered as interpolating points, where two points coincide. Because the two coincident points in (62a) are zero, it may be more convenient to determine the coefficients φ_0, φ_1, φ_2 in the constitutive equation (58) instead of (57). Thus, we use the uni-axial creep law

$$d = K(3/2)^n(\sigma')^n \quad (63)$$

instead of (56b), where $\sigma' = 2\sigma/3$. Because of (62b), i.e. $X_{II} = X_{III} \equiv$ $\equiv -\sigma/3$, and (63), we find from (43a) the coefficient

$$a_0 = Y_I \equiv K(3/2)^n(\sigma')^n = K\sigma^n . \quad (64a)$$

Furthermore, because of (62b), (64a), and $Y_{II} = -\nu d = -\nu K\sigma^n$, we find from (43b) the coefficient

$$a_1 = (1+\nu)K\sigma^{n-1} . \quad (64b)$$

The derivative f'_{II} at the coincident points $X_{II} = X_{III}$ can be determined in the following way. From (63) we derive:

$$f' \equiv \partial d/\partial\sigma' = n(3/2)^n(\sigma')^{n-1} = nd/\sigma' \quad (65a)$$

or

$$f'_{II} = nd_{II}/\sigma'_{II} . \quad (65b)$$

From (62b,c) we read

$$\sigma'_{II} = -\sigma/3 \quad \text{and} \quad d_{II} = -\nu d_I \equiv -\nu d = -\nu K\sigma^n \ ,$$

so that (65b) can be written as

$$f'_{II} = 3\nu n K\sigma^{n-1} \ . \tag{65c}$$

Considering (62b) and (64b), we calculate from (43c) the coefficient:

$$a_2 = (1 + \nu - 3\nu n)K\sigma^{n-2} \ . \tag{64c}$$

Inserting (64a,b,c) in (22a,b,c), we finally determine the scalar functions ($\psi \equiv \varphi$):

$$\varphi_0 = \frac{1}{9}(1 - 8\nu + 6\nu n)K\sigma^n \tag{66a}$$

$$\varphi_1 = \frac{2}{3}(1 + \nu + \frac{3}{2}\nu n)K\sigma^{n-1} \tag{66b}$$

$$\varphi_2 = (1 + \nu - 3\nu n)K\sigma^{n-2} \ . \tag{66c}$$

In order to control the result (66a,b,c) we make the following tests.

For the <u>uni-axial case</u> ($i = j = 1$) the constitutive equation (58) leads to

$$d = \varphi_0 + 2\varphi_1\sigma/3 + 4\varphi_2\sigma^2/9 \ , \tag{67}$$

and this equation immediately leads to NORTON's power law (56b), if we insert the result (66a,b,c).

The <u>transverse contraction ratio</u> defined as

$$\nu := -(d_{22}/d_{11})_V = -(d_{33}/d_{11})_V \tag{68}$$

can be calculated from (58):

$$\nu = -(\varphi_0 - \frac{1}{3}\varphi_1\sigma + \frac{1}{9}\varphi_2\sigma^2)/(\varphi_0 + \frac{2}{3}\varphi_1\sigma + \frac{4}{9}\varphi_2\sigma^2) \ . \tag{69}$$

With the result (66a,b,c) we find the identity $\nu = \nu$.

The condition (59) for <u>imcompresible</u> behaviour leads with (62b) and (66a,b,c) to

$$(1 - 8\nu + 6\nu n)K\sigma^n = -2(1 + \nu - 3\nu n)K\sigma^n \ ,$$

from which we immediately read: $\nu = 1/2$.

The <u>volume change</u> is obtained from (58) by forming the trace ($i = j = k$):

$$d_{kk} = 3\varphi_0 + \varphi_2\sigma'^{(2)}_{kk} = 3\varphi_0 + 2\varphi_2 J'_2 \ . \tag{70a}$$

For uni-axial loading (62b) we find with (66a,c) the trivial result:

$$d_{kk} = (1 - 2\nu)K\sigma^n \equiv (1 - 2\nu)d \tag{70b}$$

which is identical with the trace of (62c).

The <u>rate of dissipation of creep energy</u>

$$\dot{D} = \sigma_{ij}d_{ji} \tag{71}$$

can be expressed through

$$\dot{D} = J_1\varphi_0 + 2J'_2\varphi_1 + (3J'_3 + 2J_1J'_2/3)\varphi_2, \tag{72}$$

if the constitutive equation (58) is inserted into (71). For the uni-axial case (62a,b,c) with $J_1 \equiv \sigma$, $J'_2 \equiv \sigma^2/3$, $J'_3 \equiv 2\sigma^2/27$ we find from (72) and (66a,b,c)

$$\dot{D} = K\sigma^{n+1} \equiv \sigma d \ . \tag{73}$$

Thus, the hypothesis

$$\dot{D} = \sigma_{ij}d_{ji} \equiv \sigma d \tag{74}$$

can be used in order to determine the equivalent stress σ as a function of the stress invariants [see equations (79) to (82)].

For incompressible behaviour (59) the rate of dissipation of creep energy (72) is reduced to the simple relation

$$\dot{D} = 2J_2'\varphi_1 + 3J_3'\varphi_2 \tag{75}$$

in which only deviator invariants are contained. Uni-axial load (62b) leads from (75) to (73), if we insert (66b,c) with $\nu = 1/2$ into (75).

Assuming incompressibility (59) and neglecting tensorial nonlinearity ($\varphi_2 = 0 \Rightarrow a_2 = 0$, $\varphi_0 = 0$, and $\varphi_1 = a_1$), we find from (58) the simplified constitutive equation

$$d_{ij} = a_1\sigma_{ij}' \tag{76a}$$

which can be written as

$$d_{ij} = \frac{3}{2}K\sigma^{n-1}\sigma_{ij}' , \tag{76b}$$

if we use equation (64b) with $\nu = 1/2$. The result (76b) is identical to the constitutive equation proposed by LECKIE and HAYHURST in [15]. If we insert the MISES equivalent stress $\sigma = \sqrt{3J_2'}$ into (76b), we can find the constitutive equation

$$d_{ij} = \frac{3}{2}K(3J_2')^{\frac{n-1}{2}}\sigma_{ij}' \tag{76c}$$

used by ODQUIST and HULT in [16].

Inserting the deviator $\sigma_{ij}' = \sigma_{ij} - J_1\delta_{ij}/3$ and its square

$$\sigma'^{(2)}_{ij} = \sigma^{(2)}_{ij} - 2J_1\sigma_{ij}/3 + J_1^2\delta_{ij}/9 \tag{77}$$

into equation (58), one can find a constitutive equation of the form (57) with the scalar coefficients

$$\varphi_0^* = \varphi_0 - J_1\varphi_1/3 + J_1^2\varphi_2/9 \ , \tag{78a}$$

$$\varphi^* = \varphi_1 - 2J_1\varphi_2/3 \ , \tag{78b}$$

$$\varphi^* \equiv \varphi_2 \ . \tag{78c}$$

As mentioned above one can determine the equivalent stress σ in (66a,b,c) as a function of the stress invariants, if we use the hypothesis (74) in connection with (58), (72), (66a,b,c), and (56b). The result is:

$$\sigma^3 + A\sigma^2 + B\sigma + C = 0 \tag{79}$$

where the abbreviations

$$A \equiv - (1 - 8\nu + 6\nu n)\ J_1/9 \ , \tag{80a}$$

$$B \equiv - 4(1 + \nu + 3\nu n/2)\ J_2'/3 \ , \tag{80b}$$

$$C \equiv - (1 + \nu - 3\nu n)(3J_3' + 2J_1J_2'/3) \tag{80c}$$

are used. Thus, the skalar coefficients (66a,b,c) or alternatively (78a, b,c) are functions of irreducible invariants and of experimental data:

$$\varphi_0 = \varphi_0\ (J_1,J_2',J_3';K,n,\nu),\ \ldots,\ \varphi_2 = \varphi_2\ (J_1,J_2',J_3';K,n,\nu)\ . \tag{81a,b,c}$$

This statement is compatible with the representation theory of tensor-valued functions (47) in which the coefficients φ_α are scalar-valued functions of the integrity basis.

For incompressible behaviour equation (79) is reduced to

$$\sigma^3 - (2+n)\, J_2'\sigma + 9\,(n-1)\, J_3'/2 = 0 \tag{82}$$

because of (75) and $\nu = 1/2$. In this case the equivalent stress σ in (82) is influenced by the deviator invariants J_2', J_3' and the parameter n.

At the end of this paragraph two further examples should be mentioned. First, the tensorial generalization of the uni-axial stress-strain relation (55a) can be found in the alternative forms:

$$\sigma_{ij} = \sigma(\psi_0^*\delta_{ij} + \psi_1^*\varepsilon_{ij} + \psi_2^*\varepsilon_{ij}^{(2)}) \ ,$$

$$\sigma_{ij} = \sigma(\psi_0\delta_{ij} + \psi_1\varepsilon_{ij}' + \psi_2\varepsilon_{ij}'^{(2)}) \ ,$$

where [similar to (78a,b,c)] the relations

$$\psi^* = \psi_0 - I_1\psi_1/3 + I_1^2\psi_2/9 \ ,$$

$$\psi_1^* = \psi_1 - 2I_1\psi_2/3 \ ,$$

$$\psi_2^* = \psi_2$$

with $I_1 \equiv \varepsilon_{kk}$ and the scalar functions

$$\psi_0 = (1 - 2I_2'\psi_2)/3 \ ,$$

$$\psi_1 = [C + 2/(1 + \nu)]/2\varepsilon \ ,$$

$$\psi_2 = [1/(1 + \nu) - C]/(1 + \nu)^2\varepsilon^2$$

with $2I_2' \equiv \varepsilon_{ij}'\varepsilon_{ji}'$ and

$$C \equiv \frac{1}{2}\frac{9+4\nu}{(1+\nu)^2}\left[1 \pm \sqrt{1-\frac{8(1+\nu)}{(9+4\nu)^2}}\right]$$

are used.

Another example discussed in [17] is the tensorial generalization of the RAMBERG-OSGOOD stress-strain relation [18]

$$\varepsilon = \sigma/E + k(\sigma/E)^n .$$

By using the interpolation method we can find the tensorial representation

$$\varepsilon_{ij} = \varphi_0\delta_{ij} + \varphi_1\sigma'_{ij} + \varphi_2\sigma'^{(2)}_{ij} ,$$

which is similar to (58) and where the scalar functions

$$\varphi_0 = \frac{1}{3}(1-n)\frac{\sigma}{E} + \frac{1}{9}(1-8\nu+6\nu n)\left[\sigma/E+k(\sigma/E)^n\right] ,$$

$$\varphi_1 = \frac{1-n}{3E} + \frac{2(1+\nu)/3+\nu n}{E}\left[1+k(\sigma/E)^{n-1}\right] ,$$

$$\varphi_2 = \frac{1}{E\sigma}\left\{(1+\nu-3\nu n)\left[1+k(\sigma/E)^{n-1}\right] - \frac{3}{2}(1-n)\right\}$$

have been derived in [17]. The alternative representation

$$\varepsilon_{ij} = \varphi_0^*\delta_{ij} + \varphi_1^*\sigma_{ij} + \varphi_2^*\sigma^{(2)}_{ij}$$

can be achieved, if we make use of the relations (78a,b,c).

8. SEPARATION OF TENSOR VARIABLES

In the following the uni-axial stress-strain-rate relation

$$d = K\sigma^n D^m \quad \text{with} \quad D := 1/(1-\omega) \tag{83}$$

should be generalized to multi-axial states of stress, where ω is the parameter of damage [3,4]. This can be done by representing (83) as a tensor-valued function

$$d_{ij} = f_{ij}(\underset{\sim}{\sigma},\underset{\sim}{D}) = \frac{1}{2}\sum_{\nu,\mu=0}^{2} \psi_{(\nu,\mu)}\left[\sigma_{ik}^{(\nu)}D_{kj}^{(\mu)} + D_{ik}^{(\mu)}\sigma_{kj}^{(\nu)}\right] . \tag{84}$$

Now, the main problem is: to determine the nine scalar coefficients $\psi_{(\nu,\mu)}$ as functions of the integrity basis and experimental data. To solve this problem we suggest the following method which may be useful for practical applications. A representation with the same tensor generators as contained in (84) can be found by seperating the two tensor variables $\underset{\sim}{\sigma}$ and $\underset{\sim}{D}$ in the following way:

$$d_{ij} = f_{ij}(\underset{\sim}{\sigma},\underset{\sim}{D}) = \frac{1}{2}(X_{ik}Y_{kj} + Y_{ik}X_{kj}) , \tag{85}$$

where the isotropic tensor functions

$$\left.\begin{aligned} X_{ij} &= X_{ij}(\underset{\sim}{\sigma}) = \varphi_0^*\delta_{ij} + \varphi_1^*\sigma_{ij} + \varphi_2^*\sigma_{ij}^{(2)} \\ \varphi_\nu^* &= \varphi_\nu^*(\mathrm{tr}\underset{\sim}{\sigma}^\lambda) = \varphi_\nu^*(\sigma_I,\sigma_{II},\sigma_{III}) , \end{aligned}\right\} \tag{86}$$

$$\left.\begin{aligned} Y_{ij} &= Y_{ij}(\underset{\sim}{D}) = \Phi_0\delta_{ij} + \Phi_1 D_{ij} + \Phi_2 D_{ij}^{(2)} \\ \Phi_\mu &= \Phi_\mu(\mathrm{tr}\underset{\sim}{D}^\lambda) = \Phi_\mu(D_I,D_{II},D_{III}) \end{aligned}\right\} \tag{87}$$

($\mu,\nu = 0,1,2$ and $\lambda = 1,2,3$) are used.

Thus, we find the representation (84) with the scalar coefficients

$$\psi^*_{(\nu,\mu)} = \varphi^*_\nu \, \Phi_\mu , \qquad \mu,\nu = 0,1,2, \tag{88}$$

where the scalars φ^*_ν are determined in (78a,b,c). Similarly to (15) the coefficients Φ_μ can be found by solving the system of linear equations:

$$\left.\begin{aligned} \Phi_0 + \Phi_1 D_I + \Phi_2 D_I^2 &= (D_I)^{m_I} \\ \Phi_0 + \Phi_1 D_{II} + \Phi_2 D_{II}^2 &= (D_{II})^{m_{II}} \\ \Phi_0 + \Phi_1 D_{III} + \Phi_2 D_{III}^2 &= (D_{III})^{m_{III}} . \end{aligned}\right\} \tag{89}$$

The exponents $m_I, \dots, m_{III}$ in (89) are determined by using the creep law (83) in tests on specimens cut along the mutually perpendicular directions x_1, x_2, x_3.

Because of $D_{ij} := (\delta_{ij} - \omega_{ij})^{(-1)}$ and $\omega_{ij} = \mathrm{diag}\{(1-\alpha), (1-\beta), (1-\gamma)\}$, the principal values in (89) can be expressed through [14]:

$$D_I \equiv 1/\alpha , \qquad D_{II} \equiv 1/\beta , \qquad D_{III} \equiv 1/\gamma , \tag{90}$$

where α,β,γ are fractions which represent the net cross-section elements of CAUCHY's tetrahedron perpendicular to the coordinate axes [3].
In the case of two equal parameters, for instance $\alpha \neq \beta = \gamma$, the scalars Φ_μ can be determined by using the interpolation method described in section 4.

Instead of (86) we can use the isotropic tensor function

$$X_{ij} = X_{ij}(\underset{\sim}{\sigma}') = \varphi_0 \delta_{ij} + \varphi_1 \sigma'_{ij} + \varphi_2 \sigma'^{(2)}_{ij} \tag{91}$$

and find the representation

$$d_{ij} = \frac{1}{2} \sum_{\nu,\mu=0}^{2} \varphi_\nu \Phi_\mu \left[\sigma'^{(\nu)}_{ik} D^{(\mu)}_{kj} + D^{(\mu)}_{ik} \sigma'^{(\nu)}_{kj}\right] , \tag{92}$$

where the scalar coefficients φ_ν are determined in (66a,b,c) and the

Φ_μ are taken from (89).

The scalar coefficients $\psi_{(\nu,\mu)} \equiv \varphi_\nu \Phi_\mu$ in (92) must be functions of the integrity basis

$$\left.\begin{array}{llll} J_1 \equiv \sigma_{kk} , & J_2' \equiv \sigma_{ij}'\sigma_{ji}'/2 , & J_3' \equiv \sigma_{ij}'\sigma_{jk}'\sigma_{ki}'/3 , & \\ L_1 \equiv D_{kk} , & L_2 \equiv D_{kk}^{(2)} , & L_3 \equiv D_{kk}^{(3)} , & \Omega_1' \equiv \sigma_{ij}'D_{ji} , \\ \Omega_2' \equiv \sigma_{ij}'^{(2)}D_{ji} , & \Omega_3' \equiv \sigma_{ij}'D_{ji}^{(2)} , & \Omega_4' \equiv \sigma_{ij}'^{(2)}D_{ji}^{(2)} , & \end{array}\right\} \qquad (93)$$

and experimental data. To show this we can start from (74) and find similarly to (79) the cubic equation

$$\sigma^3 + A^*\sigma^2 + B^*\sigma + C^* = 0 , \qquad (94)$$

if we insert equations (85), (87), (91), (66a,b,c) into (74). In (94) the following abbreviations are used:

$$A^* \equiv -\frac{1}{9}(1 - 8\nu + 6\nu n)\left[\Phi_0 J_1 + \Phi_1\left(\Omega_1' + \frac{1}{3}J_1L_1\right) + \Phi_2\left(\Omega_3' + \frac{1}{3}J_1L_2\right)\right]/D^m , \qquad (95a)$$

$$B^* \equiv -\frac{2}{3}\left(1 + \nu + \frac{3}{2}\nu n\right)\left[2\Phi_0 J_2' + \Phi_1\left(\Omega_2' + \frac{1}{3}J_1\Omega_1'\right) + \Phi_2\left(\Omega_4' + \frac{1}{3}J_1\Omega_3'\right)\right]/D^m , \qquad (95b)$$

$$C^* \equiv -(1 + \nu - 3\nu n)\left[3\Phi_0\left(J_3' + \frac{2}{9}J_1J_2'\right) + \Phi_1\left(J_2'\Omega_1' + J_3'L_1 + \frac{1}{3}J_1\Omega_2'\right) + \right.$$
$$\left. + \Phi_2\left(J_2'\Omega_3' + J_3'L_2 + \frac{1}{3}J_1\Omega_4'\right)\right]/D^m , \qquad (95c)$$

$$D \equiv (D_I D_{II} D_{III})^{1/3}, \quad m \equiv (m_I + m_{II} + m_{III})/3 . \qquad (95d,e)$$

We see, that the elements of the integrity basis (93) and experimental data are contained in (95a,b,c,d,e). Thus, the coefficients $\psi_{(\nu,\mu)} \equiv \equiv \varphi_\nu \Phi_\mu$ in (92) are scalar functions of the integrity basis and experimental data:

$$\psi_{(\nu,\mu)}=\psi_{(\nu,\mu)}(J_1,J_2',J_3';L_1,L_2,L_3;\Omega_1',\ldots,\Omega_4';\nu,n,K;m_I,m_{II},m_{III};D_I,D_{II},D_{III}). \tag{96}$$

In the case of damage (ω_{ij}) and initial anisotropy (A_{ijkl}) the constitutive equations discussed in [3,5,14] are tensor-valued functions like (46) which can be represented as (47), where the scalar functions φ_α can be determined in a similar way as described above.

REFERENCES

1. BETTEN, J., Interpolation Methods of Tensor Functions, Fourth International Conference on Mathematical Modelling, ETH-Zürich (Switzerland) 1983, published in the proceedings: Mathematical Modelling in Science and Technology (eds. X.J.R. Avula, R.E. Kalman, A.I. Liapis and E.Y. Rodin), Pergamon Press, New York /.../ Frankfurt 1984, 52-57.
2. BETTEN, J., Integrity Basis for a Second-Order and a Fourth-Order Tensor, International J. Math. & Math. Sci. 5 (1982): 87-96.
3. BETTEN, J., Damage Tensors in Continuum Mechanics, Euromech Colloquium 147 on "Damage Mechanics", Paris VI, Cachan 1981, published in: Journal de Mécanique théorique et appliquée 2 (1983): 13-32.
4. BETTEN, J., Net-Stress Analysis in Creep Mechanics, Ingenieur-Archiv 52 (1982): 405-419.
5. BETTEN, J., Elastizitäts- und Plastizitätslehre, Vieweg-Verlag, Braunschweig/Wiesbaden 1985.
6. BETTEN, J., Elementare Tensorrechnung für Ingenieure, Vieweg-Verlag, Braunschweig 1977.
7. FITZGERALD, J.E., A tensorial Hencky Measure of Strain and Strain Rate for finite Deformations, J. Appl. Phys. 51 (1980): 5111-5115.

8. RICHTER, H., Verzerrungstensor, Verzerrungsdeviator und Spannungstensor bei endlichen Formänderungen, Z. Angew. Math. Mech. (ZAMM) 29 (1949): 65-75.

9. SEDOV, L.I., Foundations of the Non-Linear Mechanics of Continua, Pergamon Press, Oxford /.../ Frankfurt 1966.

10. LEHMANN, Th., Einige Bemerkungen zu einer allgemeinen Klasse von Stoffgesetzen für große elasto-plastische Formänderungen, Ingenieur-Archiv 41 (1972): 297-310.

11. BETTEN, J., On the Creep-Behaviour of an elastic-plastic thick-walled circular cylindrical Tube subjected to internal Pressure, International Symposium on Mechanics of inelastic Media and Structures, Warsaw 1978, published in: Mechanics of inelastic Media and Structures (eds. O. Mahrenholtz and A. Sawczuk), Polish Academy of Science, Warszawa/Poznań 1982, 51-72.

12. BETTEN, J., Zur Kriechaufweitung zylindrischer Hochdruckbehälter, Rheol. Acta 19 (1980): 517-524.

13. BETTEN, J., Zum Traglastverfahren bei nichtlinearem Stoffgesetz, Ingenieur-Archiv 44 (1975): 199-207.

14. BETTEN, J., Applications of Tensor Functions to the Formulation of Constitutive Equations involving Damage and Initial Anisotropy, IUTAM-Symposium on Mechanics of Damage and Fatigue, Haifa and Tel Aviv 1985, to be published in the proceedings (eds. S.R. Bodner and Z. Hashin).

15. LECKIE; F.A. and HAYHURST, D.R., Constitutive Equations for Creep Rupture, Acta Metallurgica 25 (1977): 1059-1070.

16. ODQUIST, F.K.G. and HULT, J., Kriechfestigkeit metallischer Werkstoffe, Springer-Verlag, Berlin/Göttingen/Heidelberg 1962.

17. BETTEN, J., Tensorielle Verallgemeinerung einachsiger Stoffgesetze, Z. Angew. Math. Mech. (ZAMM) 66 (1986), in press.

18. RAMBERG, W. and OSGOOD, W.R., Description of Stress-Strain Curves by three Parameters, NACA Technical Note No. 902, July 1943.

Chapter 14

TENSOR FUNCTION THEORY AND CLASSICAL PLASTIC POTENTIAL

J. Betten
Technical University Aachen, F.R. Germany

1. INTRODUCTION

In the theory of elasticity an elastic potential (strain-energy function W) is assumed, from which the constitutive equations can be derived by using the relation $\sigma_{ij} = \partial W/\partial \varepsilon_{ij}$, where $\underset{\sim}{\sigma}$ and $\underset{\sim}{\varepsilon}$ are appropriately defined stress and strain tensors, respectively. In the isotropic special case, when the elastic constitutive equation can be represented as an isotropic tensor function

$$\sigma_{ij} = \varphi_0\delta_{ij} + \varphi_1\varepsilon_{ij} + \varphi_2\varepsilon_{ij}^{(2)}, \tag{1}$$

the elastic potential is a scalar-valued function only of the strain tensor and can be represented in the form $W = W(S_1, S_2, S_3)$, where S_1, S_2, S_3 are the basic invariants of the strain tensor (finite or infinitesimal strain tensor). In [1] it has been shown in detail that the scalar coefficients in (1) can be expressed through the elastic potential:

$$\varphi_0 \equiv \partial W/\partial S_1 , \qquad \varphi_1 \equiv 2\partial W/\partial S_2 , \qquad \varphi_2 \equiv 3\partial W/\partial S_3 . \qquad \text{(2a,b,c)}$$

Eliminating the elastic potential, one can find the following sufficient and necessary conditions

$$2\partial\varphi_0/\partial S_2 = \partial\varphi_1/\partial S_1 , \quad 3\partial\varphi_1/\partial S_3 = 2\partial\varphi_2/\partial S_2 , \quad 3\partial\varphi_0/\partial S_3 = \partial\varphi_2/\partial S_1 , \qquad (3)$$

i.e., the elastic potential is "compatible" with the tensor function theory (1), if the conditions (3) have been fulfiled [1].

In the classical theory of plasticity a plastic potential F is assumed, and the flow rule

$$d^p\varepsilon_{ij} = d\Lambda(\partial F/\partial\sigma_{ij}) \qquad (4)$$

is used in order to find the constitutive equations, where $^p\varepsilon_{ij}$ are the components of the plastic part of the strain tensor $\underset{\sim}{\varepsilon}$, and the factor $d\Lambda$ is LAGRANGE's multiplier.

In the following the classical plastic potential theory is compared with the tensor function theory. It will be shown, that the former theory is compatible with the latter, if the material is isotropic, and if additional conditions are fulfiled.

However, for anisotropic materials the plastic potential theory only furnishes restricted forms of constitutive equations, even if a general plastic potential has been assumed [2]. Consequently, the classical flow rule (4) must be modified for anisotropic solids. In the following, appropiate modifications are discussed and resulting conditions of "compatibility" are derived.

2. ISOTROPY

For isotropic materials the plastic potential $F = F(\sigma_{ij})$ can be expressed as a single-valued function $F = F[S_\nu(\underset{\sim}{\sigma})]$ of the irreducible invariants $S_\nu \equiv \mathrm{tr}\,\underset{\sim}{\sigma}^\nu$, $\nu = 1,2,3$, of the stress tensor. Thus, considering the form $F = F(S_1,S_2,S_3)$ and using the classical flow rule (4), one immediately obtains the constitutive equation

$$d^p\varepsilon_{ij} = \left[(\partial F/\partial S_1)\delta_{ij} + 2(\partial F/\partial S_2)\sigma_{ij} + 3(\partial F/\partial S_3)\sigma_{ij}^{(2)}\right]d\Lambda \,. \qquad (5)$$

Instead of (5), one can represent the tensor $d^p\varepsilon_{ij}$ as a symmetric tensor-valued function of one second-order argument tensor:

$$d^p\varepsilon_{ij} = f_{ij}(\sigma_{pq}) = \varphi_0\delta_{ij} + \varphi_1\sigma_{ij} + \varphi_2\sigma_{ij}^{(2)}, \qquad (6)$$

where φ_0, φ_1, φ_2 are scalar-valued functions of the integrity basis S_1, S_2, S_3.

Comparing the "minimum polynomial representation" (6) with the result (5) based upon the plastic potential theory, one can find the following identities

$$\varphi_0 \equiv (\partial F/\partial S_1)d\Lambda \,, \qquad \varphi_1 \equiv 2(\partial F/\partial S_2)d\Lambda \,, \qquad \varphi_2 \equiv 3(\partial F/\partial S_3)d\Lambda \,. \quad (7)$$

By eliminating the plastic potential F in (7) we can find additional restrictions imposed on the scalar functions φ_0, φ_1, and φ_2, if the existence of a plastic potential is assumed:

$$2\,\partial\varphi_0/\partial S_2 = \partial\varphi_1/\partial S_1 \,, \quad 3\partial\varphi_0/\partial S_3 = \partial\varphi_2/\partial S_1 \,, \quad 3\partial\varphi_1/\partial S_3 = 2\partial\varphi_2/\partial S_2 \,. \quad (8)$$

Thus, in the isotropic special case the plastic potential theory is compatible with the tensor function theory, if the conditions of "compatibility" in (8) have been fulfiled.

For example, the requirements in (8) can be satisfied, if one assumes:

$$\varphi_0 = \varphi_0(S_1) \,, \qquad \varphi_1 = \varphi_1(S_2) \,, \qquad \varphi_2 = \varphi_2(S_3) \,. \qquad (9)$$

Then, because of (7), the plastic potential can be represented as

$$F = F(S_1,S_2,S_3) = g_1(S_1) + g_2(S_2) + g_3(S_3) \; , \tag{10}$$

where g_1, g_2, and g_3 are arbitrary functions of the invariants S_1, S_2, and S_3, respectively.

The conditions of "compatibility" in (8) can be found in the following way. The identities (7) can be written in the form

$$\partial F/\partial S_1 \equiv \varphi_0 L, \qquad \partial F/\partial S_2 \equiv \varphi_1 L/2 \; , \qquad \partial F/\partial S_3 \equiv \varphi_2 L/3 \; , \tag{11}$$

where

$$L \equiv 1/d\Lambda = L(S_1,S_2,S_3) \tag{12}$$

is normally a scalar-valued function of the integrity basis. The plastic potential F in (11) can be eliminated, if we consider the $\binom{3}{2}$ equalities

$$\frac{\partial^2 F}{\partial S_1 \partial S_2} = \frac{\partial^2 F}{\partial S_2 \partial S_1} \; , \quad \frac{\partial^2 F}{\partial S_1 \partial S_3} = \frac{\partial^2 F}{\partial S_3 \partial S_1} \; , \quad \frac{\partial^2 F}{\partial S_2 \partial S_3} = \frac{\partial^2 F}{\partial S_3 \partial S_2} \tag{13}$$

i.e., assuming the second partial derivatives of the plastic potential are continuous. Hence, from (11) and (13) we find a system of linear equations in the partial derivatives $\partial L/\partial S_\lambda \equiv L,_\lambda$ with $\lambda = 1,2,3$:

$$\left.\begin{aligned} \varphi_1 L,_1 - 2\varphi_0 L,_2 + \quad 0 \quad &= [2(\varphi_0),_2 - (\varphi_1),_1]L \\ \varphi_2 L,_1 + \quad 0 \quad - 3\varphi_0 L,_3 &= [3(\varphi_0),_3 - (\varphi_2),_1]L \\ 0 \quad + 2\varphi_2 L,_2 - 3\varphi_1 L,_3 &= [3(\varphi_1),_3 - 2(\varphi_2),_2]L \; . \end{aligned}\right\} \tag{14}$$

On the right hand side in (14) the abbreviations $(\varphi_0),_2 \equiv \partial\varphi_0/\partial S_2, \ldots, (\varphi_2),_2 \equiv \partial\varphi_2/\partial S_2$ are used. Because the coefficient determinant in (14) vanishes,

$$\begin{vmatrix} \varphi_1 & -2\varphi_0 & 0 \\ \varphi_2 & 0 & -3\varphi_0 \\ 0 & 2\varphi_2 & -3\varphi_1 \end{vmatrix} = 0 , \qquad (15)$$

one can assume, that the right hand side of the system (14) is also equal to zero. It is possible to find the three conditions of "compatibility" in (8). Note, that these conditions are only sufficient. But if we use CRAMER's rule, we find the following sufficient and necessary condition

$$[3(\varphi_1),_3 - 2(\varphi_2),_2]\varphi_0 + [(\varphi_2),_1 - 3(\varphi_0),_3]\varphi_1 + [2(\varphi_0),_2 - (\varphi_1),_1]\varphi_2 = 0, \qquad (16)$$

which contains (8) as a special case. As has already been mentioned before, the conditions in (8) are only sufficient, whereas the similar conditions (3) in elasticity are both sufficient and necessary.

For more complicated examples than the above mentioned the plastic potential theory is not compatible with the tensor function theory. For instance, MURAKAMI and SAWCZUK found in [3], that for the model of prestrained plastic solids the flow theory of classical plasticity only furnishes restricted forms of constitutive equations. In [4] LEHMANN proposed an extended form of the classical flow rule (4) with respect to experimental results and based upon thermodynamical considerations. In [5] oriented solids are considered and an appropiate modification of the classical flow rule (4) is discussed.

3. ORIENTED SOLIDS

In order to describe yielding and failure of oriented solids BOEHLER and SAWCZUK [6,7] use the tensor generator $\underset{\sim}{A} = \underset{\sim}{v} \otimes \underset{\sim}{v}$, where the vector $\underset{\sim}{v}$ specifies a privileged direction. Then the constitutive equation can be represented by the minimum polynomial

$$d^{p}\varepsilon_{ij} = f_{ij}(\sigma_{pq}, A_{pq}) = \frac{1}{2} \sum_{\lambda,\nu=0}^{2} \varphi_{(\lambda,\nu)} (M_{ij}^{[\lambda,\nu]} + M_{ji}^{[\lambda,\nu]}) \tag{17}$$

where $[\lambda,\nu]$ several symmetric tensor generators are formed by matrix products of the forms

$$M_{ij}^{[\lambda,\nu]} \equiv \sigma_{ik}^{(\lambda)} A_{kj}^{(\nu)}, \qquad \lambda,\nu = 0,1,2 . \tag{18}$$

The $\varphi_{(\lambda,\nu)}$'s in (17) are nine scalar functions of the integrity basis, the elements of which can be found by forming all the irreducible traces of the matrix products (18):

$$M_{rr}^{[\lambda,\nu]} \equiv \sigma_{pq}^{(\lambda)} A_{qp}^{(\nu)}, \text{ where } \lambda,\nu = 1,2 \text{ and for } \begin{array}{l} \lambda = 0 \Rightarrow \nu = 1,2,3 \\ \nu = 0 \Rightarrow \lambda = 1,2,3, \end{array} \tag{19a}$$

which can be written as:

$$\left. \begin{array}{l} S_{\lambda} \equiv \mathrm{tr}\, \underset{\sim}{\sigma}^{\lambda} , \ T_{\nu} \equiv \mathrm{tr}\, \underset{\sim}{A}^{\nu} ; \ \lambda,\nu = 1,2,3 ; \\ \Omega_{1} \equiv \mathrm{tr}\, \underset{\sim}{\sigma}\underset{\sim}{A} , \ \Omega_{2} \equiv \mathrm{tr}\, \underset{\sim}{\sigma}\underset{\sim}{A}^{2}, \ \Omega_{3} \equiv \mathrm{tr}\, \underset{\sim}{A}\underset{\sim}{\sigma}^{2} , \ \Omega_{4} \equiv \mathrm{tr}\, \underset{\sim}{\sigma}^{2}\underset{\sim}{A}^{2} . \end{array} \right\} \tag{19b}$$

From the ten invariants in (19) only the seven stress-dependent invariants are essential for the plastic potential:

$$F = F(M_{rr}^{[\lambda,\nu]}) , \text{ where } \lambda,\nu = 1,2 \text{ but for } \nu = 0 \Rightarrow \lambda = 1,2,3 . \tag{20}$$

Thus, by using the flow rule (4), one can find the constitutive equation

$$\left. \begin{array}{l} d^{p}\varepsilon_{ij} = d\Lambda \displaystyle\sum_{\lambda,\nu} (\partial F/\partial M_{rr}^{[\lambda,\nu]}) Q_{ijpq}^{[\lambda]} A_{qp}^{[\nu]}, \\ \qquad \text{where } \lambda,\nu = 1,2 \text{ and for } \nu = 0 \Rightarrow \lambda = 1,2,3 . \end{array} \right\} \tag{21}$$

In (21) the λ several fourth-rank tensors $\underset{\sim}{Q}^{[\lambda]}$ are defined as

$$Q^{[\lambda]}_{pqij} \equiv \partial\sigma^{(\lambda)}_{pq} / \partial\sigma_{ij} = \sum_{\alpha=0}^{\lambda=1} [\sigma^{(\alpha)}_{pi}\sigma^{(\lambda-1-\alpha)}_{qj} + \sigma^{(\alpha)}_{pj}\sigma^{(\lambda-1-\alpha)}_{qi}]/2 \tag{22}$$

and have the following properties of symmetry:

$$Q^{[\lambda]}_{pqij} = Q^{[\lambda]}_{qpij} = Q^{[\lambda]}_{pqji} = Q^{[\lambda]}_{ijpq} \ . \tag{23}$$

One can see, that because of (22) the value λ cannot be equal to zero. Therefore, only the seven stress-dependent invariants of the integrity basis (19) are relevant for the plastic potential (20). Furthermore, by comparing the constitutive equations (17) and (21), one can see, that the minimum polynomial (17) consists of nine tensor generators, whereas (21) has only seven terms, i.e., the plastic potential theory furnishes only restricted forms of constitutive equations, even if a general plastic potential is assumed in an anisotropic case.

It may be more useful for practical applications to represent the constitutive equation (17) in a canonical form

$$d^P\varepsilon_{ij} = {}^0H_{ijkl}\delta_{kl} + {}^1H_{ijkl}\sigma_{kl} + {}^2H_{ijkl}\sigma^{(2)}_{kl} , \tag{24}$$

where the fourth-order tensor-valued functions ${}^0\underset{\sim}{H}$, ${}^1\underset{\sim}{H}$, and ${}^2\underset{\sim}{H}$ are defined in the following way:

$${}^0H_{ijkl} \equiv \varphi_{(0,0)}m^{(0)}_{ijkl} + \varphi_{(0,1)}m_{ijkl} + \varphi_{(0,2)}m^{[2]}_{ijkl} \tag{25a}$$

$${}^1H_{ijkl} \equiv \varphi_{(1,0)}m^{(0)}_{ijkl} + \varphi_{(1,1)}m_{ijkl} + \varphi_{(1,2)}m^{[2]}_{ijkl} \tag{25b}$$

$${}^2H_{ijkl} \equiv \varphi_{(2,0)}m^{(0)}_{ijkl} + \varphi_{(2,1)}m_{ijkl} + \varphi_{(2,2)}m^{[2]}_{ijkl} \tag{25c}$$

having the symmetric tensors $\underset{\sim}{m}^{[\nu]}$, $\nu = 0,1,2$, of rank four:

$$m^{[\nu]}_{ijkl} \equiv (A^{(\nu)}_{ik}\delta_{jl} + A^{(\nu)}_{il}\delta_{jk} + \delta_{ik}A^{(\nu)}_{jl} + \delta_{il}A^{(\nu)}_{jk})/4 \ . \tag{26}$$

Especially, for $\nu = 0$ one finds in (26) the zero power tensor of rank four:

$$m^{[0]}_{ijkl} \equiv m^{(0)}_{ijkl} \equiv m_{ijpq} m^{(-1)}_{pqkl} = (\delta_{ik}\delta_{jl} + \delta_{il}\delta_{jk})/2 = m^{(-1)}_{ijpq} m_{pqkl} \ . \tag{27}$$

The representation found in the canonical form (24) consists of three terms, which are the contributions of zero, first, and second-orders in the stress tensor $\underset{\sim}{\sigma}$, influenced by the functions ${}^{0}\underset{\sim}{H}$, ${}^{1}\underset{\sim}{H}$, and ${}^{2}\underset{\sim}{H}$, respectively. In finding these fourth-order tensor-valued functions, the identities

$$\{\tfrac{1}{2}(\underset{\sim}{X}^{\lambda}\underset{\sim}{Y}^{\nu} + \underset{\sim}{Y}^{\nu}\underset{\sim}{X}^{\lambda})\}_{ij} \equiv \eta^{[\nu]}_{ijkl} X^{(\lambda)}_{kl} \equiv \xi^{[\lambda]}_{ijkl} Y^{(\nu)}_{kl} \ ,$$

$$\xi^{[\lambda]}_{ijkl} \equiv (X^{(\lambda)}_{ik}\delta_{jl} + X^{(\lambda)}_{il}\delta_{jk} + \delta_{ik}X^{(\lambda)}_{jl} + \delta_{il}X^{(\lambda)}_{jk})/4 \ ,$$

$$\eta^{[\nu]}_{ijkl} \equiv (Y^{(\nu)}_{ik}\delta_{jl} + Y^{(\nu)}_{il}\delta_{jk} + \delta_{ik}Y^{(\nu)}_{jl} + \delta_{il}Y^{(\nu)}_{jk})/4$$

have been used [5], where $\underset{\sim}{X}$ and $\underset{\sim}{Y}$ are two symmetric second-order tensors. Similar identities for non-symmetric tensors were introduced in [8].

Similarly to (24) one can represent the constitutive equation (21), derived from the plastic potential theory, in a canonical form:

$$d^{p}\varepsilon_{ij} = {}^{0}h_{ijkl}\delta_{kl} + {}^{1}h_{ijkl}\sigma_{kl} + {}^{2}h_{ijkl}\sigma^{(2)}_{kl} \ , \tag{28}$$

where the fourth-order tensor-valued functions ${}^{0}\underset{\sim}{h}$, ${}^{1}\underset{\sim}{h}$, and ${}^{2}\underset{\sim}{h}$ are defined by using (26) in the following way:

$${}^{0}h_{ijkl} \equiv \left[(\partial F/\partial S_1) m^{(0)}_{ijkl} + (\partial F/\partial \Omega_1) m_{ijkl} + (\partial F/\partial \Omega_2) m^{[2]}_{ijkl}\right] d\Lambda \tag{29a}$$

$${}^{1}h_{ijkl} \equiv 2\left[(\partial F/\partial S_2) m^{(0)}_{ijkl} + (\partial F/\partial \Omega_3) m_{ijkl} + (\partial F/\partial \Omega_4) m^{[2]}_{ijkl}\right] d\Lambda \tag{29b}$$

$${}^{2}h_{ijkl} \equiv 3\left[(\partial F/\partial S_3) m^{(0)}_{ijkl} + \quad 0 \quad + \quad 0 \quad \right] d\Lambda \ . \tag{29c}$$

Without loss of generality in the case of incipient motion, the vector $\underset{\sim}{v}$ can be regarded as an unit vector in the reference configuration

[6]. Therefore, only $\nu = 0,1$ in (18) has to be taken into account. The number of tensor generators in the constitutive equation (17) is reduced from nine to six, i.e., the fourth-order tensor-valued functions (25a,b, c) can be simplified into:

$$^{0}H_{ijkl} \equiv \varphi_{(0,0)} m^{(0)}_{ijkl} + \varphi_{(0,1)} m_{ijkl} \tag{30a}$$

$$^{1}H_{ijkl} \equiv \varphi_{(1,0)} m^{(0)}_{ijkl} + \varphi_{(1,1)} m_{ijkl} \tag{30b}$$

$$^{2}H_{ijkl} \equiv \varphi_{(2,0)} m^{(0)}_{ijkl} + \varphi_{(2,1)} m_{ijkl} \; . \tag{30c}$$

Furthermore, the number of irreducible invariants (19a,b) is reduced from ten to five,

$$S_{\lambda} \equiv \mathrm{tr}\, \underset{\sim}{\sigma}^{\lambda} \;, \; \lambda = 1,2,3, \; \Omega_1 \equiv \mathrm{tr}\, \underset{\sim\sim}{\sigma A}, \; \Omega_3 \equiv \mathrm{tr}\, \underset{\sim\sim}{A\sigma}^{2} \;, \tag{31}$$

if we regard the vector $\underset{\sim}{v}$ as a unit vector.

The fourth-order tensor-valued functions (29a,b,c) resultung from the plastic potential theory, can be simplified into:

$$^{0}h_{ijkl} \equiv \left[(\partial F/\partial S_1) m^{(0)}_{ijkl} + (\partial F/\partial \Omega_1) m_{ijkl} \right] d\Lambda \tag{32a}$$

$$^{1}h_{ijkl} \equiv 2\left[(\partial F/\partial S_2) m^{(0)}_{ijkl} + (\partial F/\partial \Omega_3) m_{ijkl} \right] d\Lambda \tag{32b}$$

$$^{2}h_{ijkl} \equiv 3\left[(\partial F/\partial S_3) m^{(0)}_{ijkl} + \quad 0 \quad \right] d\Lambda \; . \tag{32c}$$

Comparing the fourth-rank tensors (29a,b,c) with the corresponding tensors (25a,b,c) or, alternatively, (32a,b,c) with (30a,b,c), one can see that the scalar functions $\varphi_{(2,1)}$ and $\varphi_{(2,2)}$ in (25c) or $\varphi_{(2,1)}$ in (30c) cannot be expressed through the plastic potential (20), i.e., in the anisotropic case the plastic potential theory with its classical flow rule (4) only furnishes restricted forms of constitutive equations even where a general plastic potential $F = F(\sigma_{ij}, A_{ij})$ is assumed. Furthermore, if one considers the functions (29a,b,c) or, alternatively (32a,b,

c) one can see, that any terms containing m_{ijkl} and $m_{ijkl}^{[2]}$ do not appear in (29c) or in (32c), i.e., the "second-order effect" in (28), characterized through (29c) or (32c), is not influenced by the anisotropy (26) of the material. Consequently, the flow rule (4) of the classical flow theory of plasticity should be modified in the anisotropic case.

The results mentioned above can also be applied to perforated materials or damaged materials, if the anisotropy tensor A_{ij} in (26) is substituted with a perforation tensor [9] or a damage tensor [8,10].

4. MODIFICATION OF THE CLASSICAL FLOW RULE

By taking the representation theory into account we can justify the following expansion of the flow rule (4):

$$d^p\varepsilon_{ij} = d\Lambda(\partial F/\partial\sigma_{ij} + \alpha m_{ijkl}\partial F/\partial A_{kl}) \quad (33)$$

Assuming that the plastic potential $F = F(\sigma_{ij}, A_{ij})$ is a scalar-valued function of the integrity basis (19a,b) and by using the modified flow rule (33), we can find a constitutive equation which is compatible with the representation theory of tensor functions (17), (24), (25a,b,c) that is: all nine scalar-valued functions $\varphi_{(\lambda,\nu)}$, $\lambda,\nu = 0,1,2$, in (17) or (25a,b,c) can be expressed through the plastic potential F:

$$\varphi_{(0,0)} \equiv d\Lambda(\partial F/\partial S_1 + 3\alpha K_2\partial F/\partial T_3) , \quad (34a)$$

$$\varphi_{(0,1)} \equiv d\Lambda(\partial F/\partial\Omega_1 + \alpha\partial F/\partial T_1 + 3\alpha K_2\partial F/\partial T_3) , \quad (34b)$$

$$\varphi_{(0,2)} \equiv d\Lambda(\partial F/\partial\Omega_2 + 2\alpha\partial F/\partial T_2 + 3\alpha K_1\partial F/\partial T_3) , \quad (34c)$$

- -

$$\varphi_{(1,0)} \equiv 2d\Lambda\ \partial F/\partial S_2 , \quad \varphi_{(1,1)} \equiv d\Lambda[\partial F/\partial\Omega_3 + (\alpha/2)\partial F/\partial\Omega_1] , \quad (34d,e)$$

$$\varphi_{(1,2)} \equiv d\Lambda(\partial F/\partial\Omega_4 + \alpha\partial F/\partial\Omega_2) , \quad (34f)$$

- -

$$\varphi_{(2,0)} \equiv 3d\Lambda \frac{\partial F}{\partial S_3}\,, \quad \varphi_{(2,1)} \equiv \frac{\alpha}{2}\, d\Lambda \frac{\partial F}{\partial \Omega_3}\,, \quad \varphi_{(2,2)} \equiv \alpha\, d\Lambda \frac{\partial F}{\partial \Omega_4}\,, \quad (34g,h,i)$$

where the abbreviations

$$K_1 \equiv T_1\,, \quad K_2 \equiv (T_2 - T_1^2)/2\,, \quad K_3 \equiv (T_1^3 - 3T_1T_2 + 2T_3)/6 \qquad (35a,b,c)$$

have been used. In this way the second-order effect in (28) is influenced through the material tensor $\underset{\sim}{A}$ because of $\alpha \neq 0$ found in (34h, i). This influence is not at all possible, if we use the tensor-valued function (29c) which results from the classical flow rule (4) which does not contain the parameter α. Thus, the plastic potential theory is compatible with the tensor function theory, if the modified flow rule (33) is used instead of (4). But among the scalar functions (34a,...,i) we can find additional relationships ("conditions of compatibility") by eliminating the plastic potential F. This can be done by closely adhering to the following procedure. First, one must find the partial derivatives $\partial F/\partial S_1, \ldots, \partial F/\partial \Omega_4$ of the plastic potential F from the system of linear equations (34a,...,i). Second, one must apply the $\binom{10}{2}$ equalities

$$\frac{\partial^2 F}{\partial S_1 \partial S_2} = \frac{\partial^2 F}{\partial S_2 \partial S_1}\,, \ldots, \frac{\partial^2 F}{\partial S_3 \partial \Omega_4} = \frac{\partial^2 F}{\partial \Omega_4 \partial S_3}\,, \ldots, \frac{\partial^2 F}{\partial \Omega_3 \partial \Omega_4} = \frac{\partial^2 F}{\partial \Omega_4 \partial \Omega_3} \qquad (36)$$

to every pair of the obove mentioned ten derivatives. Thus, in this way we have eliminated the plastic potential F and found 45 conditions of compatibility.

If the vector $\underset{\sim}{v}$ is regarded as a unit vector, then it is possible to find the following identities by first taking (30a,b,c), (31) into consideration and by then using the modified flow rule (33):

$$\varphi_{(0,0)} \equiv (\partial F/\partial S_1)d\Lambda\,, \qquad \varphi_{(0,1)} \equiv (\partial F/\partial \Omega_1)d\Lambda\,, \qquad (37a,b)$$

$$\varphi_{(1,0)} \equiv 2(\partial F/\partial S_2)d\Lambda\,, \qquad \varphi_{(1,1)} \equiv 4\varphi_{(2,1)}/\alpha + \alpha\, \varphi_{(0,1)}\,, \qquad (37c,d)$$

$$\varphi_{(2,0)} \equiv 3(\partial F/\partial S_3)d\Lambda\,, \qquad \varphi_{(2,1)} \equiv \frac{\alpha}{2}(\partial F/\partial \Omega_3)d\Lambda\,. \qquad (37e,f)$$

The results of (37a,...,f) can be used to find the partial derivatives of the plastic potential:

$$\left.\begin{aligned} &\partial F/\partial S_1 \equiv \varphi_{(0,0)} L,\ \partial F/\partial S_2 \equiv \varphi_{(1,0)} L/2,\ \partial F/\partial S_3 \equiv \varphi_{(2,0)} L/3\ ,\\ &\partial F/\partial \Omega_1 \equiv \varphi_{(0,1)} L,\ \partial F/\partial \Omega_3 \equiv 2\varphi_{(2,1)} L/\alpha . \end{aligned}\right\} \quad (38)$$

In (38)

$$L \equiv 1/d\Lambda = L(S_1, S_2, S_3, \Omega_1, \Omega_3) \quad (39)$$

is a scalar-valued function of the integrity basis (31). The reader may note the analogy between (39) and (12). The plastic potential F in (38) can be eliminated in the following way. By regarding every pair of the derivatives (38), and by using the $\binom{5}{2}$ equalities

$$\frac{\partial^2 F}{\partial S_1 \partial S_2} = \frac{\partial^2 F}{\partial S_2 \partial S_1}\ ,\ldots,\ \frac{\partial^2 F}{\partial S_2 \partial \Omega_3} = \frac{\partial^2 F}{\partial \Omega_3 \partial S_2}\ ,\ldots,\ \frac{\partial^2 F}{\partial \Omega_1 \partial \Omega_3} = \frac{\partial^2 F}{\partial \Omega_3 \partial \Omega_1}\ , \quad (40)$$

the following ten linear equations in the partial derivatives are yielded:

$$
\begin{array}{llllll}
\varphi_{(1,0)}L,_1 & -\ 2\varphi_{(0,0)}L,_2 & +\ 0 & +\ 0 & +\ 0 & = R_1L \\
\varphi_{(2,0)}L,_1 & +\ 0 & -\ 3\varphi_{(0,0)}L,_3 & +\ 0 & +\ 0 & = R_2L \\
\varphi_{(0,1)}L,_1 & +\ 0 & +\ 0 & -\ \varphi_{(0,0)}L,_4 & +\ 0 & = R_3L \\
\varphi_{(2,1)}L,_1 & +\ 0 & +\ 0 & +\ 0 & -\ \frac{\alpha}{2}\varphi_{(0,0)}L,_5 & = R_4L \\
0 & +\ 2\varphi_{(2,0)}L,_2 & -\ 3\varphi_{(1,0)}L,_3 & +\ 0 & +\ 0 & = R_5L \\
0 & +\ 2\varphi_{(0,1)}L,_2 & +\ 0 & -\ \varphi_{(1,0)}L,_4 & +\ 0 & = R_6L \\
0 & +\ 2\varphi_{(2,1)}L,_2 & +\ 0 & +\ 0 & -\ \frac{\alpha}{2}\varphi_{(1,0)}L,_5 & = R_7L \\
0 & +\ 0 & +\ 3\varphi_{(0,1)}L,_3 & -\ \varphi_{(2,0)}L,_4 & +\ 0 & = R_8L \\
0 & +\ 0 & +\ 3\varphi_{(2,1)}L,_3 & +\ 0 & -\ \frac{\alpha}{2}\varphi_{(2,0)}L,_5 & = R_9L \\
0 & +\ 0 & +\ 0 & +\ \varphi_{(2,1)}L,_4 & -\ \frac{\alpha}{2}\varphi_{(0,1)}L,_5 & = R_{10}L
\end{array}
\tag{41}
$$

where the abbreviations

$$
L,_1 \equiv \partial L/\partial s_1,\ L,_2 \equiv \partial L/\partial s_2,\ L,_3 \equiv \partial L/\partial s_3,\ L,_4 \equiv \partial L/\partial \Omega_1,\ L,_5 \equiv \partial L/\partial \Omega_3
\tag{42}
$$

and

$$R_1 \equiv 2\partial\varphi_{(0,0)}/\partial S_2 - \partial\varphi_{(1,0)}/\partial S_1, \; R_2 \equiv 3\partial\varphi_{(0,0)}/\partial S_3 - \partial\varphi_{(2,0)}/\partial S_1,$$

$$R_3 \equiv \partial\varphi_{(0,0)}/\partial\Omega_1 - \partial\varphi_{(0,1)}/\partial S_1, \; R_4 \equiv \frac{\alpha}{2}\partial\varphi_{(0,0)}/\partial\Omega_3 - \partial\varphi_{(2,1)}/\partial S_1,$$

$$R_5 \equiv 3\partial\varphi_{(1,0)}/\partial S_3 - 2\partial\varphi_{(2,0)}/\partial S_2, \; R_6 \equiv \partial\varphi_{(1,0)}/\partial\Omega_1 - 2\partial\varphi_{(0,1)}/\partial S_2,$$

$$R_7 \equiv \frac{\alpha}{2}\partial\varphi_{(1,0)}/\partial\Omega_3 - 2\partial\varphi_{(2,1)}/\partial S_2, \; R_8 \equiv \partial\varphi_{(2,0)}/\partial\Omega_1 - 3\partial\varphi_{(0,1)}/\partial S_3,$$

$$R_9 \equiv \frac{\alpha}{2}\partial\varphi_{(2,0)}/\partial\Omega_3 - 3\partial\varphi_{(2,1)}/\partial S_3, \; R_{10} \equiv \frac{\alpha}{2}\partial\varphi_{(0,1)}/\partial\Omega_3 - \partial\varphi_{(2,1)}/\partial\Omega_1 \qquad (43)$$

are introduced.

From the equations found in (41) one takes $\binom{5}{3}$ systems each of which consists of 3 linear equations. In other words one starts from (38) and combines $\binom{5}{3}$ identities. Then, by using the $\binom{5}{2}$ equalities (40), one can find 10 systems each with 3 linear equations, for instance:

$$\left.\begin{aligned} F,_1 &\equiv \varphi_{(0,0)}L \\ F,_3 &\equiv \varphi_{(2,0)}L/3 \\ F,_5 &\equiv 2\varphi_{(2,1)}L/\alpha \end{aligned}\right\} \Rightarrow \left\{\begin{aligned} &\varphi_{(2,0)}L,_1 - 3\varphi_{(0,0)}L,_3 + 0 = R_2L \\ &\varphi_{(2,1)}L,_1 + 0 - \frac{\alpha}{2}\varphi_{(0,0)}L,_5 = R_4L \\ &0 + 3\varphi_{(2,1)}L,_3 - \frac{\alpha}{2}\varphi_{(2,0)}L,_5 = R_9L. \end{aligned}\right. \qquad (44)$$

In another example we could find:

$$\left.\begin{array}{l} F_{,2} \equiv \varphi_{(1,0)}L/2 \\ F_{,4} \equiv \varphi_{(0,1)}L \\ F_{,5} \equiv 2\varphi_{(2,1)}L/\alpha \end{array}\right\} \Rightarrow \left\{\begin{array}{llll} 2\varphi_{(0,1)}L_{,2} & -\varphi_{(1,0)}L_{,4} & +\ 0 & = R_6 L \\ 2\varphi_{(2,1)}L_{,2} & +\ 0 & -\frac{\alpha}{2}\varphi_{(1,0)}L_{,5} & = R_7 L \\ 0 & +\varphi_{(2,1)}L_{,4} & -\frac{\alpha}{2}\varphi_{(0,1)}L_{,5} & = R_{10}L. \end{array}\right. \tag{45}$$

Note, that the coefficient determinants of the systems (44), (45), and all other systems, which have been found in the same way, are equal to zero:

$$\begin{vmatrix} \varphi_{(2,0)} & -3\varphi_{(0,0)} & 0 \\ \varphi_{(2,1)} & 0 & -\frac{\alpha}{2}\varphi_{(0,0)} \\ 0 & 3\varphi_{(2,1)} & -\frac{\alpha}{2}\varphi_{(2,0)} \end{vmatrix} = \ldots = \begin{vmatrix} 2\varphi_{(0,1)} & -\varphi_{(1,0)} & 0 \\ 2\varphi_{(2,1)} & 0 & -\frac{\alpha}{2}\varphi_{(1,0)} \\ 0 & \varphi_{(2,1)} & -\frac{\alpha}{2}\varphi_{(0,1)} \end{vmatrix} = 0. \tag{46}$$

Thus, if one requires that all systems be homogeneous ($R_1 = R_2 = \ldots = R_{10} = 0$), then 10 conditions of "compatibility" can be derived:

$$2\,\frac{\partial\varphi_{(0,0)}}{\partial S_2} = \frac{\partial\varphi_{(1,0)}}{\partial S_1}\,,\quad 3\,\frac{\partial\varphi_{(0,0)}}{\partial S_3} = \frac{\partial\varphi_{(2,0)}}{\partial S_1}\,,\ \ldots,\ \alpha\,\frac{\partial\varphi_{(0,1)}}{\partial\Omega_3} = \frac{\partial\varphi_{(2,1)}}{\partial\Omega_1}\,. \tag{47}$$

These conditions are fulfiled, for instance, if the following assumptions are made:

$$\left.\begin{array}{l} \varphi_{(0,0)} = \varphi_{(0,0)}(S_1),\ \varphi_{(1,0)} = \varphi_{(1,0)}(S_2),\ \varphi_{(2,0)} = \varphi_{(2,0)}(S_3), \\ \varphi_{(0,1)} = \varphi_{(0,1)}(\Omega_1),\ \varphi_{(2,1)} = \varphi_{(2,1)}(\Omega_3)\ . \end{array}\right\} \tag{48}$$

Then the plastic potential is as follows:

$$F = g_1(S_1) + g_2(S_2) + g_3(S_3) + g_4(\Omega_1) + g_5(\Omega_3) \ , \tag{49}$$

where $g_1, \ldots, g_5$ are arbitrary functions of only one invariant. The reader may note the analogy between (49) and (10).

Note, that the conditions (47) are only sufficient, but not necessary. In order to find sufficient and necessary "conditions of compatibility" CRAMER's rule should be used. After applying this rule on (44) and (45) and after taking (46) into account, one can find the following sufficient and necessary "conditions of compatibility":

$$R_2\varphi_{(2.1)} - R_4\varphi_{(2,0)} + R_9\varphi_{(0,0)} = 0 \tag{50}$$

and

$$R_6\varphi_{(2,1)} - R_7\varphi_{(0,1)} + R_{10}\varphi_{(1,0)} = 0 \ . \tag{51}$$

Continuing on in this way, the following complete set of ten sufficient and necessary "conditions of compatibility" is obtained:

$$\left.\begin{array}{l}
R_1\varphi_{(2,0)} - R_2\varphi_{(1,0)} + R_5\varphi_{(0,0)} = 0 \\
R_1\varphi_{(0,1)} - R_3\varphi_{(1,0)} + R_6\varphi_{(0,0)} = 0 \\
R_1\varphi_{(2,1)} - R_4\varphi_{(1,0)} + R_7\varphi_{(0,0)} = 0 \\
R_2\varphi_{(0,1)} - R_3\varphi_{(2,0)} + R_8\varphi_{(0,0)} = 0 \\
R_2\varphi_{(2,1)} - R_4\varphi_{(2,0)} + R_9\varphi_{(0,0)} = 0 \\
R_3\varphi_{(2,1)} - R_4\varphi_{(0,1)} + R_{10}\varphi_{(0,0)} = 0 \\
R_5\varphi_{(0,1)} - R_6\varphi_{(2,0)} + R_8\varphi_{(1,0)} = 0 \\
R_5\varphi_{(2,1)} - R_7\varphi_{(2,0)} + R_9\varphi_{(1,0)} = 0 \\
R_6\varphi_{(2,1)} - R_7\varphi_{(0,1)} + R_{10}\varphi_{(1,0)} = 0 \\
R_8\varphi_{(2,1)} - R_9\varphi_{(0,1)} + R_{10}\varphi_{(2,0)} = 0
\end{array}\right\} \tag{52}$$

In the isotropic special case the ten conditions (52) are reduced to the first one of the ten equations in (52), which is identical to (16).

The results (37a,...,f) and (52) point out, that for oriented solids the plastic potential theory can be made compatible with the representation theory of tensor functions, if the classical flow rule (4) is modified according to (33) and if, in addition, the "conditions of compatibility" in (52) have been fulfiled.

Finally, the special case of incompressibility should be considered. In such instances it is more practical to use the stress deviator $\underset{\sim}{\sigma}'$ in the integrity basis (19a,b) than to use the stress tensor $\underset{\sim}{\sigma}$. Furthermore, the modified flow rule

$$d^p\varepsilon_{ij} = d\Lambda(\partial F/\partial\sigma_{ij} + \alpha m_{\{ij\}kl}\partial F/\partial A_{kl}) \tag{53}$$

should be used instead of (33), where the fourth-order tensor

$$m_{\{ij\}kl} \equiv m_{ijkl} - m_{rrkl}\delta_{ij}/3 = m_{ijkl} - \delta_{ij}A_{kl}/3 \tag{54}$$

is deviatoric with respect to the free indices $\{ij\}$.

5. ANISOTROPY EXPRESSED THROUGH A FOURTH-RANK TENSOR

A more general case than that in (17) is described by using a material tensor of rank four with the usual conditions of symmetry:

$$A_{ijkl} = A_{jikl} = A_{ijlk} = A_{klij} . \tag{55}$$

Then the constitutive equation is a symmetric tensor-valued function

$$d^p\varepsilon_{ij} = f_{ij}(\sigma_{pq}, A_{pqrs}) , \tag{56}$$

which has to fulfill the condition of form-invariance

$$a_{ik}a_{jl}f_{kl}(\sigma_{pq}, A_{pqrs}) \equiv f_{ij}(a_{pt}a_{qu}\sigma_{tu}, a_{pt}a_{qu}a_{rv}a_{sw}A_{tuvw}) \tag{57}$$

under the transformation of the orthogonal group ($a_{ik}a_{jk} = \delta_{ij}$). Thus, a suitable representation of the constitutive equation (56) is given through a linear combination

$$d^p\varepsilon_{ij} = \sum_{\lambda,\mu,\nu} \varphi_{(\lambda,\mu,\nu)} G_{ij}^{[\lambda,\mu,\nu]} \tag{58}$$

with $[\lambda,\mu,\nu]$ several symmetric tensor generators

$$G_{ij}^{[\lambda,\mu,\nu]} = (M_{ij}^{[\lambda,\mu,\nu]} + M_{ji}^{[\lambda,\mu,\nu]})/2 \tag{59}$$

formed by matrix products of the following types

$$M_{ij}^{[\lambda,\mu,\nu]} \equiv \sigma_{ik}^{(\lambda)} A_{kjpq}^{(\mu)} \sigma_{pq}^{(\nu)}, \tag{60}$$

which fulfill the condition (57) of form-invariance.

In finding an irreducible set of tensor generators note, that, according to the HAMILTON-CAYLEY theorem, none of the exponents λ,ν of the second-order tensor $\underset{\sim}{\sigma}$ needs to be larger than 2. In generalizing the HAMILTON-CAYLEY theorem one can find in [11] that a symmetric fourth-order tensor (55) of power six and all higher powers can be expressed in terms of lower powers, that is $\mu \leqq 5$. For $\mu = 0$ the tensor $\underset{\sim}{A}$ in (60) is a spherical one. Then, on the one hand, the combinations $[\lambda,0,\nu]$ are symmetric with respect to λ,ν, and on the other hand the restriction $\lambda + \nu \leqq 2$ follows. Thus, 48 irreducible matrix products of the forms (60) can be found, which are listed in Table 1.

In formulating the plastic potential the integrity basis is used, which can be derived by forming al' irreducible traces of the matrix products (60):

$$M_{tt}^{[\lambda,\mu,\nu]} \equiv \sigma_{pq}^{(\lambda)} A_{pqrs}^{(\mu)} \sigma_{rs}^{(\nu)}, \quad \text{where } \lambda,\nu = 0,1,2 \text{ for } \mu = 1,2,\ldots,5 \text{ but for } \mu = 0 \Rightarrow \lambda + \nu \leqq 3. \tag{61}$$

These invariants are listed in Table 2.

Table 1 Matrix Products of the Forms (60) for Isotropy ($\mu = 0$) and Anisotropy ($\mu \neq 0$)

$\mu = 0$; $M_{ij}^{[\lambda,0,\nu]}$ with $\lambda + \nu \leq 2$			
$\lambda \backslash \nu$	0	1	2
0	δ_{ij}	σ_{ij}	$\sigma_{ij}^{(2)}$
1	σ_{ij}	$\sigma_{ij}^{(2)}$	
2	$\sigma_{ij}^{(2)}$		

$\mu = 1,2,\ldots,5$; $M_{ij}^{[\lambda,\mu,\nu]}$			
$\lambda \backslash \nu$	0	1	2
0	$A_{ijrr}^{(\mu)}$	$A_{ijpq}^{(\mu)}\sigma_{pq}$	$A_{ijpq}^{(\mu)}\sigma_{pq}^{(2)}$
1	$\sigma_{ik}A_{kjrr}^{(\mu)}$	$\sigma_{ik}A_{kjpq}^{(\mu)}\sigma_{pq}$	$\sigma_{ik}A_{kjpq}^{(\mu)}\sigma_{pq}^{(2)}$
2	$\sigma_{ik}^{(2)}A_{kjrr}^{(\mu)}$	$\sigma_{ik}^{(2)}A_{kjpq}^{(\mu)}\sigma_{pq}$	$\sigma_{ik}^{(2)}A_{kjpq}^{(\mu)}\sigma_{pq}^{(2)}$

Table 2 System (61) of irreducible Invariants

$M_{tt}^{[\lambda,\mu,\nu]} \equiv \sigma_{pq}^{(\lambda)}A_{pqrs}^{(\mu)}\sigma_{rs}^{(\nu)}$; $\mu = 1,2,\ldots,5$			
$\lambda \backslash \nu$	0	1	2
0	$A_{iijj}^{(\mu)}$	$A_{iirs}^{(\mu)}\sigma_{rs}$	$A_{iirs}^{(\mu)}\sigma_{rs}^{(2)}$
1		$\sigma_{pq}A_{pqrs}^{(\mu)}\sigma_{rs}$	$\sigma_{pq}A_{pqrs}^{(\mu)}\sigma_{rs}^{(2)}$
2	symmetric		$\sigma_{pq}^{(2)}A_{pqrs}^{(\mu)}\sigma_{rs}^{(2)}$
isotropy : $\mu = 0 \Rightarrow \underline{S_1 \equiv tr\,\underset{\sim}{\sigma}}, \underline{S_2 \equiv tr\,\underset{\sim}{\sigma}^2}, \underline{S_3 \equiv tr\,\underset{\sim}{\sigma}^3}$			

One can see, that the combinations $[\lambda,\mu,\nu]$ are symmetric with respect to λ,ν for all μ, i.e. in eqn (61) or in Table 2 we see 33 invariants. In addition, the invariant $A_{iijj}^{(6)}$ must be considered. Thus a set of 34 irreducible invariants is found.

The plastic potential is given in the form

$$F = F(M_{tt}^{[\lambda,\mu,\nu]}) , \tag{62}$$

and by using the classical flow rule (4) one obtains the constitutive equation

$$d^p\varepsilon_{ij} = d\Lambda \sum_{\lambda,\mu,\nu} \frac{\partial F}{\partial M_{tt}^{[\lambda\ \mu\ \nu]}} [Q_{ijpq}^{[\lambda]} A_{pqrs}^{(\mu)} \sigma_{rs}^{(\nu)} + Q_{ijpq}^{[\nu]} A_{pqrs}^{(\mu)} \sigma_{rs}^{(\lambda)}] , \tag{63}$$

where the λ several fourth-rank tensors $\underset{\sim}{Q}^{[\lambda]}$ defined as (22) are derivatives of tensor powers. One sees that only 28 invariants in the set (61) are stress-dependent. Therefore, the constitutive equation (63) contains only 28 terms, contrary to the tensor polynomial function (58) with 48 terms, that is: the classical flow rule (4) only furnishes restricted forms of the constitutive equations, even if a general plastic potential has been used.

The constitutive equations (17) or (21) are special cases and can be dealt with by inserting the special fourth-rank tensor

$$A_{ijkl} \equiv (A_{ik}A_{jl} + A_{il}A_{jk})/2 \tag{64}$$

having the tensor power

$$A_{ijkl}^{(\mu)} \equiv (A_{ik}^{(\mu)}A_{jl}^{(\mu)} + A_{il}^{(\mu)}A_{jk}^{(\mu)})/2 \tag{65}$$

in equations (58) or (63), respectively.

Finally, further material influences, which are of practical importance, should be considered, e.g. the effect of damage ($\underset{\sim}{D}$) on materials with inborn anisotropy ($\underset{\sim}{A}$). Then, we have to consider a symmetric second order tensor-valued function

$$d^p\varepsilon_{ij} = f_{ij}(\sigma_{pq}, D_{pq}, A_{pqrs}) \tag{66}$$

in three argument-tensor $\underset{\sim}{\sigma}$, $\underset{\sim}{D}$, and $\underset{\sim}{A}$ of rank 2, 2, and 4, respectively.

Different representations of (66) are discussed in [8,10].

REFERENCES

1. BETTEN, J., Elastizitäts- und Plastizitätslehre, Vieweg-Verlag, Braunschweig/Wiesbaden 1985.

2. BETTEN, J., The Classical Plastic Potential Theory in Comparison with the Tensor Function Theory, International Symposium PLASTICITY TODAY on Current Trends and Results in Plasticity, Udine 1983, published in a special Olszak Memorial volume of the Engineering Fracture Mechanics 21 (1985): 641-652.

3. MURAKAMI, S. and SAWCZUK, A., On Description of rate-independent Behaviour for prestrained Solids, Archives of Mechanics 31 (1979): 251-264.

4. LEHMANN, Th., Einige Bemerkungen zu einer allgemeinen Klasse von Stoffgesetzen für große elasto-plastische Formänderungen, Ingenieur-Archiv 41 (1972): 297-310.

5. BETTEN, J., On the Representation of the Plastic Potential of Anisotropic Solids, Colloque International du CNRS n° 319, Comportement plastique des solides anisotropes, Grenoble 1981, published in the proceedings: Plastic Behavior of Anisotropic Solids (ed. J.P. Boehler) CNRS, Paris 1985, 213-228.

6. BOEHLER, J.P. and SAWCZUK, A., Application of Representation Theorems to describe Yielding of transversely isotropic Solids, Mech. Res. Comm. 3 (1976): 277-283.

7. BOEHLER, R.P. and SAWCZUK, A., On Yielding of Oriented Solids, Acta Mechanica 27 (1977): 185-206.

8. BETTEN, J., Net-Stress Analysis in Creep Mechanics, Ingenieur-Archiv 52 (1982): 405-419.

9. LITEWKA, A. and SAWCZUK, A., A Yield Criterion for Perforated Sheets, Ingenieur-Archiv 50 (1981): 393-400.

10. BETTEN, J., Damage Tensors in Continuum Mechanics, Euromech Colloquium 147 on "Damage Mechanics", Paris VI, Cachan 1981, published in Journal de Mécanique théorique et appliquée 2 (1983): 13-32.

11. BETTEN, J., Integrity Basis for a Second-Order and a Fourth-Order Tensor, International J. Math. & Math. Sci. 5 (1982): 87-96.

MIX
Papier aus verantwortungsvollen Quellen
Paper from responsible sources
FSC® C105338

If you have any concerns about our products,
you can contact us on
ProductSafety@springernature.com

In case Publisher is established outside the EU,
the EU authorized representative is:
Springer Nature Customer Service Center GmbH
Europaplatz 3, 69115 Heidelberg, Germany

Printed by Libri Plureos GmbH
in Hamburg, Germany